A. Baklanov · B. Grisogono
Editors

Atmospheric Boundary Layers

Nature, Theory and Applications to Environmental Modelling and Security

Introduction by A. Baklanov and B. Grisogono

Previously published in journal *Boundary-Layer Meteorology*, Volume 125, No. 2

Alexander Baklanov
Danish Meteorological Institute
Copenhagen, Denmark

Branko Grisigono
Department of Geophysics
University of Zagreb
Zagreb, Croatia

Based upon papers presented at a NATO Advanced Research Workshop held in Dubrovnik, Croatia, 18–22 April, 2006.

Cover illustration: Thanks are due to S.S. Zilitinkevich and I.N. Esau for the use of Figure 3, from 'Similarity theory and calculation of turbulent fluxes at the surface for the stably stratified atmospheric boundary layer', p. 43.

ISBN 978-1-4419-2555-8 e-ISBN 978-0-387-74321-9

Printed on acid-free paper.

9 8 7 6 5 4 3 2 1

springer.com

Contents

Dedicated to Sergej S. Zilitinkevich
on the occasion of his 70th birthday

Professor Sergej S. Zilitinkevich

Atmospheric boundary layers: nature, theory and applications to environmental modelling and security

Alexander Baklanov · Branko Grisogono

This special issue presents a set of peer-reviewed papers from the NATO Advanced Research Workshop (ARW) "Atmospheric Boundary Layers: Modelling and Applications to Environmental Security", held in Dubrovnik, Croatia, 18–22 April 2006; 57 researchers from 21 countries and 4 continents participated (see the ARW web-site http://pbl-nato-arw.dmi.dk).

The principal goals of the ARW were

- to summarise and assess current knowledge on planetary boundary-layer (PBL) physics and parameterization,
- to promote the exchange of ideas and knowledge between physicists, meteorologists, and environmental modellers,
- to set a course for improving PBL parameterisations in climate, numerical weather prediction, air-quality, and emergency preparedness models.

A pleasant reason to arrange this event in April 2006 was the 70th birthday of Professor Sergej Zilitinkevich (born on 13 April 1936 in St. Petersburg, Russia). A most appropriate tribute to him is the fact that a major part of the presentations at this ARW are based on or linked with his fundamental works. His scientific biographical note by George Djolov is included in this issue. He is in perfect form, sprinkling with new ideas, and very productive scientifically.

The scientific organisation of the ARW was performed by the Organising Committee consisting of the NATO-country ARW director Alexander Baklanov (Meteorological Research Division, Danish Meteorological Institute, Copenhagen, Denmark); the Partner-country ARW director Branko Grisogono (Department of Geophysics, University of Zagreb, Croatia); Adolf Ebel (Phenish Institute for Environmental Research, University of Cologne, Germany); Sylvain M. Joffre (Finnish Meteorological Institute) and Sergej Zilitinkevich (Division of Atmospheric Sciences, University of Helsinki, Finland)—with valuable contri-

A. Baklanov (✉)
Meteorological Research, Danish Meteorological Institute, Lyngbyvej 100, Copenhagen, 2100 Denmark
e-mail: alb@dmi.dk

B. Grisogono
Geophysical Institute, Faculty of Science, University of Zagreb, Horvatovac bb, 10000 Zagreb, Croatia

Atmospheric Boundary Layers. A. Baklanov & B. Grisogono (eds.),
doi: 10.1007/978-0-387-74321-9_1,

butions from the local organisers: Iva Kavcic and Iva Grisogono (Department of Geophysics, University of Zagreb).

The programme included three introductory talks, nine key lectures, and 27 regular presentations divided into several thematic sessions. It is only natural that this issue does not cover all presentations; some of them were based on already published, submitted or still uncompleted papers.

Besides presentations, the programme included three general discussions: "Stability dependence of the turbulent Prandtl number and the critical Richardson number problem", "Turbulence closure problem", "Towards improvement of PBL schemes in operational models"; and specific discussions in working groups: (WG1) Boundary-layer physics (*L. Mahrt*); (WG2) Turbulence closure (*S.S. Zilitinkevich*); (WG3) Complex and mesoscale boundary-layer flows (*P.A. Taylor*); (WG4) Air-sea-ice interaction (*S.E. Larsen*); (WG5) Air flows within and above urban and vegetation canopies (*R. Bornstein*); (WG6) PBLs in operational models (*M.W. Rotach*); (WG7) Environmental security issues and demands from end users (*H.J.S. Fernando*). The WG chairmen (see their names in brackets) have summarised and forwarded to us the conclusions and recommendations of their groups. Below we briefly skim through working-group and general inert-group conclusions.

WG1 recommended (i) creation of a catalogue of the more extensive PBL datasets, (ii) analysing different sites with the same analysis method, and (iii) maintenance of a responsive data-base centre that can accommodate continual upgrades of datasets. More attention should be concentrated on spatial averaging of turbulent fluxes. Tower measurements can provide spatial averages, such as over a grid area, only with weak heterogeneity and the precarious assumption of Taylor's hypothesis. Most practical applications involve complex surfaces, which require more attention. The degree of failure of existing similarity theory over common complex surfaces needs to be determined to establish the likely magnitude of errors. Even modest improvements of the formulation of the flux-gradient relationships over heterogeneous surfaces would be of considerable practical use.

WG2 emphasised that traditional treatment of the turbulence energetics using solely the turbulent kinetic energy (TKE) budget equation is insufficient and causes difficulties in operational turbulence closure models. The TKE equation in combination with the down-gradient formulation for the turbulent fluxes and Kolmogorov's closure hypothesis for the eddy exchange coefficients leads to unrealistic degeneration of turbulence at Richardson numbers, Ri, exceeding some critical value, Ri_c. To Ri_c, Ri-dependent "correction coefficients" are introduced in the above formulation without physical explanation of the maintenance of turbulence in strongly stable stratification. As demonstrated at the ARW the above difficulties are caused by overlooking the turbulent potential energy (TPE) proportional to the mean squared temperature fluctuations. Together TKE and TPE comprise the turbulent total energy (TTE), whose budget equation does not include the vertical flux of buoyancy and has the form of a conservation equation securing positive TTE in any stratification. The concepts of TPE and TTE eliminate Ri_c from the turbulence closure problem and open a constructive way to create a hierarchy of energetically consistent closure schemes for operational use. The nature of organised structures in the convective PBL, namely cells and rolls in the shear-free and sheared regimes, respectively, remain not fully understood. The surface heat/mass transfer laws for the shear-free convection are obtained, but to extend the theory to sheared convection and to vertical transports within the convective zone, further observational and numerical simulation studies are needed.

WG3 identified the following mesoscale boundary-layer flows that strongly affect regional climate, weather and air quality: internal boundary layers caused by the roughness and/or thermal heterogeneity; flows over topography, such as slope winds; thermally driven features:

surface-induced convection and thunderstorms, sea and lake breezes, mountain-valley winds, urban and other heat islands, convective circulations over leads and polynias; mechanically driven flows: orographic gravity waves, downslope wind storms, retardation of frontal passages by orography or roughness; and features combing thermal and mechanical driving mechanisms, such as bora, chinook and foehn. Here the traditional computational techniques are Reynolds-averaged Navier–Stokes equation (RANS) models with varying levels of closure. TTE closures open new opportunities to improve RANS models, first of all, addressing air quality issues. Large-eddy simulation (LES) has just started to address idealised complex-terrain flows (e.g., Arctic leads, simply composed slopes, urban canopies) utilising periodic lateral boundary conditions. In view of the increasing power of supercomputers this technique will become suitable for operational modelling in the very near future.

WG4 focused on the marine atmospheric planetary boundary layer (MPBL). The smaller roughness and larger heat capacity of the ocean result in more organised convective flow patterns, such as cloud streets. Parameterization of the surface turbulent fluxes of momentum, heat and water vapour over the sea remains the key problem because of difficulties in obtaining high quality data in strong winds. Several studies revealed decreasing of the effective roughness length in very strong winds caused by the input of the sea spray into the lower surface layer. Constrained waters, coastal areas and lakes represent an even more complex problem requiring knowledge of the wind fetch and the basin depth. Modelling of the gas and particle exchange between the atmosphere and ocean still suffers from many uncertainties of both theoretical and experimental nature. For gases, an important uncertainty follows from the empirical nature of the surface exchange coefficient allowing no simple rules to account for non-stationary and heterogeneous features often found in aqueous concentration patterns and wind speeds. The mixed ice-sea surface is often highly heterogeneous with respect to the heat flux, due to the often-extreme difference in temperature between the ice surface and underlying waters. Also the drag may change strongly due to ridged borders of the ice floes. A quite successful effort has been made to understand and model the flow and fluxes for individual water openings, and also from ice-covered surfaces. However, aggregation into average surface fluxes remains strongly uncertain. Passing clouds change the radiation heat flux and thus constitute an important non-stationary effect on the MPBL. The strong wind MPBL traditionally considered as a simple neutrally stratified boundary layer exhibits novel features and, as recognised recently, should be treated as "conventionally neutral" (near-neutral close to the surface but strongly affected by the free-flow stability and stably stratified in its upper portion).

WG5 emphasised the key role of the urban boundary layer in air quality problems. Urban features essentially influence atmospheric flow and microclimate, strongly enhance atmospheric turbulence, and modify turbulent transports, dispersion and deposition of atmospheric pollutants. Considerably increased resolution in numerical weather prediction models has allowed for more realistically reproducing urban air flows and air pollution processes. This has triggered new interest in modelling and investigating experimentally specific processes essential for urban areas. Recent developments performed within the EU-funded project FUMAPEX on integrated systems for forecasting urban meteorology and air pollution and other relevant studies showed many opportunities in "urbanisation" of weather prediction, air pollution and emergency preparedness models.

WG6 and **WG3** agreed that high-resolution mesoscale RANS models and LES are complementary providing basic information on fine-scale features of air flow over complex terrain. The latter are not resolved in larger scale climate and weather prediction models and are to be parameterized through appropriate spatial averaging of turbulent fluxes (flux aggregation). This applies in particular to "hydrological heterogeneity" associated with areas with many small lakes. Currently-used flux-profile relationships determining the lower boundary conditions in both mesoscale and larger scale models need to be refined. The background Monin–Obukhov similarity theory is not applicable over complex terrain (see WG1) and does not realistically reproduce strongly stable and strongly unstable stratification regimes. Recently it has been generalised accounting for the non-local effects of the free-flow stability in stable stratification and large-scale, organised eddies in shear-free convection. Further work is needed to extend surface-layer theory to sheared convection and also to complex and sloping terrain.

WG7 considered environmental security issues and demands from end users. With growing spectra of chemical, biological and radiological (CBR) terrorism, it has become increasingly necessary to protect our ecosystems against deleterious activities of humans. The rapid rise of population in cities has created concentrated human centres that are vulnerable to extensive destruction through terrorism or by natural causes such as hurricanes, as evidenced with increasing frequency in recent decades. Increase of air pollution is another issue that is closely related to the quality of life. Given that the bulk of living entities are embedded in the PBL, it is opportune to discuss the role that boundary-layer meteorology plays in securing our environment. One of the key issues is the prediction of the pathways of CBR releases. First and foremost, suitable sensors are needed for the detection of toxins, and in the post-9/11 era such sensor technologies are rapidly advancing. What follows are in the arena of PBL modelling: optimal placement of sensor and design of sensor networks, environmental cyber-infrastructure, prediction of transport, diffusion and distribution of contaminants in the PBL, especially fast forecast models, monitoring of contaminant paths, including sensor model fusion work, long range transport and dilution, indoor air quality (air seepage through building accessories), environmental remediation, informing the authorities and providing help in emergency response.

We hope that the principal goals of the ARW will be achieved and this issue will help readers to assess the potential of recent achievements and novel ways in PBL physics and its applications. It was the unanimous opinion of all participants that the ARW has indeed made a step towards these goals, and that similar meetings, say, once in 2–3 years would strongly facilitate further progress in boundary-layer meteorology and operational environmental modelling, including the security issues.

Acknowledgements This ARW has been sponsored by the NATO Programme No. 982062 “Security through Science”. Additional supports have come from Croatian Meteorological and Hydrological Service (CMHS), Danish Meteorological Institute (DMI), Finnish Meteorological Institute (FMI), the EU Marie Curie Chair Project MEXC-CT-2003–509742 “Theory, Modelling and Role in Earth Systems” and the NORDPLUS NEIGHBOUR Project 177039/VII. We thank the NATO Administration, the CMHS Director General Ivan Čačić and the FMI Director General Pekka Plathan for making this meeting possible, productive and available to many participants. We also thank the working group leaders Larry Mahrt, Sergej Zilitinkevich, Peter Taylor, Søren Larsen, Robert Bornstein, Mathias Rotach and Joe Fernando for providing us with recommendations summarised herein.

9 July 2007

Some modern features of boundary-layer meteorology: a birthday tribute for Sergej Zilitinkevich

G. D. Djolov

Abstract The paper summarises the major scientific achievements of Sergej Zilitinkevich on the occasion of his 70th birthday.

1 Introduction

Recent urgent demands from environmental security, rational land exploitation, energy use and sustainable development have placed boundary-layer meteorology (BLM) in the forefront of modern developments in the field of atmospheric sciences. The theoretical advances, modelling tools and experimental data related to the atmospheric boundary layer are needed in a wide range of environmental models including those for climate, weather prediction, air pollution, and environmental impact assessment. Contemporary demand for high resolution modelling of complex atmosphere–hydrosphere–biosphere systems depends strongly on an accurate description of the physical processes in the turbulent planetary boundary layer (PBL).

At the workshop the participants had the pleasure of celebrating the 70th birthday of Sergej Zilitinkevich. His works contributed to many aspects of environmental physics and fluid dynamics including the theory of climate and general circulation of planetary atmospheres (Zilitinkevich and Monin 1974, 1977; Zilitinkevich 1976a, b, 1989a, b, c; Zilitinkevich et al. 1971), air–sea interaction (Zilitinkevich et al. 1978a, b, 2001), physical oceanography and hydrology (Zilitinkevich 1991a; Zilitinkevich et al. 1979, 1992; Zilitinkevich and Mironov 1992) and astrophysics (Zilitinkevich et al. 1976), not to forget his literary works (Zilitinkevich 1994, 1995).

G. D. Djolov (✉)
University of Limpopo, Private Bag 1106, Sovenga 0727, South Africa
e-mail: djolov@mweb.co.za

Atmospheric Boundary Layers. A. Baklanov & B. Grisogono (eds.),
doi: 10.1007/978-0-387-74321-9_2, © Springer Science+Business Media B.V. 2007

2 Career achievements

During his academic career in Russia, Germany and Scandinavia, the focus of Sergej's research was and still is on the atmospheric boundary layer. His work has integrated and further developed earlier outstanding achievements of the Russian, German and Scandinavian schools of the geophysical boundary layer and turbulence (A.N. Kolmogorov and A.M. Obukhov, L. Prandtl and T. von Karman, V.W. Ekman and C.G. Rossby) and has contributed significantly to the contemporary nature of these fields. His contributions have played an important role in the development of boundary-layer meteorology into a mature physical discipline, and a first book on this subject published 37 years ago (Zilitinkevich 1970) still withstands the passage of time and contains a research program relevant even today. Some of his universally accepted contributions to BLM are shortly overviewed in the present article.

Sergej Zilitinkevich's name became known worldwide in the late sixties, in connection with one of the key problems of boundary-layer meteorology: determination of the turbulent fluxes at the surface from easily available information, namely, the geostrophic wind speed and the temperature/humidity increments across the PBL. In 1967 he established the resistance and heat/mass transfer laws for both the stable and convective geophysical boundary layer in terms of what were denoted the A, B, C, D stability functions (nomenclature that became classical). Beginning from his early work (Zilitinkevich et al. 1967; Zilitinkevich and Chalikov 1968), many principal developments in this field are based on his pioneering contributions. These include an advanced version of the theory for the non-steady boundary layer (Zilitinkevich and Deardorff 1974), the asymptotic analysis and analytical solution for the A, B, C, D stability functions (Zilitinkevich 1975a), and, finally, extension of the theory to extreme stability regimes, in which semi-organised components of turbulence (overlooked in the traditional theories) play the decisive role (Zilitinkevich and Esau 2005).

Limiting ourselves to the most frequently quoted papers of Sergej Zilitinkevich, we proceed to his "directional dimensional analysis" (DD-analysis)—a novel mathematical procedure that allows one to separate motions driven by different forces (Zilitinkevich 1971, 1973). He used it to theoretically discover a new convective regime, "sheared convection" (besides the two regimes already known at the time: the buoyancy dominated "free convection" and the shear dominated "forced convection"), and to predict its basic features. In the early 1970s, this theoretical development was premature: experimental techniques were insufficiently advanced for its verification. The theory of sheared convection was convincingly confirmed by high-quality turbulence measurements only in the 1990s, and since then has become universally recognized. Important support to these works, including experimental validation, has been given by Betchov and Yaglom (1971) and Kadar and Yaglom (1990).

In his very short but very frequently quoted paper, Zilitinkevich (1972) analysed the stable boundary layer in a rotating fluid and derived the characteristic height scale $h_* \sim u_*^2 |fB_s|^{-1}$, nowadays universally accepted as the "Zilitinkevich scale" (here, u_* is the friction velocity, B_s is the surface buoyancy flux, and f is the Coriolis parameter).

In another short paper Zilitinkevich (1975b) formulated the first non-steady-turbulence equation for the depth of the evolving boundary layer often referred to as the "Zilitinkevich spin-up equation/effect". Numerous later work by many researchers was devoted to the experimental verification, generalisation and modelling applications of this theory. His own further developments of this theory are summarized in Zilitinkevich (1991b).

3 Recent work

The last 15 years or so of his research have been devoted to the fulfilment of a comprehensive multidimensional research program aimed at incorporation of boundary-layer physics into the general context of dynamic meteorology, climate physics, and operational environmental modelling. Below we note some of the principal developments, which Sergej carried out in cooperation with many groups from around the world.

A series of his studies are devoted to the extension and modification of PBL theory in order to systematically incorporate all relevant characteristics of the free atmosphere. Traditionally, the PBL is considered as driven by the horizontal pressure gradient force and affected by turbulent friction, density stratification and the Earth's rotation (characterized by u_*, B_s, and f, respectively); whereas the free atmosphere's characteristics are represented only by the values of wind and temperature at the top of the PBL (through the boundary conditions). In this context, the state of the free atmosphere, its static stability or any other properties, is considered irrelevant to the turbulent state of the atmospheric PBL. Sergej Zilitinkevich has developed an advanced, non-local theory of the neutral and stable PBL accounting for the free-atmosphere Brunt-Vaisala frequency N, baroclinic shear, $\Gamma = |\partial \mathbf{u}_g/\partial z|$ (where $\mathbf{u}_g$ is the geostrophic wind vector and z is the height), large-scale vertical velocity, w_h, and also non-steady developments of the PBL. Accordingly, instead of the traditionally recognised neutral PBL and stable PBL, we now distinguish the truly neutral ($B_s = 0$ and $N = 0$), the conventionally neutral ($B_s = 0$ and $N > 0$), the nocturnal stable ($B_s < 0$ and $N = 0$), and the long-lived stable ($B_s < 0$ and $N > 0$) PBL, which are controlled by different physical mechanisms and exhibit essentially different properties (Zilitinkevich 2002; Zilitinkevich et al. 2002, 2007a; Zilitinkevich and Baklanov 2002; Zilitinkevich and Esau 2003). Furthermore, the classical Monin–Obukhov similarity theory for the stably stratified surface layer has been revised: the classical Obukhov length scale $L = -u_*^{-3} B_s^{-1}$ is generally replaced by the composite turbulent length scale involving the rotational and the free-flow-stability scales ($L_f = u_*/f$ and $L_N = u_*/N$): $L_{\mathrm{Com}}^{-2} = L^{-2} + C' L_N^{-2} + C'' L_f^{-2}$, where C' and C'' are dimensionless constants (Zilitinkevich and Esau 2005, 2007). A dramatic example of a turbulent regime, in which the Monin–Obukhov theory fails, is the conventionally neutral PBL where $L^{-1} = 0$, and the essential length scale is L_N.

Another area of interest of Sergej Zilitinkevich is the nature of semi-organised eddies in turbulent convective flows and their role in turbulent transport. In Zilitinkevich et al. (1999, 2006) a very recent theory of the convective heat/mass transfer deserves emphasizing. As follows from perturbation analysis and large-eddy simulation (LES) of turbulent convection, semi-organized convective eddies represent complex but principally regular (or quasi-regular) flow patterns. Typically they consist of strong plumes fed by "convective winds" blowing towards plume axes, which cause shear-generated turbulence. This mechanism enhances convective heat/mass transfer and discloses an important role of the surface roughness overlooked in the classical theory. An advanced model developed on this basis is impressively confirmed by meteorological observations from numerous field campaigns and also through LES (Zilitinkevich et al. 2006). This analysis has shown that heat fluxes over rough surfaces may be up to two orders of magnitude larger than the prediction of classical theory. Furthermore, the new theory demonstrates that in shear-free convection the boundary-layer height h and the Deardorff convective velocity scale $w_* = (B_s h)^{1/3}$ must be included in the surface-layer scaling, so that the basic Monin–Obukhov similarity theory, as well as turbulence closure models expressing turbulent fluxes through local mean gradients, become insufficient. Thus after half a century of indisputable acceptance, this famous theory can no longer be considered as universal: in calm-weather convection the correct surface-layer scaling should include w_*

(in addition to u_*) and thus h; whereas in the conventionally neutral and long-lived stable PBL it should include the free-flow Brunt–Väisälä frequency N.

4 Summary

The above theoretical works of Sergey Zilitinkevich gave rise to numerous innovative developments in numerical weather prediction, climate modelling, modelling of air pollution and urban environment, optimised energy use in variable climate and weather, renewable energy planning and management, environmental impact assessment and other applied fields. Sergej's achievements in PBL physics are internationally recognised: in 2000 the European Geophysical Society presented its highest award in meteorology, the Vilhelm Bjerknes Medal, to Zilitinkevich "for his outstanding contribution to the creation of the modern theory of atmospheric turbulent boundary layers". In 2003 the EU Commission awarded him the newly established Marie Curie Chair of boundary-layer physics. Sergej has amazingly increased his scientific productivity and generation of new ideas in the last decade. His lecture at this workshop (Zilitinkevich et al. 2007b) provides a good example: his concept of the total turbulent energy (TTE) and consequences from the TTE budget equation include a theoretical solution to the critical Richardson number problem, as well as new opportunities to create a consistent turbulence closure for stably stratified flows.

References

Betchov R, Yaglom AM (1971) Comments on the theory of similarity as applied to turbulence in an unstably stratified fluid. Izv AN SSSR, Ser Fizika Atmosfery i Okeana 7:1270–1279

Kadar BA, Yaglom AM (1990) Mean fields and fluctuation moments in unstably stratified turbulent boundary layers. J Fluid Mech 212:637–662

Zilitinkevich SS (1970) Dynamics of the atmospheric boundary layer. Gidrometeoizdat. Leningrad, 290 pp

Zilitinkevich SS (1971) On turbulence and diffusion in free convection. Izv AN SSSR, Fizika Atmosfery i Okeana 7:1263–1269

Zilitinkevich SS (1972) Asymptotic formulas for the depth of the Ekman boundary layer. Izv AN SSSR, Fizika Atmosfery i Okeana 8:1086–1090

Zilitinkevich SS (1973) Shear convection. Boundary-Layer Meteorol 3:416–423

Zilitinkevich SS (1975a) Resistance laws and prediction equations for the depth of the planetary boundary layer. J Atmos Sci 32:741–752

Zilitinkevich SS (1975b) Comments on A model of the dynamics of the inversion above a convective boundary layer. J Atmos Sci 32:991–992

Zilitinkevich SS (1976a) Generation of kinetic energy of atmospheric circulation on slowly rotating planets. Doklady AN SSSR 227:1315–1318

Zilitinkevich SS (1976b) Rough estimates of some characteristics of atmospheric circulation on rotating planets. Doklady AN SSSR 228:62–65

Zilitinkevich SS (1989a) On analysis of the atmospheric general circulation on Venus. Kosmicheskije Issledovanija 27:286–291

Zilitinkevich SS (1989b) On the theory of super-rotation of the Venus atmosphere. Kosmicheskije Issledovanija 27:595–603

Zilitinkevich SS (1989c) Heat transport by the meridional circulation cell and static stability of the atmosphere on a slowly rotating planet. Kosmicheskije Issledovanija 27:932–942

Zilitinkevich SS (ed) (1991a) Modeling air–lake interaction—physical background. Springer Verlag, Berlin, 130 pp

Zilitinkevich SS (1991b) Turbulent penetrative convection. Avebury Technical, Aldershot, 149 pp

Zilitinkevich SS (1994) Hi, Professor! Zvezda (monthly literary magazine, St. Petersburg, Russia) 1:36–70

Zilitinkevich SS (1995) On meaning in fine arts. Mera (literary magazine, St. Petersburg, Russia) 1:142–151

Zilitinkevich S (2002) Third-order transport due to internal waves and non-local turbulence in the stably stratified surface layer. Quart J Roy Meteorol Soc 128:913–925

Zilitinkevich SS, Baklanov A (2002) Calculation of the height of stable boundary layers in practical applications. Boundary-Layer Meteorol 105:389–409

Zilitinkevich SS, Chalikov DV (1968) On the resistance and heat/moisture transfer laws in the interaction between the atmosphere and the underlying surface. Izv AN SSSR, Fizika Atmosfery i Okeana 4:765–772

Zilitinkevich SS, Deardorff JW (1974) Similarity theory for the planetary boundary layer of time-dependent height. J Atmos Sci 31:1449–1452

Zilitinkevich SS, Esau IN (2003) The effect of baroclinicity on the depth of neutral and stable planetary boundary layers. Quart J Roy Meteorol Soc 129:3339–3356

Zilitinkevich SS, Esau IN (2005) Resistance and heat/mass transfer laws for neutral and stable planetary boundary layers: old theory advanced and re-evaluated. Quart J Roy Meteorol Soc 131:1863–1892

Zilitinkevich S, Esau I (2007) Similarity theory and calculation of turbulent fluxes at the surface for the stably stratified atmospheric boundary layers. Boundary-Layer Meteorol, DOI: 10.1007/s10546-007-9187-4

Zilitinkevich SS, Mironov DV (1992) Theoretical model of thermocline in a freshwater basin. J Phys Oceanogr 22:988–996

Zilitinkevich SS, Monin AS (eds) (1974) Dynamics of the atmosphere of Venus. Nauka. Leningrad, 184 pp

Zilitinkevich SS, Monin AS (1977) Global interaction between the atmosphere and the ocean, Gidrometeoizdat. Leningrad. 24 pp [in English: Monin AS, Zilitinkevich SS (1977) Scale relations for global air-sea interaction. J Atmos Sci 34:1214–1223]

Zilitinkevich SS, Laikhtman DL, Monin AS (1967) Dynamics of the boundary layer in the atmosphere. Izv AN SSSR, Fizika Atmosfery i Okeana 3:297–333

Zilitinkevich SS, Monin AS, Turikov VG, Chalikov DV (1971) Numerical simulation of the circulation of the Venus atmosphere. Doklady AN SSSR 197:1291–1294

Zilitinkevich SS, Monin AS, Chalikov DV (1978a) Air–sea interaction. Wydawnictwo Polskiej Akademii Nauk, Wroclaw, 282 pp

Zilitinkevich SS, Monin AS, Chalikov DV (1978b) Air–sea interaction In: Kamenkovich VM, Monin AS (eds) Physics of the Ocean. volume 1: hydrophysics of the ocean. Nauka, Moscow, pp 208–239

Zilitinkevich SS, Chalikov DV, Resnyansky YuD (1979) Modeling the oceanic upper layer. Oceanol Acta 2:219–240

Zilitinkevich SS, Kreiman KD, Terzhevik AYu (1992) The thermal bar. J Fluid Mech 236:27–42

Zilitinkevich SS, Sunyaev RA, Shakura NI (1976) On turbulent transport of energy in accretion discs. Institute of Space Research AN SSSR. Report No. 296. 26 pp [in English: Shakura NI, Sunyaev RA, Zilitinkevich SS (1978) On the turbulent energy transport in accretion discs. Astron Astrophys 62:179–187]

Zilitinkevich SS, Gryanik VM, Lykossov VN, Mironov DV (1999) A new concept of the third-order transport and hierarchy of non-local turbulence closures for convective boundary layers. J Atmos Sci 56:3463–3477

Zilitinkevich SS, Grachev AA, Fairall CW (2001) Scaling reasoning and field data on the sea-surface roughness lengths for scalars. J Atmos Sci 58:320–325

Zilitinkevich S, Baklanov A, Rost J, Smedman A-S, Lykosov V, Calanca P (2002) Diagnostic and prognostic equations for the depth of the stably stratified Ekman boundary layer. Quart J Roy Meteorol Soc 128:25–46

Zilitinkevich SS, Hunt JCR, Grachev AA, Esau IN, Lalas DP, Akylas E, Tombrou M, Fairall CW, Fernando HJS, Baklanov A, Joffre SM (2006) The influence of large convective eddies on the surface layer turbulence. Quart J Roy Meteorol Soc 132:1423–1456

Zilitinkevich S, Esau I, Baklanov A (2007a) Further comments on the equilibrium height of neutral and stable planetary boundary layers. Quart J Roy Meteorol Soc 133:265–271

Zilitinkevich SS, Elperin T, Kleeorin N, Rogachevskii I (2007b) Energy- and flux-budget (EFB) turbulence closure model for the stably stratified flows. Part I: steady-state, homogeneous regimes. Boundary-Layer Meteorol, DOI: 10.1007/s10546-007-9189-2

Energy- and flux-budget (EFB) turbulence closure model for stably stratified flows. Part I: steady-state, homogeneous regimes

S. S. Zilitinkevich · T. Elperin · N. Kleeorin · I. Rogachevskii

Abstract We propose a new turbulence closure model based on the budget equations for the key second moments: turbulent kinetic and potential energies: TKE and TPE (comprising the turbulent total energy: TTE = TKE + TPE) and vertical turbulent fluxes of momentum and buoyancy (proportional to potential temperature). Besides the concept of TTE, we take into account the non-gradient correction to the traditional buoyancy flux formulation. The proposed model permits the existence of turbulence at any gradient Richardson number, Ri. Instead of the critical value of Richardson number separating—as is usually assumed—the turbulent and the laminar regimes, the suggested model reveals a transitional interval, $0.1 < \mathrm{Ri} < 1$, which separates two regimes of essentially different nature but both turbulent: strong turbulence at $\mathrm{Ri} \ll 1$; and weak turbulence, capable of transporting momentum but much less efficient in transporting heat, at $\mathrm{Ri} > 1$. Predictions from this model are consistent with available data from atmospheric and laboratory experiments, direct numerical simulation and large-eddy simulation.

Keywords Anisotropy · Critical Richardson number · Eddy viscosity · Heat conductivity · Kinetic, potential and total turbulent energies · Stable stratification · Turbulence closure · Turbulent fluxes · Turbulent length scale

---

S. S. Zilitinkevich (✉)
Division of Atmospheric Sciences, University of Helsinki, Helsinki, Finland
e-mail: sergej.zilitinkevich@fmi.fi

S. S. Zilitinkevich
Finnish Meteorological Institute, Helsinki, Finland

S. S. Zilitinkevich
Nansen Environmental and Remote Sensing Centre, Bjerknes Centre for Climate Research, Bergen, Norway

T. Elperin · N. Kleeorin · I. Rogachevskii
Pearlstone Center for Aeronautical Engineering Studies, Department of Mechanical Engineering, Ben-Gurion University of the Negev, Beer-Sheva, Israel

Atmospheric Boundary Layers. A. Baklanov & B. Grisogono (eds.),
doi: 10.1007/978-0-387-74321-9_3,

1 Introduction

Most of practically-used turbulence closure models are based on the concept of downgradient transport. Accordingly the models express turbulent fluxes of momentum and scalars as products of the mean gradient of the transported property and the corresponding turbulent transport coefficient (eddy viscosity, K_M, heat conductivity, K_H, or diffusivity, K_D). Following Kolmogorov (1941), turbulent transport coefficients are taken to be proportional to the turbulent velocity scale, u_T, and length scale, l_T:

$$K_M \sim K_H \sim K_D \sim u_T l_T. \tag{1}$$

Usually u_T^2 is identified with the turbulent kinetic energy (TKE) per unit mass, E_K, and is calculated from the TKE budget equation using the Kolmogorov closure for the TKE dissipation rate:

$$\varepsilon_K \sim E_K/t_T, \tag{2}$$

where $t_T \sim l_T/u_T$ is the turbulent dissipation time scale. This approach is justified when it is applied to neutral stability flows, where l_T can be taken to be proportional to the distance from the nearest wall.

However, this method encounters difficulties in stratified flows (both stable and unstable). The turbulent Prandtl number $\mathrm{Pr}_T = K_M/K_H$ exhibits essential dependence on the stratification and cannot be considered as constant. Furthermore, as follows from the budget equations for the vertical turbulent fluxes, the velocity scale u_T, which characterises the vertical turbulent transport, is determined as the root mean square (r.m.s.) vertical velocity $u_T \sim \sqrt{E_z}$ (where E_z is the energy of the vertical velocity fluctuations). In neutral stratification $E_z \sim E_K$, which is why the traditional equation $u_T \sim \sqrt{E_K}$ holds true. However, in strongly stable stratification this equation is insufficiently accurate because of the stability dependence of the anisotropy of turbulence $A_z \equiv E_z/E_K$, e.g., A_z generally decreases with increasing stability.

To reflect the effect of stratification, the turbulent length scales for the momentum, l_{TM}, and heat, l_{TH}, are taken to be unequal. As a result, the above-described closure scheme (formulated by Kolmogorov for neutral stratification and well-grounded only in this case) loses its constructiveness: the unsolved part of the problem is merely displaced from $\{K_M, K_H\}$ to $\{l_{TM}, l_{TH}\}$. In that case, the TKE budget equation becomes insufficient to determine additional unknown parameters.

Numerous alternative turbulence closures have been formulated using the budget equations for other turbulent parameters (in addition to the TKE) together with heuristic hypotheses and empirical relationships. However no consensus has been reached (see overviews by Weng and Taylor 2003; Umlauf and Burchard 2005).

In this study we analyse the effects of density stratification on turbulent energies and vertical turbulent fluxes in stably stratified atmospheric (or oceanic) boundary layers, in which the horizontal variations of the mean velocity and temperature are much weaker than the vertical variations. The proposed theory provides realistic stability dependencies of the turbulent Prandtl number, the vertical anisotropy, and the vertical turbulent length scale. Our work is presented in meteorological terms, but all the results can be easily reformulated in terms of water currents in oceans or lakes. In this case buoyancy is expressed through temperature and salinity instead of temperature and humidity.

We consider a minimal set of the budget equations for the second-order moments, namely equations for the vertical fluxes of buoyancy (proportional to the potential temperature) and momentum, the TKE and the turbulent potential energy (TPE), proportional to the mean squared potential temperature fluctuation. In these equations we account for some commonly neglected effects but leave a more detailed treatment of the third-order transports and the pressure–velocity correlations for future analysis. In particular, we advance the familiar "return to isotropy" model in order to more realistically determine the stability dependence of the vertical anisotropy, A_z. We also take into account a non-gradient correction to the traditional, downgradient formulation for the turbulent flux of potential temperature. This approach allows us to derive a reasonably simple turbulence closure scheme including realistic energy budgets and stability dependence of Pr_T.

We consider the total (kinetic + potential) turbulent energy (TTE), derive the TTE budget equation, and demonstrate that the TTE in stably stratified sheared flows does not completely decay even in very strong static stability. This conclusion, which is deduced from the general equations independently of the concrete formulation for the turbulent length scale, refutes the widely accepted concept of the critical Richardson number.

For the reader's convenience we recall that the gradient Richardson number, Ri, is defined as the squared ratio of the Brunt–Väisälä frequency, N, to the velocity shear, S:

$$\mathrm{Ri} = \left(\frac{N}{S}\right)^2, \tag{3a}$$

$$S^2 = \left(\frac{\partial U}{\partial z}\right)^2 + \left(\frac{\partial V}{\partial z}\right)^2, \tag{3b}$$

$$N^2 = \beta \frac{\partial \Theta}{\partial z}, \tag{3c}$$

where z is the vertical coordinate, U and V are the mean velocity components along the horizontal axes x and y, Θ is the mean potential temperature, $\beta = g/T_0$ is the buoyancy parameter, $g = 9.81\,\mathrm{m\,s^{-1}}$ is the acceleration due to gravity, and T_0 is a reference absolute temperature. As originally proposed by Richardson (1920), the Richardson number quantifies the effect of static stability on turbulence. Subsequent researches in the theory of stably stratified turbulent flows focussed on the question whether or not stationary turbulence can be maintained by the velocity shear at very large Richardson numbers.

A widely accepted opinion is that turbulence decays when Ri exceeds some critical value, $\mathrm{Ri_c}$ (with the frequently quoted estimate of $\mathrm{Ri_c} = 0.25$). However, the concept of a critical Ri was neither rigorously derived from basic physical principles nor demonstrated empirically; indeed, it contradicts long standing experimental evidence.

It is worth emphasizing that turbulence closure models based on the straightforward application of the TKE budget equation and Kolmogorov's closure hypotheses, Eqs. (1) and (2), imply the existence of Ri_c. In practical atmospheric modelling these closures are not acceptable. In particular, they lead to unrealistic decoupling of the atmosphere from the underlying surface when the Richardson number in the surface layer exceeds Ri_c. Since the milestone study of Mellor and Yamada (1974), in order to prevent the undesirable appearance of Ri_c, turbulence closures in practical use have been equipped with correction coefficients specifying the ratios $K_M(u_T l_T)^{-1}$ and $K_H(u_T l_T)^{-1}$ in the form of different single-valued functions of Ri. Usually these functions are not derived in the context of the closure in use, but are either determined empirically or taken from independent theories. Using this approach, modellers ignore the fact that corrections could be inconsistent with the formalism of the basic closure model.

2 Reynolds equations and budget equations for second moments

We consider atmospheric flows in which typical variations of the mean wind velocity $\mathbf{U} = (U_1, U_2, U_3) = (U, V, W)$ and potential temperature Θ (or virtual potential temperature involving specific humidity) in the vertical direction [along the x_3 (or z) axis] are much larger than in the horizontal direction [along the x_1, x_2 (or x, y) axes], so that the terms proportional to their horizontal gradients in the budget equations for turbulent statistics can be neglected. Θ is defined as $\Theta = T(P_0/P)^{1-1/\gamma}$, where T is the absolute temperature, P is the pressure, P_0 is its reference value, and $\gamma = c_p/c_v = 1.41$ is the ratio of specific heats.

We also assume that the vertical scale of motions (which is limited to the height scale of the atmosphere or the ocean, $H \sim 10^4$ m) is much smaller than the horizontal scale, so that the mean flow vertical velocity is typically much smaller than the horizontal velocity. In this context, to close the Reynolds equations we need only the vertical component, F_z, of the potential temperature flux, F_i, and two components of the Reynolds stresses, τ_{ij}, that represent the vertical turbulent flux of momentum: τ_{13} and τ_{23}.

The mean flow is determined by the momentum equations:

$$\frac{\mathrm{D}U_1}{\mathrm{D}t} = fU_2 - \frac{1}{\rho_0}\frac{\partial P}{\partial x} - \frac{\partial \tau_{13}}{\partial z}, \tag{4}$$

$$\frac{\mathrm{D}U_2}{\mathrm{D}t} = -fU_1 - \frac{1}{\rho_0}\frac{\partial P}{\partial y} - \frac{\partial \tau_{23}}{\partial z}, \tag{5}$$

and the thermodynamic energy equation:

$$\frac{\mathrm{D}\Theta}{\mathrm{D}t} = -\frac{\partial F_z}{\partial z} + J, \tag{6}$$

where $\mathrm{D}/\mathrm{D}t = \partial/\partial t + U_k \partial/\partial x_k$, $\tau_{ij} = \langle u_i u_j \rangle$, $F_i = \langle u_i \theta \rangle$, t is the time, $f = 2\Omega \sin\varphi$, with Ω_i the earth's rotation vector parallel to the polar axis ($|\Omega_i| \equiv \Omega = 0.76 \times 10^{-4}\,\mathrm{s}^{-1}$), φ is the latitude, ρ_0 is the mean density, J is the heating/cooling rate ($J = 0$ in adiabatic processes), P is the mean pressure, $\mathbf{u} = (u_1, u_2, u_3) = (u, v, w)$ and θ are the velocity and potential–temperature fluctuations. The angle brackets denote the ensemble average [see Holton (2004) or Kraus and Businger (1994)].

The budget equations for the TKE, $E_K = \frac{1}{2}\langle u_i u_i \rangle$, the "energy" of the potential temperature fluctuations, $E_\theta = \frac{1}{2}\langle \theta^2 \rangle$, the potential temperature flux, $F_i = \langle u_i\, \theta \rangle$ [with the vertical component $F_3 = F_z = \langle w\theta \rangle$], and the Reynolds stress $\tau_{ij} = \langle u_i u_j \rangle$ [with the components $\tau_{i3} = \langle u_i w \rangle$ ($i = 1, 2$) representing the vertical flux of momentum] read (see, e.g., Kaimal and Finnigan (1994), Kurbatsky (2000) and Cheng et al. (2002)):

$$\frac{\mathrm{D}E_K}{\mathrm{D}t} + \nabla \cdot \mathbf{\Phi}_K = -\tau_{ij}\frac{\partial U_i}{\partial x_j} + \beta F_z - \varepsilon_K \tag{7a}$$

or approximately

$$\frac{\mathrm{D}E_K}{\mathrm{D}t} + \frac{\partial \Phi_K}{\partial z} \approx -\tau_{i3}\frac{\partial U_i}{\partial z} + \beta F_z - \varepsilon_K, \tag{7b}$$

$$\frac{\mathrm{D}E_\theta}{\mathrm{D}t} + \nabla \cdot \mathbf{\Phi}_\theta = -F_z\frac{\partial \Theta}{\partial z} - \varepsilon_\theta, \tag{8a}$$

or approximately

$$\frac{\mathrm{D}E_\theta}{\mathrm{D}t} + \frac{\partial \Phi_\theta}{\partial z} = -F_z \frac{\partial \Theta}{\partial z} - \varepsilon_\theta, \tag{8b}$$

$$\frac{\mathrm{D}F_i}{\mathrm{D}t} + \frac{\partial}{\partial x_j}\Phi_{ij}^{(F)} = \beta_i \left\langle \theta^2 \right\rangle + \frac{1}{\rho_0}\left\langle \theta \nabla_i p \right\rangle - \tau_{ij}\frac{\partial \Theta}{\partial z}\delta_{j3} - F_j \frac{\partial U_i}{\partial x_j} - \varepsilon_i^{(F)}, \tag{9a}$$

and for $F_3 = F_z$

$$\begin{aligned} \tfrac{\mathrm{D}F_z}{\mathrm{D}t} + \tfrac{\partial}{\partial z}\Phi_F &= \beta \left\langle \theta^2 \right\rangle + \tfrac{1}{\rho_0}\left\langle \theta \tfrac{\partial}{\partial z} p \right\rangle - \left\langle w^2 \right\rangle \tfrac{\partial \Theta}{\partial z} - \varepsilon_z^{(F)} \\ &\approx C_\theta \beta \left\langle \theta^2 \right\rangle - \left\langle w^2 \right\rangle \tfrac{\partial \Theta}{\partial z} - \varepsilon_z^{(F)}, \end{aligned} \tag{9b}$$

$$\frac{\mathrm{D}\tau_{ij}}{\mathrm{D}t} + \frac{\partial}{\partial x_k}\Phi_{ijk}^{(\tau)} = -\tau_{ik}\frac{\partial U_j}{\partial x_k} - \tau_{jk}\frac{\partial U_i}{\partial x_k} + \left[\beta(F_j \delta_{i3} + F_i \delta_{j3}) + Q_{ij} - \varepsilon_{ij}^{(\tau)}\right] \tag{10a}$$

and for τ_{i3} (i = 1, 2)

$$\frac{\mathrm{D}\tau_{i3}}{\mathrm{D}t} + \frac{\partial}{\partial z}\Phi_i^{(\tau)} = -\left\langle w^2 \right\rangle \frac{\partial U_i}{\partial z} - \left[-\beta F_i - Q_{i3} + \varepsilon_{i3}^{(\tau)}\right] \approx -\langle w^2 \rangle \frac{\partial U_i}{\partial z} - \varepsilon_{i3}, \tag{10b}$$

where $\beta_i = \beta e_i$ and $\mathbf{e}$ is the vertical unit vector, $F_i = \langle u_i \, \theta \rangle$ $(i = 1, 2)$ are the horizontal fluxes of potential temperature, $-\tau_{ij}\partial U_i/\partial x_j$ is the TKE production rate, and δ_{ij} is the unit tensor (δ_{ij}= 1 for $i = j$ and δ_{ij}= 0 for $i \neq j$).

Here, $\mathbf{\Phi}_K$, $\mathbf{\Phi}_\theta$, etc. are the third-order moments representing the turbulent transports of the TKE and the "energy" of potential temperature fluctuations:

$$\mathbf{\Phi}_K = \tfrac{1}{\rho_0}\left\langle p\, \mathbf{u} \right\rangle + \tfrac{1}{2}\left\langle u^2\, \mathbf{u} \right\rangle, \quad \text{that is } \Phi_K = \tfrac{1}{\rho_0}\left\langle p\, w \right\rangle + \tfrac{1}{2}\left\langle u^2\, w \right\rangle, \tag{11a}$$

$$\mathbf{\Phi}_\theta = \tfrac{1}{2}\left\langle \theta^2\, \mathbf{u} \right\rangle, \quad \text{that is } \Phi_\theta = \tfrac{1}{2}\left\langle \theta^2\, w \right\rangle, \tag{11b}$$

and the turbulent transports of the fluxes of potential temperature and momentum:

$$\Phi_{ij}^{(F)} = \frac{1}{2\rho_0}\left\langle p\, \theta \right\rangle \delta_{ij} + \left\langle u_i\, u_j\, \theta \right\rangle, \quad \Phi_{33}^{(F)} = \Phi_F = \frac{1}{2\rho_0}\left\langle p\, \theta \right\rangle + \left\langle w^2\, \theta \right\rangle, \tag{12}$$

$$\Phi_{ijk}^{(\tau)} = \left\langle u_i u_j u_k \right\rangle + \frac{1}{\rho_0}\left(\left\langle p u_i \right\rangle \delta_{jk} + \left\langle p u_j \right\rangle \delta_{ik}\right), \tag{13a}$$

$$\Phi_{i33}^{(\tau)} = \Phi_i^{(\tau)} = \left\langle u_i w^2 \right\rangle + \frac{1}{\rho_0}\left\langle p u_i \right\rangle, \ (i = 1, 2). \tag{13b}$$

Q_{ij} are correlations between the fluctuations of pressure, p, and the velocity shears:

$$Q_{ij} = \frac{1}{\rho_0}\left\langle p \left(\frac{\partial u_i}{\partial x_j} + \frac{\partial u_j}{\partial x_i} \right) \right\rangle. \tag{14}$$

In the above equations, ε_k, $\varepsilon_{ij}^{(\tau)}$, ε_θ and $\varepsilon_i^{(F)}$ are operators including the molecular transport coefficients:

$$\varepsilon_K = \nu \left\langle \frac{\partial u_i}{\partial x_k}\frac{\partial u_i}{\partial x_k} \right\rangle, \quad \varepsilon_{ij}^{(\tau)} = 2\nu \left\langle \frac{\partial u_i}{\partial x_k}\frac{\partial u_j}{\partial x_k} \right\rangle, \tag{15a}$$

$$\varepsilon_\theta = -\kappa \left\langle \theta\, \Delta\, \theta \right\rangle, \quad \varepsilon_i^{(F)} = -\kappa \left(\left\langle u_i\, \Delta\, \theta \right\rangle + \mathrm{Pr} \left\langle \theta\, \Delta\, u_i \right\rangle \right), \tag{15b}$$

where ν is the kinematic viscosity, κ is the temperature conductivity, and $\Pr = \nu/\kappa$ is the Prandtl number. Of these terms, ε_K, $\varepsilon_{ii}^{(\tau)}$ (that is the diagonal elements $\varepsilon_{11}^{(\tau)}$, $\varepsilon_{22}^{(\tau)}$, $\varepsilon_{33}^{(\tau)}$), ε_θ and $\varepsilon_i^{(F)}$ are essentially positive and represent the dissipation rates for E_K, τ_{ii}, E_θ and $F_i^{(F)}$, respectively. Following Kolmogorov (1941), they are taken to be proportional to the ratios of the dissipating statistical moment to the turbulent dissipation time scale, t_T:

$$\varepsilon_K = \frac{E_K}{C_K t_T}, \quad \varepsilon_{ii}^{(\tau)} = \frac{\tau_{ii}}{C_K t_T}, \quad \varepsilon_\theta = \frac{E_\theta}{C_P t_T}, \quad \varepsilon_i^{(F)} = \frac{F_i}{C_F t_T}, \tag{16}$$

where C_K, C_P and C_F are dimensionless constants.

The physical mechanisms of dissipation of the non-diagonal components of the Reynolds stress, $\tau_{ij} (i \neq j)$, are more complicated. The terms $\varepsilon_{ij}^{(\tau)} = 2\nu \left\langle \frac{\partial u_i}{\partial x_k} \frac{\partial u_j}{\partial x_k} \right\rangle$ in Eq. (10b) are comparatively small and are not even necessarily positive, whereas the dissipative role is to a large extent performed by the pressure-shear correlations and the horizontal turbulent transport of the potential temperature. Moreover, our analysis does not account for the vertical transport of momentum (that is for the contribution to τ_{i3}) due to internal gravity waves [see, e.g., Sect. 9.4 in Holton (2004)]. Leaving the detailed analyses of the τ_{i3} budget for future work, we now introduce the following "effective dissipation rate" for the Reynolds stress:

$$\varepsilon_{i3(\text{eff})} \equiv \varepsilon_{i3}^{(\tau)} - \beta F_i - Q_{i3} + (\text{unaccounted factors}), \quad i = 1, 2; \tag{17}$$

and apply to it the Kolmogorov closure hypothesis whereby $\varepsilon_{i3(\text{eff})} \sim \tau_{i3}/t_\tau$, and t_τ is an "effective dissipation time scale" [the term $\varepsilon_{i3}^{(\tau)}$ is estimated as $\varepsilon_{i3}^{(\tau)} \sim O(Re^{-1/2})$ and can be neglected]. Accounting for the difference between t_τ and the Kolmogorov dissipation time scale, t_T [see Eq. (16)], our effective dissipation rates become

$$\varepsilon_{i3(\text{eff})} = \frac{\tau_{i3}}{\Psi_\tau t_T}, \tag{18}$$

where $\Psi_\tau = t_\tau / t_T$ is an empirical dimensionless coefficient. There are no grounds *a prior* to assume that this coefficient is constant. Coefficient Ψ_τ can depend on the static stability but is neither zero nor infinite, and it is also conceivable that this stability dependence is monotonic.

In further analysis we employ the approximate version of Eq. (9b). As shown in Appendix A, the second term on the r.h.s. of Eq. (9b), namely $\rho_0^{-1} \langle \theta \partial p/\partial z \rangle$, is essentially negative and scales as $\beta \langle \theta^2 \rangle$. On these grounds, in its approximate version the sum $\beta \langle \theta^2 \rangle + \rho_0^{-1} \langle \theta \partial p/\partial z \rangle$ is replaced by $C_\theta \beta \langle \theta^2 \rangle$, where $C_\theta < 1$ is an empirical dimensionless constant.

3 Turbulent energies

We first consider the concept of turbulent potential energy (TPE). Using the state equation and the hydrostatic equation, the density and the buoyancy in the atmosphere are expressed through potential temperature, θ, and specific humidity, q (in the ocean, through θ and salinity, s). These variables are adiabatic invariants that are conserved in the vertically displaced portions of fluid, so that the density is also conserved. This allows us to determine density fluctuations, $\rho' = (\partial \rho / \partial z) \delta z$, and the fluctuations of potential energy per unit mass:

$$\delta E_P = \frac{g}{\rho_0} \int_z^{z+\delta z} \rho' \, dz = \frac{1}{2} \frac{b'^2}{N^2}. \tag{19}$$

Let us consider the thermally stratified atmosphere, where the buoyancy, b, is expressed through the potential temperature, $b = \beta\theta$. Consequently the TPE is proportional to the energy of the potential temperature fluctuations:

$$E_P = \left(\frac{\beta}{N}\right)^2 E_\theta = \frac{1}{2}\left(\frac{\beta}{N}\right)^2 \langle\theta^2\rangle. \tag{20}$$

Then by multiplying Eq. (8b) by $(\beta/N)^2 = (\partial\Theta/\partial z)^{-1}$ and assuming that N changes only slowly compared to turbulent variations we arrive at the following TPE budget equation[1]:

$$\frac{\mathrm{D}E_P}{\mathrm{D}t} + \frac{\partial}{\partial z}\Phi_P = -\beta F_z - \varepsilon_P = -\beta F_z - \frac{E_P}{C_P t_T}, \tag{21}$$

where $\Phi_P = (\beta/N)^2\Phi_\theta$ and $\varepsilon_P = (\beta/N)^2\varepsilon_\theta$. The term βF_z appears in Eqs. (7b) and (21) with opposite signs and describes the energy exchange between TKE and TPE.

The sum of the TKE and TPE is simply the total turbulent energy (TTE):

$$E = E_K + E_P = \frac{1}{2}\left(\langle\mathbf{u}^2\rangle + \left(\frac{\beta}{N}\right)^2\langle\theta^2\rangle\right), \tag{22}$$

and the TTE budget equation is immediately derived by summing up Eqs. (7b) and (21). Generally speaking, the time-scale constants C_K and C_P in Eq. (16), which characterise the kinetic and the potential energy dissipation rates, can differ. Here, for simplicity, we use $C_K = C_P$. Then the TTE budget equation becomes

$$\frac{DE}{Dt} + \frac{\partial}{\partial z}\Phi_T = -\tau_{i3}\frac{\partial U_i}{\partial z} - \frac{E}{C_K t_T}, \tag{23}$$

where $\Phi_T = \Phi_K + \Phi_P$ is the TTE vertical flux.

In the steady state, Eq. (23) reduces to a simple balance between the TTE production = τS (where $\tau^2 = \tau_{13}^2 + \tau_{23}^2$) and the TTE dissipation $\sim E t_T^{-1}$, which yields $E \sim \tau S t_T$. In Section 5 we demonstrate that for a very large Ri the ratios τ/E, E_K/E and E_z/E_K tend to become non-zero constants. In that case estimating t_T through the turbulent length scale, l_z, as $t_T \sim l_z E_z^{-1/2} \sim l_z E^{-1/2}$ yields an asymptotic large-Ri estimate, $E \sim (l_z S)^2 > 0$. This reasoning does not allow the existence of the critical Richardson number.

As a matter of interest, traditional analyses of the turbulent energy have been basically limited to using TKE budget, Eq. (7b). Equation (8b) for the squared potential temperature fluctuations, although it is well-known for decades, has been ignored in the operationally used turbulent closure models. Only rather recently, E_θ has been treated in terms of the TPE, see Dalaudier and Sidi (1987), Hunt et al. (1988), Canuto and Minotti (1993), Schumann and Gerz (1995), Hanazaki and Hunt (1996,2004), Keller and van Atta (2000), Stretch et al. (2001), Canuto et al. (2001), Cheng et al. (2002), Luyten et al. (2002, p. 257), Jin et al. (2003), Umlauf (2005) and Rehmann and Hwang (2005). Zilitinkevich (2002) employed the TKE and the TPE budget equations on equal terms to derive an energetically consistent turbulent closure model, avoiding the traditional hypothesis $K_H \sim K_M \sim E_K t_T$ (which leads to a dead end, at least in stable stratification). All three budgets, for TKE, TPE and TTE have been considered by Canuto and Minotti (1993) and Elperin et al. (2002).

[1] Alternatively the TPE budget equation can be derived from the equation for the fluctuation of buoyancy, b, namely, by multiplying this equation by bN^{-2}, and then applying statistical averaging. It follows then that Eq. (21) holds true independently of the assumption that N changes slowly.

4 Local model for the steady-state, homogeneous regime

4.1 Anisotropy of turbulence

In this section we consider the equilibrium turbulence regime and neglect the third-order transport terms, so that the left-hand sides (l.h.s.) in all budget equations become zero. We limit our analysis to boundary-layer type flows, in which the horizontal gradients of the mean velocity and temperature are negligibly small. For these conditions the TKE production rate becomes

$$\Pi = -\boldsymbol{\tau} \cdot \frac{\partial \mathbf{U}}{\partial z} = \tau S, \tag{24}$$

where $\boldsymbol{\tau} = (\tau_{xz}, \tau_{yz}, 0)$, and $\tau \equiv |\boldsymbol{\tau}|$. It goes without saying that Π is determined differently in other types of turbulent flows, in particular in the wave boundary layer below the ocean surface or in the capping inversion layer above the long-lived atmospheric stable boundary layer, where the TKE is at least partially produced by the breaking of surface waves in water or internal gravity waves in the atmosphere. Note that in the laboratory conditions these mechanisms are similar to the oscillating-grid generation of turbulence rather than to turbulence generation by shear.

Taking $C_P = C_K$ [see discussion of Eq. (23) in Sect. 3], Eqs. (19)–(23) yield the following expressions for the turbulent energies:

$$E = C_K t_T \Pi, \tag{25a}$$
$$E_P = -C_K t_T \beta F_z = E \mathrm{Ri}_f, \tag{25b}$$
$$E_K = C_K t_T (\Pi + \beta F_z) \equiv C_K t_T \Pi (1 - \mathrm{Ri}_f) = E(1 - \mathrm{Ri}_f), \tag{25c}$$

where Ri_f is the familiar flux Richardson number defined as the ratio of the TKE consumption needed for overtaking buoyancy forces to the TKE production by the velocity shear:

$$\mathrm{Ri}_f \equiv -\frac{\beta F_z}{\Pi} = \frac{\mathrm{Ri}}{\mathrm{Pr}_T} = \frac{E_P}{E}. \tag{26}$$

The above analysis implies that Ri_f is then the ratio of TPE to TTE, a fact that has been overlooked until recently.[2] Ri_f is equal to zero in neutral stratification, monotonically increases with increasing stability, but obviously cannot exceed unity. Hence, for very strong static stability (at $\mathrm{Ri} \to \infty$) it must approach a non-zero, positive limit, $\mathrm{Ri}_f^\infty < 1$. This conclusion by no means supports the existence of the critical gradient Richardson number. Indeed, Ri_f is an internal parameter that is controlled by turbulence in contrast to $\mathrm{Ri} = (\beta \partial \Theta / \partial z)/(\partial \mathbf{U}/\partial z)^2$, which is an "external" parameter that characterises the mean flow.

It is worth recalling that the key parameter characterising vertical turbulent transports is the TKE of the vertical velocity fluctuations, $E_z = \frac{1}{2}\langle w^2 \rangle$, rather than the full TKE. In order to determine E_z, we need to consider all three budget equations (10a) for the diagonal Reynolds stresses, $\tau_{11} = 2E_1 = 2E_x = \langle u^2 \rangle$, $\tau_{22} = 2E_2 = 2E_y = \langle v^2 \rangle$ and $\tau_{33} = 2E_3 = 2E_z = \langle w^2 \rangle$. In the steady state these budget equations become

$$\frac{E_i}{C_K t_T} = -\tau_{i3} \frac{\partial U_i}{\partial z} + \frac{1}{2} Q_{ii}, \quad i = 1, 2, \tag{27a}$$
$$\frac{E_z}{C_K t_T} = \frac{E_3}{C_K t_T} = \beta F_z + \frac{1}{2} Q_{33}. \tag{27b}$$

[2] Taking into account that C_P and C_K can differ, Ri_f is proportional rather than equal to E_P/E.

The sum of the pressure-velocity shear correlation terms, $\sum Q_{ii} = \sum \rho_0^{-1} \langle p\partial u_i/\partial x_i\rangle$, is zero because of the continuity equation, $\sum \partial u_i/\partial x_i = 0$. Hence, they are neither productive nor dissipative; they simply describe the conversion of the energy of "rich" components into the energy of "poorer" components.

In order to determine Q_{11}, Q_{22} and Q_{33}, we generalize the familiar "return-to-isotropy" hypothesis as follows:

$$Q_{11} = -\frac{2C_r}{3C_K t_T}(3E_1 - E_K\Psi_1), \tag{28a}$$

$$Q_{22} = -\frac{2C_r}{3C_K t_T}(3E_2 - E_K\Psi_2), \tag{28b}$$

$$Q_{33} = -\frac{2C_r}{3C_K t_T}(3E_3 - E_K\Psi_3). \tag{28c}$$

Here, C_r and Ψ_i (i = 1, 2, 3) are dimensionless empirical coefficients; C_r accounts for the difference between the relaxation-time and the dissipation-time scales (as a first approximation, we take these two time scales to be proportional, $t_r \sim t_T$, so that $C_r = t_r/t_T =$ constant); Ψ_i govern redistribution of TKE between the components. When $\Psi_i = 1$ the above relations reduce to their original form (Rotta 1951) and are known to be a good approximation for neutrally stratified flows. In stable stratification, we need to leave room for their possible stability dependence. As a first approximation, we assume

$$\Psi_i = 1 + C_i \mathrm{Ri}_f, \quad i = 1, 2, 3, \tag{29}$$

where Ri_f is the flux Richardson number, and C_i are empirical constants. Their sum must be zero, $C_1 + C_2 + C_3 = 0$, in order to satisfy the condition $\sum Q_{ii} = 0$ (which is needed to guarantee that $E_K = E_1 + E_2 + E_3$). Linear functions of Ri_f on the r.h.s. of Eq. (29) are taken as simple approximations providing the only possible (from the physical point of view) finite, non-zero limits: $\Psi_i = 1$ at Ri = 0, and $\Psi_i \to 1 + C_i \mathrm{Ri}_f^{\infty}$ at $\mathrm{Ri} \to \infty$.

Because the energy exchange between the horizontal components of TKE, E_1 and E_2, is not directly affected by the stable stratification, we take the first two energy-exchange constants to be equal, $C_1 = C_2$. Then, the condition $C_1 + C_2 + C_3 = 0$ implies that only one of the three constants is independent and $C_1 = C_2 = -\frac{1}{2}C_3$.

Equations (27)–(28) yield

$$E_i = \frac{C_r}{3(1+C_r)}E_K\Psi_i - \frac{C_K}{1+C_r}t_T\tau_{i3}\frac{\partial U_i}{\partial z}, \quad i = 1, 2, \tag{30a}$$

$$E_z = \frac{C_r}{3(1+C_r)}E_K\Psi_3 + \frac{C_K}{1+C_r}t_T\beta F_z. \tag{30b}$$

In the plain-parallel neutral boundary layer with $\mathbf{U} = (U,\ 0, 0)$, Eqs. (30a) and (30b) reduce to

$$\frac{E_x}{E_K} = \frac{3+C_r}{3(1+C_r)}, \tag{31a}$$

$$\frac{E_y}{E_K} = \frac{E_z}{E_K} = \frac{C_r}{3(1+C_r)}. \tag{31b}$$

Given the vertical component of TKE, E_z, the turbulent dissipation time scale, $t_T = l_T E_K^{-1/2}$, can alternatively be expressed through the vertical turbulent length scale l_z:

$$t_T = \frac{l_z}{E_z^{1/2}}. \tag{32}$$

Then eliminating t_T from Eq. (25c) and Eq. (30b), and substituting Eq. (29) for Ψ_3 yields

$$E_z = \left[\frac{C_K C_r \Psi_3}{3(1+C_r)}\left(\Pi + \left(\frac{3}{C_r\Psi_3}+1\right)\beta F_z\right) l_z\right]^{2/3}, \quad (33a)$$

$$\Psi_3 = 1 + C_3 \mathrm{Ri}_f. \quad (33b)$$

This formulation recovers the traditional return-to-isotropy formulation when $C_3 = 0$.

In order to close the system, the horizontal components of the TKE, E_x and E_y, are not required. We leave the discussion of these components to a separate paper, in which our closure is extended to passive scalars and applied to turbulent diffusion.

4.2 Vertical turbulent fluxes of momentum and potential temperature

Of the non-diagonal Reynolds stresses we consider only those representing the vertical fluxes of momentum $\tau_{13} = \tau_{xz} = \langle uw \rangle$ and $\tau_{23} = \tau_{yz} = \langle vw \rangle$ which are needed to close the momentum equations (4)–(5) and are determined by Eq. (10b). In the steady state, using Eqs. (17)–(18) for the effective Reynolds-stress dissipation rate, we obtain the following relation for the non-diagonal Reynolds stresses:

$$\tau_{i3} = -2\Psi_\tau E_z^{1/2} l_z \frac{\partial U_i}{\partial z}. \quad (34)$$

Likewise, of the three components of the potential-temperature flux, we consider only the vertical flux $F_3 = F_z$ that is needed to close the thermodynamic energy Equation (6). The vertical flux F_z is determined by Eq. (9b). Taking $\beta E_\theta = (N^2/\beta)E_P = -C_K N^2 l_z F_z / E_z^{1/2}$ [after Eqs. (25b) and (32)], the steady-state version of Eq. (9b) becomes

$$F_z = -\frac{2C_F E_z^{1/2} l_z}{1 + 2C_\theta C_F C_K (N l_z)^2 E_z^{-1}}\left(\frac{\partial \Theta}{\partial z}\right). \quad (35)$$

Substituting here $N^2 = \beta \partial\Theta/\partial z$ shows that F_z depends on $\partial\Theta/\partial z$ weaker than linearly and at $\partial\Theta/\partial z \to \infty$ tends to a finite limit:

$$F_{z,\max} = -\frac{E_z^{3/2}}{C_\theta C_K \beta l_z}. \quad (36)$$

It follows then that F_z in a turbulent flow cannot be considered as a given external parameter. This conclusion is consistent with our reasoning in Sect. 4.1 that the flux Richardson number $\mathrm{Ri}_f = -\beta F_z (\tau S)^{-1}$ is an internal parameter of turbulence that cannot be arbitrarily prescribed. According to Eq. (36), the maximum value of the buoyancy flux βF_z, in the strong stability limit, is proportional to the dissipation rate, $E_z^{3/2} l_z^{-1}$, of the energy of vertical velocity fluctuations.[3]

Equations (34) and (35) allow us to determine the eddy viscosity and conductivity:

$$K_M \equiv \frac{-\tau_{i3}}{\partial U_i/\partial z} = 2\Psi_\tau E_z^{1/2} l_z, \quad (37a)$$

$$K_H \equiv \frac{-F_z}{\partial\Theta/\partial z} = \frac{2C_F E_z^{1/2} l_z}{1 + 2C_\theta C_F C_K (N l_z)^2 E_z^{-1}}. \quad (37b)$$

[3] A principally similar analysis of the budget equation for F_z has been performed by Cheng et al. (2002). Their Eq. (15i) implies the same maximum value of F_z as our Eq. (36). It worth noting that Eq. (35) imposes an upper limit on the downward heat flux in the deep ocean (which is known to be a controlling factor of the rate of the global warming).

Consequently, the Kolmogorov closure hypothesis applied to the effective Reynolds-stress dissipation rate, Eqs. (17)–(18), yields the eddy-viscosity formulation, Eq. (37a), basically similar to the traditional formulation, Eq. (1), whereas Eq. (37b) for eddy conductivity differs essentially from this formulation.

It may appear that our derivation of Eq. (37a) essentially depends on the hypothetical concept of the effective dissipation rate, Eqs. (17)–(18). Actually we employ this merely for the reader's convenience, to avoid overly complex derivations. Principally, the same result, namely the downgradient momentum-flux formulation equivalent to Eqs. (34) and (37a), follows from analyses of the budget equations for the Reynolds stresses in the k-space using the familiar "τ-approximation" (see, e.g., Elperin et al. 2002, 2006).

Recall now that Ψ_τ is a dimensionless, non-zero, bounded coefficient that can only monotonically depend on the static stability [see Eqs. (17)–(18) and their discussion in Sect. 2]. Let us approximate the stability dependence of Ψ_τ by the following linear function of the flux Richardson number, Ri_f:

$$\Psi_\tau = C_{\tau 1} + C_{\tau 2}\mathrm{Ri}_f, \tag{38}$$

where $C_{\tau 1}$ and $C_{\tau 2}$ are dimensionless constants to be determined empirically. Equation (38) provides the only physically meaningful, finite, non-zero limits, namely, $\Psi_\tau = C_{\tau 1}$ at $\mathrm{Ri} = 0$, and $\Psi_\tau \to C_{\tau 1} + C_{\tau 2}\mathrm{Ri}_f^\infty$ at $\mathrm{Ri} \to \infty$ [cf. our argument in support of Eq. (29)].

4.3 Turbulent Prandtl number and other dimensionless parameters

The system of Eq. (33a)–(35), although unclosed until we determine the vertical turbulent length scale l_z, reveals a "partial invariance" with respect to l_z and allows determining the turbulent Prandtl number, Pr_T, the flux Richardson number, Ri_f, and other dimensionless characteristics of turbulence in the form of universal functions of the gradient Richardson number, Ri. Obviously such universality is relevant only to the steady-state homogeneous regime. In non-steady, heterogeneous regimes, all these characteristics are not single-valued functions of Ri.

Recalling that $\Pi = K_M S^2$ and $\mathrm{Ri}_f \equiv -\beta F_z/\Pi$, Eqs. (33a) and (37a) give

$$\frac{E_z}{(Sl_z)^2} = \Psi(\mathrm{Ri}_f) \equiv \frac{2C_K C_r \Psi_3 \Psi_\tau}{3(1+C_r)}\left[1 - \left(\frac{3}{C_r\Psi_3} + 1\right)\mathrm{Ri}_f\right], \tag{39}$$

where Ψ_3 and Ψ_τ are linear functions of Ri_f given by Eqs. (33b) and (38). Then dividing K_M [determined by Eq. (37a)] by K_H [determined by Eq. (37b)] and expressing E_z through Eq. (39) yields the following surprisingly simple expressions:

$$\mathrm{Pr}_T \equiv \frac{K_M}{K_H} = \frac{\mathrm{Ri}}{\mathrm{Ri}_f} = \frac{\Psi_\tau}{C_F} + \frac{3(1+C_r)C_\theta}{C_r\Psi_3}\mathrm{Ri}\left[1 - \left(\frac{3}{C_r\Psi_3} + 1\right)\mathrm{Ri}_f\right]^{-1}, \tag{40}$$

and

$$\frac{1}{\mathrm{Ri}} = \frac{C_F\Psi_\tau^{-1}}{\mathrm{Ri}_f} - \frac{3C_F(1+C_r)C_\theta\Psi_\tau^{-1}}{C_r\Psi_3(1-\mathrm{Ri}_f) - 3\mathrm{Ri}_f}, \tag{41}$$

which do not include l_z. Equation (41) together with Eqs. (33b) and (38) specify Ri as a single-valued, monotonically increasing function of Ri_f determined in the interval $0 < \mathrm{Ri}_f < \mathrm{Ri}_f^\infty$, where Ri_f^∞ is given by Eq. (45). Therefore, the inverse function, namely,

$$\mathrm{Ri}_f = \Phi(\mathrm{Ri}), \tag{42}$$

is a monotonically increasing function of Ri, changing from 0 at $\mathrm{Ri} = 0$ to Ri_f^∞ at $R_i \to \infty$.

According to the above equations, the Ri dependencies of Ri_f and Pr_T (which is also a monotonically increasing function of Ri) are characterised by the following asymptotic limits:

$$\mathrm{Pr}_T \approx \frac{\Psi_\tau^{(0)}}{C_F} + \left(\frac{3C_\theta(1+C_r)}{C_r} + \frac{C_{\tau 2}}{C_F}\right)\mathrm{Ri} \quad \to \mathrm{Pr}_T^{(0)} = \frac{\Psi_\tau^{(0)}}{C_F}, \tag{43a}$$

$$\mathrm{Ri}_f \approx \frac{C_F}{\Psi_\tau^{(0)}}\mathrm{Ri} \quad \text{at} \quad Ri \ll 1, \tag{43b}$$

$$\mathrm{Pr}_T \approx \frac{1}{\mathrm{Ri}_f^\infty}\mathrm{Ri}, \tag{44a}$$

$$\mathrm{Ri}_f \to \mathrm{Ri}_f^\infty \quad \text{at} \quad \mathrm{Ri} \gg 1, \tag{44b}$$

where

$$\mathrm{Ri}_f^\infty = \frac{C_r \Psi_3^\infty}{C_r \Psi_3^\infty + 3[1 + C_\theta(1+C_r)]}, \tag{45}$$

and the superscripts "(0)" and "∞" mean "at Ri = 0" and "at Ri$\to \infty$", respectively.

Equations (33a)–(35) allow us to determine, besides Pr_T, three other dimensionless parameters the vertical anisotropy of turbulence:

$$A_z \equiv \frac{E_z}{E_K} = \frac{C_r \Psi_3}{3(1+C_r)}\left[1 - \left(\frac{3}{C_r \Psi_3} + 1\right)\mathrm{Ri}_f\right](1-\mathrm{Ri}_f)^{-1}, \tag{46}$$

the squared ratio of the turbulent flux of momentum to the TKE (which characterises the correlation between vertical and horizontal velocity fluctuations):

$$\left(\frac{\tau}{E_K}\right)^2 = \frac{2\Psi_\tau A_z}{C_K(1-\mathrm{Ri}_f)}, \tag{47}$$

and the ratio of the squared vertical flux of potential temperature to the product of the TKE and the "energy" of the potential temperature fluctuations:

$$\frac{F_z^2}{E_K E_\theta} = \frac{2\Psi_\tau A_z}{C_K\, \mathrm{Pr}_T}. \tag{48}$$

Equations (41), (46)–(48) determine the Ri dependencies of A_z, $\tau^2 E_K^{-2}$ and $F_z^2 (E_K E_\theta)^{-2}$, which are characterised by the following asymptotic limits:

$$A_z \to A_z^{(0)} = \frac{C_r}{3(1+C_r)}, \tag{49a}$$

$$\left(\frac{\tau}{E_K}\right)^2 \to \frac{2\Psi_\tau^{(0)} A_z^{(0)}}{C_K}, \tag{49b}$$

$$\frac{F_z^2}{E_K E_\theta} \to \frac{2C_F A_z^{(0)}}{C_K} \quad \text{at} \quad \mathrm{Ri} \ll 1, \tag{49c}$$

$$A_z \to A_z^\infty = C_\theta Ri_f^\infty (1 - Ri_f^\infty)^{-1}, \tag{50a}$$

$$\left(\frac{\tau}{E_K}\right)^2 \to \frac{2\Psi_\tau^\infty A_z^\infty}{C_K(1 - \mathrm{Ri}_f^\infty)}, \tag{50b}$$

$$\frac{F_z^2}{E_K E_\theta} \to \frac{2\Psi_\tau^\infty A_z^\infty}{C_K \, \mathrm{Pr}_T^\infty} \quad \text{at} \quad \mathrm{Ri} \gg 1. \tag{50c}$$

It must be noted that the turbulent velocity scale in Eqs. (34)–(37a) is $\sqrt{E_z}$ rather than $\sqrt{E_K}$. However, in a number of currently used turbulence closure models the stability dependence of $A_z = E_z/E_K$ is neglected and $\sqrt{E_K}$ is taken as an ultimate velocity scale to characterise the vertical turbulent transports. This is done unfortunately without serious theoretical or experimental grounds. On the contrary, Eq. (46) implies an essential Ri dependence of A_z, which is in agreement with currently available data [see Mauritsen and Svensson (2007) and our data analysis in Sect. 5 below].

4.4 Vertical turbulent length scale

Two basic factors impose limits on the vertical turbulent length scale, l_z, in geophysical flows: the height over the surface (the geometric limit) and the stable stratification.

In neutral stratification, l_z is restricted by the geometric limit[4]:

$$l_z \sim z. \tag{51}$$

For the strong stable stratification limit, different formulations have been proposed. Monin and Obukhov (1954) proposed the following length scale widely used in boundary-layer meteorology:

$$L \equiv \frac{\tau^{3/2}}{-\beta F_z} = \frac{\tau^{1/2}}{S\mathrm{Ri}_f}. \tag{52}$$

Our local closure model is consistent with this limit: Eqs. (34), (39) and (52) yield

$$l_z = \frac{\mathrm{Ri}_f}{(2\Psi_\tau)^{1/2}\Psi^{1/4}} L. \tag{53}$$

Furthermore any interpolation formula for l_z linking the limits $l_z \sim z$ and $l_z \sim L$ should have the form

$$l_z = z\Psi_l(\mathrm{Ri}_f), \tag{54}$$

where Ψ_l is a function of Ri_f.

Well-known alternatives to L are the Ozmidov scale: $\varepsilon_K^{1/2} N^{3/2}$ (Ozmidov 1990); the local energy balance scale: $E_z^{1/2} N^{-1}$ (e.g., Table 3 in Cuxart et al. 2006); and the shear sheltering scale: $E_z^{1/2} S^{-1}$ (Hunt et al. 1985, 1988). Using our local closure equations (Sect. 4.1–4.3) the ratio of each of these scales to L can be expressed through a corresponding function of Ri_f. Hence, any interpolation linking the neutral stratification limit, $l_z \sim z$, with all the above limits will still have the same form as Eq. (54).

[4] In rotating fluids, the direct effect of the angular velocity, Ω, on turbulent eddies is characterised by the rotational limit, $E_z^{1/2}/\Omega$. In geophysical, stably stratified flows it has only a secondary importance. We leave the discussion of this effect for future work.

Consequently, Eq. (54) represents a general formulation for the vertical turbulent length scale in the steady-state, homogeneous, stably stratified flows. In other words, the stability dependence of l_z is fully characterised by the universal function $\Psi_l(\mathrm{Ri}_f)$. This function should satisfy the following physical requirements: in neutral stratification it attains the maximum value, $\Psi_l(0) = 1$ [the omitted empirical constant combines with the coefficients $C_K = C_P, C_F$ and Ψ_τ in Eqs. (16) and (18)], and with increasing Ri_f it should monotonically decrease. Finally, at $\mathrm{Ri}_f \to \mathrm{Ri}_f^\infty$ this function should tend to zero [otherwise Eq. (33a) would give $E_z > 0$ at $\mathrm{Ri}_f \to \mathrm{Ri}_f^\infty$, which is physically senseless].

We propose a simple approximation to satisfy these requirements: $\Psi_l = \left(1 - \mathrm{Ri}_f/\mathrm{Ri}_f^\infty\right)^n$, where n is a positive constant. Using an empirical value of $n = 4/3$ (see the next Section) we arrive at the following relation for l_z:

$$l_z = z\left(1 - \frac{\mathrm{Ri}_f}{\mathrm{Ri}_f^\infty}\right)^{4/3}. \tag{55}$$

Obviously, in non-steady, heterogeneous regimes l_z should be determined through a prognostic equation accounting for its advection and temporal evolution.

5 Comparison of the local model with experimental and numerically simulated data

To determine the empirical dimensionless constants $C_r, C_K, C_F, C_\theta, C_{\tau1}, C_{\tau2}, C_3$ and n we compare results from the local closure model presented in Sect. 4 with experimental, large-eddy simulation (LES) and direct numerical simulation (DNS) data.

As mentioned earlier, the local model is applied to homogeneous turbulence and does not include transports of turbulent energies and turbulent fluxes. At the same time practically all currently available data represent vertically (in a number of cases, both vertically and horizontally) heterogeneous flows, in which the above transports are more or less pronounced. In these conditions, fundamental dimensionless parameters of turbulence, such as Pr_T, Ri_f, $(\tau/E_K)^2$, $F_z^2/(E_K E_\theta)$ and A_z, can be only approximately considered as universal functions of Ri. Mauritsen and Svensson (2007) have demonstrated quite reasonable Ri dependencies of the above parameters based on datasets obtained in several recent field campaigns. To reduce inevitable deviations from universality and to more accurately determine empirical constants, we now more carefully select data and rule out those that represent strongly heterogeneous regimes.

Figures 1a, b show the turbulent Prandtl number, Pr_T, and flux Richardson number, $\mathrm{Ri}_f = \mathrm{Ri}/\mathrm{Pr}_T$, versus the gradient Richardson number, Ri. They demonstrate reasonable agreement between data from atmospheric and laboratory experiments, LES and DNS. Data for $\mathrm{Ri} \to 0$ in Fig. 1 are consistent with the commonly accepted empirical estimate of $\mathrm{Pr}_T^{(0)} \equiv \mathrm{Pr}_T|_{\mathrm{Ri}\to 0} = 0.8$ [see data collected by Churchill (2002) and Foken (2006) and the theoretical analysis of Elperin et al. (1996)]. Figure 1b clearly demonstrates that Ri_f at large Ri levels off, allowing an estimate of its limiting value, $\mathrm{Ri}_f^\infty = 0.2$.

Figure 2 shows Ri dependencies of the dimensionless turbulent fluxes: (a) $\hat{\tau}^2 \equiv (\tau/E_K)^2$ and (b) $\hat{F}_z^2 \equiv F_z^2/(E_K E_\theta)$. It is long recognised [see, e.g., Sect. 5.3 and 8.5 in Monin and Yaglom (1971)] that in neutral stratification, atmospheric observations give more variable and generally smaller values of these dimensionless turbulent fluxes than laboratory experiments. This is not surprising because measured values of the TKE, E_K, in the atmosphere are contaminated with low-frequency velocity fluctuations caused by the interaction of the airflow

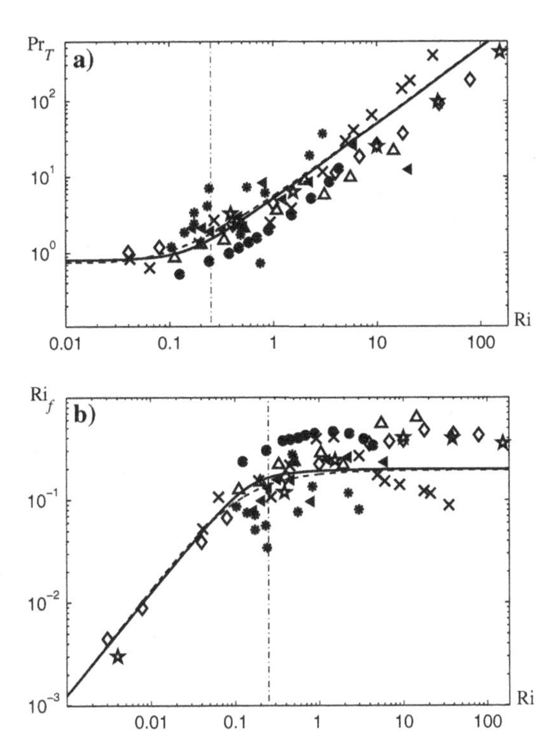

Fig. 1 Ri dependences of (**a**) turbulent Prandtl number, $\mathrm{Pr}_T = K_M/K_H$, and (**b**) flux Richardson number, $\mathrm{Ri}_f = -\beta F_z(\tau S)^{-1}$, based on meteorological observations: slanting black triangles (Kondo et al. 1978), snowflakes (Bertin et al. 1997); laboratory experiments: black circles (Strang and Fernando 2001), slanting crosses (Rehmann and Koseff 2004), diamonds (Ohya 2001); LES: triangles (new data provided by Igor Esau); DNS: five-pointed stars (Stretch et al. 2001). Solid lines show our model for homogeneous turbulence; dashed line, analytical approximations after Eq. (64)

with surface heterogeneities. These low-frequency fluctuations, however, should not be confused with shear-generated turbulence. Therefore, to validate our turbulence closure model it is only natural to use data on $\hat{\tau}^2$ obtained from laboratory experiments and/or numerical simulations. Relying on these data presented in Fig. 2a, we obtain $(\tau/E_K)^{(0)} = 0.326$ for $\mathrm{Ri} \ll 1$; and $(\tau/E_K)^{\infty} = 0.18$ for $\mathrm{Ri} \gg 1$ [the superscripts "(0)" and "∞" mean "at Ri = 0" and "at Ri$\rightarrow \infty$"]. These estimates are consistent with the conditions $(\hat{\tau}^2)^{(0)}/(\hat{F}_z^2)^{(0)} = \mathrm{Pr}_T^{(0)} = 0.8$, and $(\hat{F}_z^2)^{\infty} = 0$ that follow from Eqs. (47)–(48). Furthermore, Fig. 3, which shows the Ri dependencies of the re-normalised fluxes, (a) $\hat{\tau}^2/(\hat{\tau}^2)^{(0)}$ and (b) $\hat{F}_z^2/(\hat{F}_z^2)^{(0)}$, reveals essential similarity in the shape of these dependencies based on atmospheric, laboratory and LES data, and provides additional support to our analysis.

Data on the vertical anisotropy of turbulence, $A_z = E_z/E_K$, are shown in Fig. 4. These data are quite ambiguous and need to be analysed carefully. For neutral stratification, we adopt the estimate of $A_z^{(0)} = 0.25$ based on precise results from laboratory experiments (Agrawal et al. 2004) and DNS (Moser et al. 1999). These data are now commonly accepted and have been shown to be consistent with independent data on the wall-layer turbulence (L'vov et al. 2006). Current and previous atmospheric data (e.g., those shown in Fig. 75 in Monin and Yaglom 1971) yield smaller values of $A_z^{(0)}$, but, as already mentioned, they overestimate the horizontal TKE and, consequently, underestimate A_z, especially in neutral stratification. Such overestimating arises from meandering of atmospheric boundary-layer flow caused by non-uniform features of the earth's surface (hills, houses, groups of trees, etc.). At the same time, very large values of Ri in currently available experiments and numerical simulations are relevant to turbulent flows above the boundary layer, where the TKE of local origin (controlled by the local Ri) is often small compared to the TKE transported from

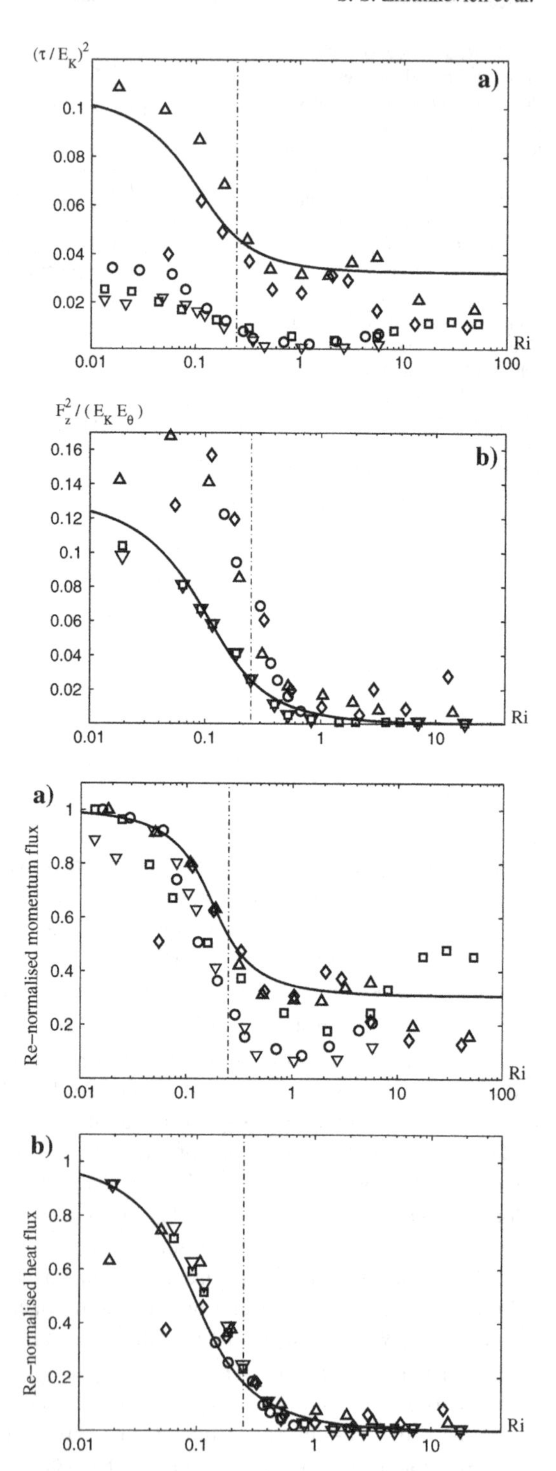

Fig. 2 Same as in Fig. 1 but for the squared dimensionless turbulent fluxes of (**a**) momentum, $\hat{\tau}^2 = (\tau/E_K)^2$, and (b) potential temperature, $\hat{F}_z^2 = F_z^2/(E_K E_\theta)$, based on laboratory experiments: diamonds (Ohya 2001) and LES: triangles (new data provided by Igor Esau); and meteorological observations: squares [CME = Carbon in the Mountains Experiment, Mahrt and Vickers (2005)], circles [SHEBA = Surface Heat Budget of the Arctic Ocean, Uttal et al. (2002)] and overturned triangles [CASES-99 = Cooperative Atmosphere-Surface Exchange Study, Poulos et al. (2002), Banta et al. (2002)]

Fig. 3 Same as in Fig. 2 but for re-normalised turbulent fluxes: (**a**) $\hat{\tau}^2/(\hat{\tau}^2)^{(0)}$ and (**b**) $\hat{F}_z^2/(\hat{F}_z^2)^{(0)}$, where the superscript (0) indicates mean values at Ri = 0 [hence $\hat{\tau}^2/(\hat{\tau}^2)^{(0)}$ and $\hat{F}_z^2/(\hat{F}_z^2)^{(0)}$ equal 1 at Ri = 0]

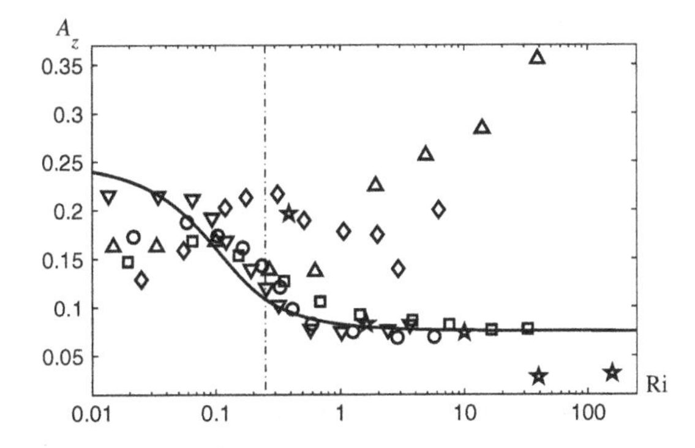

Fig. 4 Same as in Figure 2 but for the vertical anisotropy of turbulence, $A_z = E_z/E_K$, on addition of DNS data of Stretch et al. (2001) shown by five-pointed stars

the lower, strong-shear layers. It is not surprising that the spread of data on A_z versus Ri is quite large. Nonetheless, atmospheric data characterise A_z as a monotonically decreasing function of Ri and allow at least an approximate estimate of its lower limit: $A_z^{\infty}= 0.075$.

Below we use the estimates of $A_z^{(0)}$, $(\tau^2 E_K^{-2})_{\mathrm{Ri}=0}$, $\mathrm{Pr}_T^{(0)}$, Ri_f^{∞}, A_z^{∞}, and $(\tau^2 E_K^{-2})_{\mathrm{Ri}=\infty}$ to determine our empirical constants.

We start with data for neutral stratification. The empirical estimate of $A_z^{(0)} = 0.25$ yields

$$C_r = 3A_z^{(0)}(1 - 3A_z^{(0)})^{-1} = 3. \tag{56}$$

Then we combine Eq. (25c) for E_K with Eq. (32) for t_T and consider the logarithmic boundary layer, in which $l_z = z, \tau = \tau|_{z=0} \equiv u_*^2$ and $S = u_*(kz)^{-1}$(u_* is the friction velocity and k is the von Karman constant) to obtain

$$C_K = k(A_z^{(0)})^{1/2}\left(\frac{E_K}{\tau}\right)^{3/2}_{\mathrm{Ri}=0} = 1.08. \tag{57}$$

This estimate is based on the well-determined empirical value of $k \approx 0.4$, and the above values of $(\tau/E_K)^{(0)} = 0.326$ and $A_z^{(0)} = 0.25$. Then taking $C_K = 1.08$, $\mathrm{Pr}_T^{(0)} = 0.8$ and using Eqs. (43a) and (47) we obtain

$$C_{\tau 1} = \frac{C_K}{2A_z^{(0)}}\left(\frac{E_K}{\tau}\right)^{-2}_{\mathrm{Ri}=0} = 0.228, \tag{58}$$

$$C_F = C_{\tau 1}/\mathrm{Pr}_T^{(0)} = 0.285. \tag{59}$$

Taking $C_r= 3$, $A_z^{\infty}= 0.075$, $\mathrm{Ri}_f^{\infty}= 0.2$ and using Eq. (46) we obtain

$$\Psi_3^{\infty} = \frac{A_z^{\infty}}{A_z^{(0)}} + \frac{3\mathrm{Ri}_f^{\infty}}{C_r(1 - \mathrm{Ri}_f^{\infty})} = 0.55; \quad C_3 = \frac{1}{\mathrm{Ri}_f^{\infty}}\left(\Psi_3^{\infty} - 1\right) = -2.25. \tag{60}$$

The constants C_1 and C_2 determine only the energy exchange between the horizontal velocity components and do not affect any other aspects of our closure model. Taking them equal (based on symmetry reasons) and recalling that $C_1 + C_2 + C_3 = 0$ yields

$$C_1 = C_2 = -\frac{1}{2}C_3 = 1.125. \tag{61}$$

Taking $C_K = 1.08$, $\mathrm{Ri}_f^\infty = 0.2$, $A_z^\infty = 0.075$ and $(\tau/E_K)^\infty = 0.18$, Eq. (50b) gives

$$\Psi_\tau^\infty = \frac{C_K\left[(\tau/E_K)^\infty\right]^2(1-\mathrm{Ri}_f^\infty)}{2A_z^\infty} = 0.187;\quad C_{\tau 2} = \frac{1}{\mathrm{Ri}_f^\infty}\left(\Psi_\tau^\infty - C_{\tau 1}\right) = -0.208 \quad (62)$$

Then C_θ is determined from Eq. (45) written in the strong stability limit:

$$C_\theta = \frac{1}{1+C_r}\left[\frac{C_r \Psi_3^\infty}{3}\left(\frac{1}{\mathrm{Ri}_f^\infty} - 1\right) - 1\right] = 0.3. \quad (63)$$

Using the above values of dimensionless constants, the function $\mathrm{Ri}_f = \Phi(\mathrm{Ri})$ determined by Eqs. (41)–(42) is shown in Fig. 1b by the solid line. For practical use we propose the following explicit approximation of this function (with 5% accuracy):

$$\mathrm{Ri}_f = \Phi(\mathrm{Ri}) \approx 1.25\mathrm{Ri}\frac{(1\ +\ 36\mathrm{Ri})^{1.7}}{(1\ +\ 19\mathrm{Ri})^{2.7}}, \quad (64)$$

which is shown in Fig. 1b by the dashed line.

In the above estimates we did not use data on the dimensionless heat flux $\hat{F}_z^2 \equiv F_z^2/(E_K E_\theta)$ shown in Fig. 2b and 3b. The good correspondence between data and the theoretical curves in these figures serves as an empirical confirmation to our model.

The last empirical constant to be determined is the exponent n in Eq. (55). We eliminate l_z from Eqs. (53) and (54) to obtain

$$\frac{z}{L} = \frac{\mathrm{Ri}_f}{(2\Psi_\tau)^{1/2}\Psi^{1/4}\Psi_l}, \quad (65)$$

where Ψ_τ and Ψ are functions of Ri_f as specified by Eqs. (38) and (39). Given the dependence $\Psi_l(\mathrm{Ri}_f)$, the Eqs. (41) and (65) allow us to determine Ri_f and Ri as single-valued functions of z/L. *Vice versa*, given, e.g., the dependence $\mathrm{Ri}(z/L)$, Ψ_l can be determined as a single-valued function of Ri_f.

We can apply this analysis to deduce $\Psi_l(\mathrm{Ri}_f)$ from the empirical dependence of Ri on z/L obtained by Zilitinkevich and Esau (2007) using LES DATABASE64 (Esau 2004; Beare et al. 2006; Esau and Zilitinkevich 2006) and data from the field campaign SHEBA (Uttal et al. 2002). In Fig. 5, we present the above LES data together with our approximation based on Eq. (55). The exponent $n = 4/3$ is obtained from the best fit of the theoretical curve to all these data.

Strictly speaking, the suggested local, algebraic closure model is applicable only to homogeneous flows, in particular, to the nocturnal stable atmospheric boundary layer (ABL) of depth h, where non-local vertical turbulent transport plays a comparatively minor role, whereas $\boldsymbol{\tau}$ and F_z are reasonably accurately represented by universal functions of z/h (see, e.g., Fig. 1 in Zilitinkevich and Esau 2005); or with more confidence to the lower 10% of the ABL, the so-called surface layer, where τ and F_z can be taken depth-constant: $\tau \approx \tau|_{z=0} = u_*^2$ and $F_z \approx F_z|_{z=0} = F_*$.

As mentioned earlier, Eqs. (64) and (65) determine Ri_f and Ri as single-valued functions of z/L:

$$\mathrm{Ri}_f = \Phi_{\mathrm{Rif}}\left(\frac{z}{L}\right), \quad (66a)$$

$$\mathrm{Ri} = \Phi_{\mathrm{Ri}}\left(\frac{z}{L}\right). \quad (66b)$$

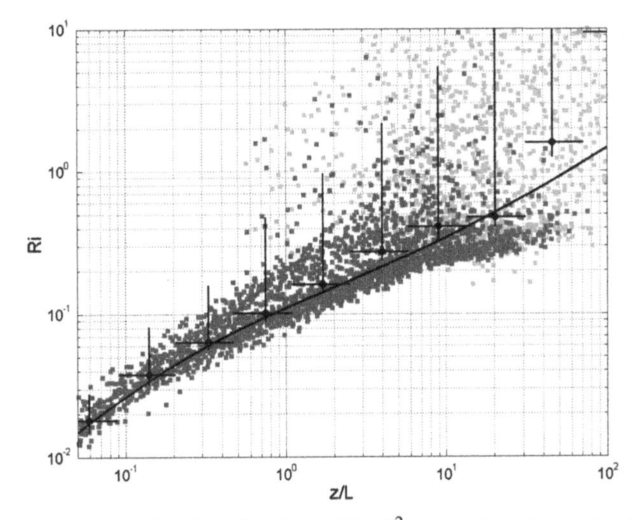

Fig. 5 Gradient Richardson number, $\mathrm{Ri} = \beta(\partial\Theta/\partial z)(\partial U/\partial z)^{-2}$, versus dimensionless height z/L in the nocturnal atmospheric boundary layer (ABL). Dark- and light-grey points show LES data within and above the ABL, respectively; heavy black points with error bars are bin-averaged values of Ri [from Fig. 3 of Zilitinkevich and Esau (2007)]. Solid line is calculated after Eqs. (41), (55) and (65) with $n = 4/3$

Consequently, our model applied to the steady-state, homogeneous regime in the surface layer, is consistent with the similarity theory of Monin and Obukhov (1954). Given τ and F_z, this model allows us to determine z/L dependencies of all the dimensionless parameters considered above, as well as the familiar similarity-theory functions specifying mean velocity and temperature profiles:

$$\Phi_M \equiv \frac{kz}{\tau^{1/2}}\left(\frac{\partial U}{\partial z}\right) = \frac{k}{\mathrm{Ri}_f}\left(\frac{z}{L}\right) = \frac{k}{\Phi_{\mathrm{Rif}}(z/L)}\left(\frac{z}{L}\right), \tag{67a}$$

$$\Phi_H \equiv \frac{k_T z \tau^{1/2}}{-F_z}\left(\frac{\partial \Theta}{\partial z}\right) \equiv k_T \frac{\mathrm{Pr}_T}{\mathrm{Ri}_f}\left(\frac{z}{L}\right) \equiv k_T \frac{\mathrm{Ri}}{\mathrm{Ri}_f^2}\left(\frac{z}{L}\right)$$
$$= k_T \frac{\Phi_{\mathrm{Ri}}(z/L)}{\Phi^2_{\mathrm{Rif}}(z/L)}\left(\frac{z}{L}\right), \tag{67b}$$

where k is the von Karman constant expressed through our constants by Eq. (57) and $k_T = k/\mathrm{Pr}_T^{(0)}$. At $\mathrm{Ri} \ll 1$, Eqs. (66a, b) reduce to $\mathrm{Ri}_f \approx kz/L$ and $\mathrm{Ri} \approx \mathrm{Pr}_T^{(0)}\, kz/L$ while Eqs. (67a, b) recover the familiar wall-layer formulation. $\Phi_M(z/L)$ and $\Phi_H(z/L)$ calculated according to our model are shown in Figure 6 together with LES data from Zilitinkevich and Esau (2007).

In contrast to the commonly accepted paradigm that both Φ_M and Φ_H depend on z/L linearly, LES data and our solution show different asymptotic behaviours, namely, linear for Φ_M and stronger than linear for Φ_H. This result deserves discussion. Indeed, the traditional formulation, $\Phi_M, \Phi_H \sim z/L$ at $z/L \gg 1$, implies that Pr_T levels off (rather than increases) with increase of z/L and, as a consequence, that surface-layer turbulence decays when Ri exceeds a critical value, $\mathrm{Ri}_c \approx 0.25$. However, as demonstrated in Sects. 1 and 3 this conclusion is erroneous.

Note that the linear dependences, $\Phi_M \sim \Phi_H \sim z/L$, were traditionally derived from the heuristic "z-less stratification" concept, which postulates that the distance from the surface,

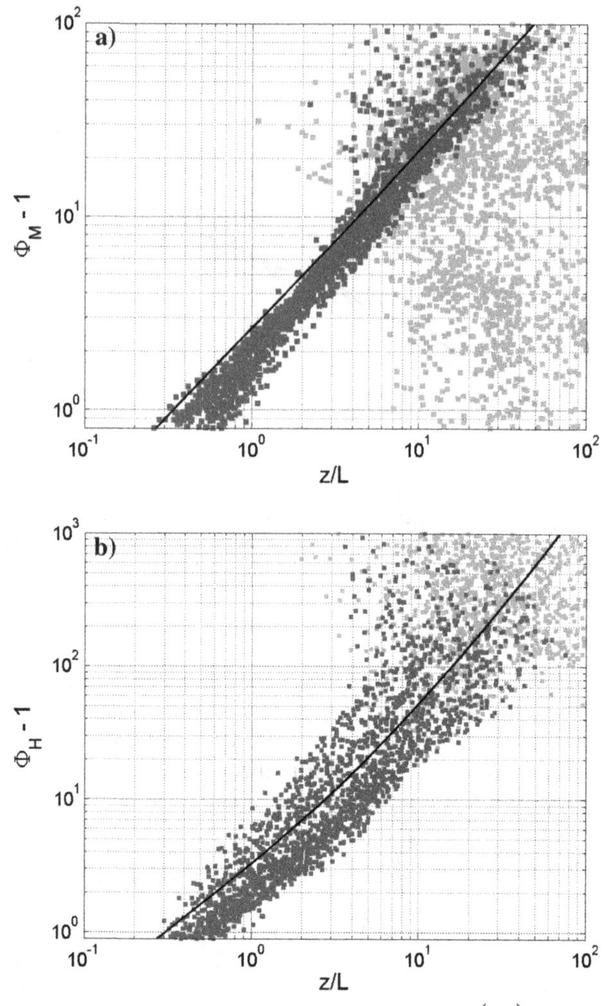

Fig. 6 Dimensionless vertical gradients of (**a**) mean velocity, $\Phi_M = \frac{kz}{\tau^{1/2}}\left(\frac{\partial U}{\partial z}\right)$, and (**b**) potential temperature, $\Phi_H = \frac{k_T z \tau^{1/2}}{-F_z}\left(\frac{\partial \Theta}{\partial z}\right)$, versus z/L, based on our local closure model [solid lines plotted after Eq. (5.13a, b)] compared to the same LES data as in Fig. 5 [from Figs. 1 and 2 of Zilitinkevich and Esau (2007)]

z, does not appear in the set of parameters that characterise the vertical turbulent length scale in sufficiently strong static stability ($z/L \gg 1$). Without this assumption the linear asymptote for Φ_H loses ground while for Φ_M it holds true. Indeed, the existence of a finite upper limit for the flux Richardson number $\mathrm{Ri}_f \to \mathrm{Ri}_f^\infty$ at $z/L \to \infty$ immediately yields the asymptotic relation:

$$\Phi_M \approx C_U \frac{z}{L} \text{ at } \frac{z}{L} \gg 1, \tag{68}$$

where $C_U = (\mathrm{Ri}_f^\infty)^{-1} \approx 5$.

It is important to note that the algebraic closure model presented in Sect. 4, is applicable only to homogeneous turbulence regimes. Therefore it probably serves as a reasonable approximation for the nocturnal ABL separated from the free flow by the neutrally stratified residual layer, but not for the conventionally neutral and the long-lived stable ABLs, which develop against the stably stratified free flow and exhibit essentially non-local features, such as the distant effect of the free-flow stability on the surface-layer turbulence (see Zilitinkevich 2002; Zilitinkevich and Esau 2005). In order to reproduce these types of ABL realistically, an adequate turbulence closure model should take into account the non-local transports.

6 Summary and conclusions

The structure of the most widely used turbulence closure models for neutrally and stably stratified geophysical flows follows Kolmogorov (1941): vertical turbulent fluxes are assumed to be downgradient; the turbulent exchange coefficients, namely, the eddy viscosity, K_M, conductivity, K_H, and diffusivity, K_D, are assumed to be proportional to the turbulent length scale, l_T, and the turbulent velocity scale, u_T, which in turn is taken to be proportional to the square root of the TKE, $E_K^{1/2}$, so that $K_{\{M,H,D\}} \sim E_K^{1/2} l_T$; and E_K is determined solely from the TKE budget equation. Kolmogorov developed this formulation for neutral stratification, where it provides quite a good approximation. However, when applied to essentially stable stratification Kolmogorov's model predicts that TKE decays at Richardson numbers exceeding a critical value, Ri_c (close to 0.25), which contradicts experimental evidence. To avoid this drawback, modern closure models modify the original Kolmogorov formulation assuming $K_{\{M,H,D\}} = f_{\{M,H,D\}}(\mathrm{Ri}) E_K^{1/2} l_z$, where stability functions $f_{\{M,H,D\}}(\mathrm{Ri})$ are determined either theoretically or empirically. Given these functions, it remains to determine l_T and then, apparently, the closure problem is solved.

Such a conclusion, however, is premature. The concepts of the downgradient turbulent transport and the turbulent exchange coefficients, as well as the relationships $K_{\{M,H,D\}} = f_{\{M,H,D\}}(\mathrm{Ri}) E_K^{1/2} l_T$ are consistent with the flux-budget equations only in comparatively simple particular cases relevant to the homogeneous regime of turbulence. Only in these cases the turbulent exchange coefficients can be rigorously defined, in contrast to the turbulent fluxes that represent clearly defined, measurable parameters, governed by the flux-budget equations. It is therefore preferable to rely on the flux-budget equations rather than to formulate hypotheses about virtual exchange coefficients.

Furthermore, the TKE budget equation does not fully characterise turbulent energy transformations, not to mention that the vertical turbulent transports are controlled by the energy of vertical velocity fluctuations, E_z, rather than E_K.

In this study we do not follow the above "main stream" approach, and instead of solely using the TKE budget equation, we employ the budget equations for turbulent potential energy (TPE) and turbulent total energy (TTE = TKE + TPE), which guarantees maintenance of turbulence by the velocity shear in any stratification.

Furthermore, we do not accept *a priori* the concept of downgradient turbulent transports (implying universal existence of turbulent exchange coefficients). Instead, we use the budget equations for key turbulent fluxes and derive (rather than postulate) formulations for the exchange coefficients, when it is physically grounded as in the steady-state homogeneous regime.

In the budget equation for the vertical flux of potential temperature we take into account a crucially important mechanism: generation of the countergradient flux due to the buoyancy effect of potential-temperature fluctuations (compensated, but only partially so, by the correlation between the potential-temperature and the pressure-gradient fluctuations). We show that this is an important mechanism responsible for the principle difference between the heat and the momentum transfer.

To determine the energy of the vertical velocity fluctuations, we modify the traditional return-to-isotropy formulation accounting for the effect of stratification on the redistribution of the TKE between horizontal and vertical velocity components. We then derive a simple algebraic version of an energetically consistent closure model for the steady-state, homogeneous regime, and verify it against available experimental, LES and DNS data.

As seen from Figs. 1–4 showing Ri dependencies of the turbulent Prandtl number, $\mathrm{Pr}_T = K_M/K_H$, the flux Richardson number, Ri_f, dimensionless turbulent fluxes, $(\tau/E_K)^2$ and $F_z^2(E_K E_\theta)^{-1}$, and anisotropy of turbulence, $A_z = E_z/E_K$, our model, in compliance with the majority of data, reveals the existence of two essentially different regimes of turbulence separated by a comparatively narrow interval of Ri around a threshold value of $\mathrm{Ri} \approx 0.25$ (shown in the figures by the vertical dashed lines). On both sides of the transitional interval, $0.1 < \mathrm{Ri} < 1$, the ratios $(\tau/E_K)^2$ and $F_z^2(E_K E_\theta)^{-1}$ approach plateaus corresponding to the very high efficiency of the turbulent transfer at $\mathrm{Ri} < 0.1$, and to the strongly different efficiencies of the momentum transfer (which is still pronounced) and heat transfer (which is very weak) at $\mathrm{Ri} > 1$.

It is hardly incidental that the above threshold coincides with the critical Richardson number, Ri_c, derived from the classical perturbation analyses. These analyses have demonstrated that the infinitesimal perturbations grow exponentially at $\mathrm{Ri} < \mathrm{Ri}_c$ but do not grow at $\mathrm{Ri} > \mathrm{Ri}_c$ when, as we understand now, the onset of turbulent events requires finite perturbations. Consequently, the transitional interval, $0.1 < \mathrm{Ri} < 1$, indeed separates two essentially different regimes: strong turbulence at $\mathrm{Ri} < 0.1$ and weak turbulence at $\mathrm{Ri} > 1$, but do not separate the turbulent and the laminar regimes as traditionally assumed.

What we presented in this study is just a first step towards developing consistent and practically useful turbulence closure models based on a minimal set of equations, which indispensably includes the TTE budget equation and does not imply the existence of the critical Richardson number. Two other recent studies follow this approach. Mauritsen et al. (2007) have developed a simple closure model employing the TTE budget equation and empirical Ri dependences of $(\tau/E_K)^2$ and $F_z^2(E_K E_\theta)^{-1}$ (similar to those shown in our Fig. 2–3). L'vov, Procaccia and Rudenko (Private communication) perform detailed analyses of the budget equations for the Reynolds stresses in the turbulent boundary layer (relevant to the strong turbulence regime) taking into consideration the dissipative effect of the horizontal heat flux explicitly, in contrast to our "effective-dissipation approximation".

As already mentioned, the present study is limited to the local, algebraic closure model that is applicable to the steady-state, homogeneous turbulence regime. Generalised versions of this model, based on the same physical analyses but accounting for the third-order transports (Φ_K, Φ_P, Φ_F and $\Phi_{\{1,2\}}^{(\tau)}$) will be presented in forthcoming papers.

Our data analysis provides only a plausible first verification rather than a comprehensive validation of the proposed model. Special efforts are needed to extend our data analysis using additional field, laboratory and numerically simulated data (e.g., Rohr et al., 1988; Shih et al. 2000). In future work, particular attention should also be paid to direct verification of our approximations, such as those for the term $\rho_0^{-1}\langle\theta\partial p/\partial z\rangle$ taken to be proportional to $\beta\langle\theta^2\rangle$ in Eq. (9b), and for the term $\varepsilon_{i3(\mathrm{eff})} \equiv \varepsilon_{i3}^{(\tau)} - \beta F_i - Q_{i3}$ assumed to be proportional to τ_{i3}/t_T in Eq. (10b).

In the present state, our closure model does not account for vertical transports arising from internal waves. The dual nature of fluctuations representing both turbulence and waves in stratified flows has been suggested, e.g., by Jacobitz et al. (2005). The role of waves and the need for their inclusion in the context of turbulence closure models has been discussed, e.g., by Jin et al. (2003) and Baumert and Peters (2005). Direct account of the wave-driven transports of momentum and both kinetic and potential energies is a topic of our current research.

Acknowledgements We thank Vittorio Canuto, Igor Esau, Victor L'vov, Thorsten Mauritsen and Gunilla Svensson for discussions and Igor Esau for his valuable contribution to data analysis and for making Figures 5 and 6. This work has been supported by EU Marie Curie Chair Project MEXC-CT-2003-509742, EU Project FUMAPEX EVK4-CT-2002-00097, ARO Project W911NF-05-1-0055, Carl-Gustaf Rossby International Meteorological Institute in Stockholm, German-Israeli Project Cooperation (DIP) administrated by the Federal Ministry of Education and Research (BMBF), Israel Science Foundation governed by the Israeli Academy of Science, Israeli Universities Budget Planning Committee (VATAT) and Israeli Atomic Energy Commission.

Appendix A: The pressure term in the budget equation for turbulent flux of potential temperature

The approximation used in Sect. 2:

$$\beta\left\langle\theta^2\right\rangle+\frac{1}{\rho_0}\left\langle\theta\frac{\partial}{\partial z}p\right\rangle=C_\theta\beta\left\langle\theta^2\right\rangle \tag{69}$$

with C_θ = constant < 1 can be justified as follows. Taking the divergence of the momentum equation we arrive at the following equation

$$\frac{1}{\rho_0}\Delta p=-\beta\frac{\partial}{\partial z}\theta. \tag{70}$$

Applying the inverse Laplacian to Eq. (70) yields

$$\frac{1}{\rho_0}p=\beta\Delta^{-1}\left(\frac{\partial\theta}{\partial z}\right),\quad\text{and}\quad\frac{1}{\rho_0}\left\langle\theta\frac{\partial}{\partial z}p\right\rangle=-\beta\left\langle\theta\Delta^{-1}\frac{\partial^2}{\partial z^2}\theta\right\rangle. \tag{71}$$

We employ the following scaling estimate:

$$\frac{\left\langle\theta\Delta^{-1}\left(\frac{\partial^2\theta}{\partial z^2}\right)\right\rangle}{\left\langle\theta^2\right\rangle}\approx\left(1+\alpha^{-1}\right)\left(1-\frac{\arctan\sqrt{\alpha}}{\sqrt{\alpha}}\right), \tag{72}$$

where $\alpha=l_\perp^2/l_z^2-1$, l_z and $l_\perp$ are the correlation lengths of the correlation function $\langle\theta(t,\mathbf{x}_1)\theta(t,\mathbf{x}_2)\rangle$ in the vertical and the horizontal directions.

Equations (71) and (72) yield

$$\frac{\frac{1}{\rho_0}\left\langle\theta\frac{\partial p}{\partial z}\right\rangle}{\left\langle\theta^2\right\rangle}\approx-\begin{cases}\frac{1}{3}\left(1+\frac{2}{5}\alpha\right) & \text{in the thermal isotropy limit } (\alpha\ll 1)\\ 1-\frac{\pi}{2\sqrt{\alpha}} & \text{in the infinite thermal anisotropy limit } (\alpha\gg 1).\end{cases} \tag{73}$$

Consequently, the coefficient C_θ= {1 + [r.h.s. of Eq. (73)]} turns into 2/3 in the thermal isotropy limit (corresponding to neutral stratification) and vanishes in an imaginary case of the infinite thermal anisotropy. Our empirical estimate, Eq. (63), of C_θ= 0.3 is a reasonable compromise between these two extremes.

References

Agrawal A, Djenidi L, Antobia RA (2004) URL http://in3.dem.ist.utl.pt/lxlaser2004/pdf/paper_28_1.pdf
Banta RM, Newsom RK, Lundquist JK, Pichugina YL, Coulter RL, Mahrt L (2002) Nocturnal low-level jet characteristics over Kansas during CASES-99. Boundary-Layer Meteorol 105:221–252
Baumert HZ, Peters H (2005) A novel two equation turbulence closure model for high Reynolds numbers. Part A: homogeneous, non-rotating, stratified shear layers. In: Baumert HZ, Simpson JH, Sündermann J (eds) Marine turbulence. theory, observations and models. Cambridge University Press, pp 14–30
Beare RJ, MacVean MK, Holtslag AAM, Cuxart J, Esau I, Golaz JC, Jimenez MA, Khairoudinov M, Kosovic B, Lewellen D, Lund TS, Lundquist JK, McCabe A, Moene AF, Noh Y, Raasch S, Sullivan P (2006) An intercomparison of large eddy simulations of the stable boundary layer. Boundary-Layer Meteorol 118:247–272
Bertin F, Barat J, Wilson R (1997) Energy dissipation rates, eddy diffusivity, and the Prandtl number: an in situ experimental approach and its consequences on radar estimate of turbulent parameters. Radio Sci 32:791–804
Canuto VM, Minotti F (1993) Stratified turbulence in the atmosphere and oceans: a new sub-grid model. J Atmos Sci 50:1925–1935
Canuto VM, Howard A, Cheng Y, Dubovikov MS (2001) Ocean turbulence. Part I: one-point closure model - momentum and heat vertical diffusivities. J Phys Oceanogr 31:1413–1426
Cheng Y, Canuto VM, Howard AM (2002) An improved model for the turbulent PBL. J Atmosph Sci 59:1550–1565
Churchill SW (2002) A reinterpretation of the turbulent Prandtl number. Ind Eng Chem Res 41:6393–6401
Cuxart J, 23 co-authors (2006) Single-column model intercomparison for a stably stratified atmospheric boundary layer. Boundary-Layer Meteorol 118:273–303
Dalaudier F, Sidi C (1987) Evidence and interpretation of a spectral gap in the turbulent atmospheric temperature spectra. J Atmos Sci 44:3121–3126
Elperin T, Kleeorin N, Rogachevskii I (1996) Isotropic and anisotropic spectra of passive scalar fluctuations in turbulent fluid flow. Phys Rev E 53:3431–3441
Elperin T, Kleeorin N, Rogachevskii I, Zilitinkevich S (2002) Formation of large-scale semi-organized structures in turbulent convection. Phys Rev E 66:066305 (1–15)
Elperin T, Kleeorin N, Rogachevskii I, Zilitinkevich S (2006) Turbulence and coherent structures in geophysical convection. Boundary-Layer Meteorol 119:449–472
Esau IN, Zilitinkevich SS (2006) Universal dependences between turbulent and mean flow parameters in stably and neutrally stratified planetary boundary layers. Nonlin Proce Geophys 13:135–144
Foken T (2006) 50 years of the Monin–Obukhov similarity theory. Boundary-Layer Meteorol 119:431–447
Hanazaki H, Hunt JCR (1996) Linear processes in unsteady stably stratified turbulence. J Fluid Mech 318:303–337
Hanazaki H, Hunt JCR (2004) Structure of unsteady stably stratified turbulence with mean shear. J Fluid Mech 507:1–42
Hunt JCR, Kaimal JC, Gaynor JE (1985) Some observations of turbulence in stable layers. Quart J Roy Meteorol Soc 111:793–815
Hunt JCR, Stretch DD, Britter RE (1988) Length scales in stably stratified turbulent flows and their use in turbulence models. In: Puttock JS (ed) Proc. I.M.A. Conference on "Stably Stratified Flow and Dense Gas Dispersion" Clarendon Press, pp 285–322
Holton JR (2004) An introduction to dynamic meteorology. Academic Press, New York 535 pp
Jacobitz FG, Rogers MM, Ferziger JH (2005) Waves in stably stratified turbulent flow. J Turbulence 6:1–12
Jin LH, So RMC, Gatski TB (2003) Equilibrium states of turbulent homogeneous buoyant flows. J Fluid Mech 482:207–233
Kaimal JC, Finnigan JJ (1994) Atmospheric boundary layer flows. Oxford University Press, New York, 289 pp
Keller K, Van Atta CW (2000) An experimental investigation of the vertical temperature structure of homogeneous stratified shear turbulence. J Fluid Mech 425:1–29
Kraus EB, Businger JA (1994) Atmosphere-ocean interaction. Oxford University Press, Oxford 362 pp
Kolmogorov AN (1941) Energy dissipation in locally isotropic turbulence. Doklady AN SSSR 32(1):19–21
Kondo J, Kanechika O, Yasuda N (1978) Heat and momentum transfer under strong stability in the atmospheric surface layer. J Atmos Sci 35:1012–1021
Kurbatsky AF (2000) Lectures on turbulence. Novosibirsk State University Press, Novosibirsk
Luyten PJ, Carniel S, Umgiesser G (2002) Validation of turbulence closure parameterisations for stably stratified flows using the PROVESS turbulence measurements in the North Sea. J Sea Research 47:239–267

L'vov VS, Pomyalov A, Procaccia I, Zilitinkevich SS (2006) Phenomenology of wall bounded Newtonian turbulence. Phys Rev E 73:016303, 1–13
Mahrt L, Vickers D (2005) Boundary layer adjustment over small-scale changes of surface heat flux. Boundary-Layer Meteorol 116:313–330
Mauritsen T, Svensson G (2007) Observations of stably stratified shear-driven atmospheric turbulence at low and high Richardson numbers. J Atmos Sci 64:645–655
Mauritsen T, Svensson G, Zilitinkevich SS, Esau E, Enger L, Grisogono B (2007) A total turbulent energy closure model for neutrally and stably stratified atmospheric boundary layers. J Atmos Sci (In press)
Mellor GL, Yamada T (1974) A hierarchy of turbulence closure models for planetary boundary layers. J Atmos Sci 31:1791–1806
Moser RG, Kim J, Mansour NN (1999) Direct numerical simulation of turbulent channel flow up to Re = 590. Phys Fluids 11:943–945
Monin AS, Obukhov AM (1954) Main characteristics of the turbulent mixing in the atmospheric surface layer. Trudy Geophys Inst AN SSSR 24(151):153–187
Monin AS, Yaglom AM (1971) Statistical fluid mechanics. Vol 1. MIT Press, Cambridge, Massachusetts, 769 pp
Ohya Y (2001) Wind-tunnel study of atmospheric stable boundary layers over a rough surface. Boundary-Layer Meteorol 98:57–82
Ozmidov RV (1990) Diffusion of contaminants in the ocean. Kluwer Academic Publishers, Dordrecht, The Netherlands, 283 pp
Poulos GS, Blumen W, Fritts DC, Lundquist JK, Sun J, Burns SP, Nappo C, Banta R, Newsom R, Cuxart J, Terradellas E, Balsley B, Jensen M (2002) CASES-99: a comprehensive investigation of the stable nocturnal boundary layer. Bull Amer Meteorol Soc 83:555–581
Rehmann CR, Hwang JH (2005) Small-scale structure of strongly stratified turbulence. J Phys Oceanogr 32:154–164
Rehmann CR, Koseff JR (2004) Mean potential energy change in stratified grid turbulence Dynamics Atmospheres Oceans. 37:271–294
Richardson LF (1920) The supply of energy from and to atmospheric eddies. Proc Roy Soc London A 97:354–373
Rohr JJ, Itsweire EC, Helland KN, Van Atta CW (1988) Growth and decay of turbulence in a stably stratified shear flow. J Fluid Mech 195:77–111
Rotta JC (1951) Statistische theorie nichthomogener turbulenz. Z Physik 129:547–572
Schumann U, Gerz T (1995) Turbulent mixing in stably stratified sheared flows. J App Meteorol 34:33–48
Shih LH, Koseff JR, Ferziger JH, Rehmann CR (2000) Scaling and parameterization of stratified homogeneous turbulent shear flow. J Fluid Mech 412:1–20
Strang EJ, Fernando HJS (2001) Vertical mixing and transports through a stratified shear layer. J Phys Oceanogr 31:2026–2048
Stretch DD, Rottman JW, Nomura KK, Venayagamoorthy SK (2001) Transient mixing events in stably stratified turbulence, In: 14th Australasian Fluid Mechanics Conference:Adelaide, Australia, 10–14 December 2001
Umlauf L (2005) Modelling the effects of horizontal and vertical shear in stratified turbulent flows. Deep-Sea Res 52:1181–1201
Umlauf L, Burchard H (2005) Second-order turbulence closure models for geophysical boundary layers. A review of recent work. Continental Shelf Res 25:725–827
Uttal T, Curry JA, McPhee MG, Perovich DK, 24 other co-authors (2002) Surface Heat Budget of the Arctic Ocean. Bull Amer Meteorol Soc 83:255–276
Weng W, Taylor P (2003) On modelling the one-dimensional Atmospheric Boundary Layer. Boundary-Layer Meteoro, present volume 107:371–400
Zilitinkevich S (2002) Third-order transport due to internal waves and non-local turbulence in the stably stratified surface layer. Quart J Roy Meteorol Soc 128:913–925
Zilitinkevich SS, Esau IN (2005) Resistance and heat/mass transfer laws for neutral and stable planetary boundary layers: old theory advanced and re-evaluated. Quart J Roy Meteorol Soc 131:1863–1892
Zilitinkevich SS, Esau IN (2007) Similarity theory and calculation of turbulent fluxes at the surface for the stably stratified atmospheric boundary layers. Boundary-Layer Meteorol (In press, present volume)
Zilitinkevich SS, Perov VL, King JC (2002) Near-surface turbulent fluxes in stable stratification: calculation techniques for use in general circulation models. Quart J Roy Meteorol Soc 128:1571–1587

Similarity theory and calculation of turbulent fluxes at the surface for the stably stratified atmospheric boundary layer

Sergej S. Zilitinkevich · Igor N. Esau

Abstract In this paper we revise the similarity theory for the stably stratified atmospheric boundary layer (ABL), formulate analytical approximations for the wind velocity and potential temperature profiles over the entire ABL, validate them against large-eddy simulation and observational data, and develop an improved surface flux calculation technique for use in operational models.

Keywords Monin–Obukhov similarity theory · Planetary boundary layer · Prandtl number · Richardson number · Stable stratification · Surface fluxes in atmospheric models · Surface layer

1 Introduction

Parameterisation of turbulence in atmospheric models comprises two basic problems:

- turbulence closure—to calculate vertical turbulent fluxes, first of all, the fluxes of momentum and potential temperature: $\vec{\tau}$ and F_θ through the mean gradients: $d\vec{U}/dz$ and $d\Theta/dz$ (where z is the height, $\vec{U}$ and Θ are the mean wind speed and potential temperature);
- flux–profile relationships—to calculate the fluxes at the earth's surface: $\tau_* = \tau|_{z=0}$ and $F_* = F_\theta|_{z=0}$ through the mean wind speed $U_1 = U|_{z=z_1}$ and potential temperature $\Theta_1 = \Theta|_{z=z_1}$ at a given level, z_1, above the surface.

S. S. Zilitinkevich (✉)
Division of Atmospheric Sciences, University of Helsinki, Helsinki, Finland
e-mail: sergej.zilitinkevich@fmi.fi

I. N. Esau · S. S. Zilitinkevich
Nansen Environmental and Remote Sensing Centre/Bjerknes Centre for Climate Research, Bergen, Norway

S. S. Zilitinkevich
Finnish Meteorological Institute, Helsinki, Finland

Atmospheric Boundary Layers. A. Baklanov & B. Grisogono (eds.),
doi: 10.1007/978-0-387-74321-9_4,

We focus on the flux–profile relationships for stable and neutral stratifications. At first sight, these could be obtained numerically using an adequate turbulence-closure model. However, this way is too computationally expensive: the mean gradients close to the surface are very sharp, which requires very high resolution, not to mention that the adequate closure for strongly stable stratification can hardly be considered as a fully understood, easy problem. Hence the practically sound problem is to analytically express the surface fluxes τ_* and F_* through $U_1 = U|_{z=z_1}$ and $\Theta_1 = \Theta|_{z=z_1}$ available in numerical models (and similarly for the fluxes of humidity and other scalars). In numerical weather prediction (NWP) and climate models, the lowest computational level is usually taken $z_1 \approx 30\,\mathrm{m}$ (see Ayotte et al. 1996; Tjernstrom 2004).

In neutral or near-neutral stratification the solution to the above problem is given by the logarithmic wall law:

$$\frac{dU}{dz} = \frac{\tau^{1/2}}{kz}, \tag{1a}$$

$$\frac{d\Theta}{dz} = \frac{-F_\theta}{k_T \tau^{1/2} z}, \tag{1b}$$

$$U = \frac{\tau^{1/2}}{k} \ln \frac{z}{z_{0u}}, \tag{1c}$$

$$\Theta = \Theta_s + \frac{-F_\theta}{k_T \tau^{1/2}} \ln \frac{z}{z_{0T}}, \tag{1d}$$

$$\Theta_0 + \frac{-F_\theta}{k_T \tau^{1/2}} \ln \frac{z}{z_{0u}}, \tag{1e}$$

where k and k_T are the von Karman constants, z_{0u} and z_{0T} are the roughness lengths for momentum and heat, Θ_s is the potential temperature at the surface, and Θ_0 is the aerodynamic surface potential temperature, that is the value of $\Theta(z)$ extrapolated logarithmically down to the level $z = z_{0u}$ [determination of the difference $\Theta_0 - \Theta_s = k_T^{-1}(-F_\theta \tau^{-1/2}) \ln(z_{0u}/z_{0T})$ comprises an independent problem; see, e.g., Zilitinkevich et al. (2001, 2002)]. As follows from Eq. 1, $\tau_1^{1/2} = kU_1(\ln z/z_{0u})^{-1}$ and $F_{\theta 1} = -kk_T U_1(\Theta_1 - \Theta_0)(\ln z/z_{0u})^{-2}$. The turbulent fluxes τ_1 and $F_{\theta 1}$ at the level $z = z_1$ can be identified with the surface fluxes: $\tau_1 = \tau_*$ and $F_{\theta 1} = F_*$, provided that z_1 is much less then the height, h, of the atmospheric boundary layer (ABL). In neutral stratification, a typical value of h is a few hundred metres, so that the requirement $z_1 \approx 30\,\mathrm{m} \ll h$ is satisfied.

In stable stratification, the problem becomes more complicated. Its commonly accepted solution is based, firstly, on the assumption that the level z_1 belongs to the surface layer [that is the lowest one tenth of the ABL, where the turbulent fluxes do not diverge considerably from their surface values: $\tau \approx \tau_*$ and $F_\theta \approx F_*$] and, secondly, on the Monin–Obukhov (MO) similarity theory for surface-layer turbulence (Monin and Obukhov 1954).

The MO theory states that the turbulent regime in the stratified surface layer is fully characterised by the turbulent fluxes, $\tau \approx \tau_* = u_*^2$ (where u_* is the friction velocity) and $F_\theta \approx F_*$, and the buoyancy parameter, $\beta = g/T_0$ (where g is the acceleration of gravity, and T_0 is a reference value of absolute temperature), which determine the familiar Obukhov length scale

$$L = \frac{\tau^{3/2}}{-\beta F_\theta}, \tag{2}$$

whereas the velocity and potential temperature gradients are expressed through universal functions, Φ_M and Φ_H, of the dimensionless height $\xi = z/L$:

$$\frac{kz}{\tau^{1/2}}\frac{dU}{dz} = \Phi_M(\xi), \tag{3a}$$

$$\frac{k_T \tau^{1/2} z}{-F_\theta}\frac{d\Theta}{dz} = \Phi_H(\xi). \tag{3b}$$

From the requirement of consistency with the wall law for neutral stratification, Eq. 1, it follows that $\Phi_M = \Phi_H \to 1$ at $\xi \ll 1$. The asymptotic behaviour of Φ_M and Φ_H in strongly stable stratification (at $\xi \gg 1$) is traditionally derived from the concept of z-less stratification, which states that at $z \gg L$ the distance above the surface, z, no longer affects turbulence. If so, the velocity- and temperature-gradient formulations should become independent of z, which immediately suggests the linear asymptotes: $\Phi_M \sim \Phi_H \sim \xi$. The linear interpolation between the neutral and the strong stability limits gives

$$\Phi_M = 1 + C_{U1}\xi, \tag{4a}$$

$$\Phi_H = 1 + C_{\Theta 1}\xi, \tag{4b}$$

where C_{U1} and $C_{\Theta 1}$ are empirical dimensionless constants.

The above analysis is usually considered as relevant only to the surface layer. However, the basic statement of the MO similarity theory, namely, that surface-layer turbulence is fully characterised by τ, F_θ and β, is applicable to locally generated turbulence in a more general context. Nieuwstadt (1984) was probably the first who extended the MO theory by substituting the height-dependent τ and F_θ for the height-constant τ_* and F_*, and demonstrated its successful application to the entire nocturnal stable ABL. In the present paper we employ this extended version of the MO theory.

In the surface layer, substituting Eq. 4 for Φ_M and Φ_H into Eq. 3 and integrating over z, yields the log-linear approximation:

$$U = \frac{u_*}{k}\left(\ln\frac{z}{z_{u0}} + C_{U1}\frac{z}{L_s}\right), \tag{5a}$$

$$\Theta - \Theta_0 = \frac{-F_*}{k_T u_*}\left(\ln\frac{z}{z_{u0}} + C_{\Theta 1}\frac{z}{L_s}\right), \tag{5b}$$

where $L_s = u_*^3(-\beta F_*)^{-1}$.

Since the late 1950s, Eqs. 3–5 have been compared with experimental data in numerous works that basically gave estimates of C_{U1} close to 2 and $C_{\Theta 1}$ also close to 2 but with a wider spread (see overview by Högström 1996; Yague et al., 2006). Experimentalists often admitted that for Θ the log-linear formulation is worse than for U (e.g., the above reference) but no commonly accepted alternative formulations were derived from physical grounds. Esau and Byrkjedal (2007) analysed data from large-eddy simulations (LES) and disclosed that the coefficient $C_{\Theta 1}$ in Eq. 4b is not a constant but increases with increasing z/L.

According to Eqs. 3–4 the Richardson number, $\mathrm{Ri} \equiv \beta(d\Theta/dz)(dU/dz)^{-2}$, monotonically increases with increasing z/L, and at $z/L \to \infty$ achieves its maximum value: $\mathrm{Ri}_c = k^2 C_{\Theta 1} k_T^{-1} C_{U1}^{-2}$. In other words, Eq. 4 is not applicable to $\mathrm{Ri} > \mathrm{Ri}_c$. This conclusion is consistent with the critical Richardson number concept, universally accepted at the time when the MO theory and Eqs. 3–5 were formulated.

However, as recognised recently, the concept of the critical Ri contradicts both experimental evidence and analysis of the turbulent kinetic and potential energy budgets (see Zilitinkevich et al. 2007b). This conclusion is by no means new. Long ago it has been understood that turbulent closures or surface flux schemes implying the critical Ri lead to erroneous conclusions, such as unrealistic decoupling of air flows from the underlying surface in all cases when Ri> Ri_c. It is not surprising that modellers do not use Eq. 4 as well as other formulations of similar type, even though they are supported by experimental data. Instead, operational modellers develop their own flux–profile relationships, free of the critical Ri, and evaluate them indirectly—fitting the model results to the available observational data. Different points of view of experimentalists and operational modellers on the flux–profile relationships have factually caused two nearly independent lines of inquiry in this field (see discussion in Zilitinkevich et al. 2002).

One more point deserves emphasising. Currently used flux-calculation schemes identify the turbulent fluxes calculated at the level z_1 with the surface fluxes. However, in strongly stable stratification, especially in the long-lived stable ABL, the ABL height, h, quite often reduces to a few dozen metres[1] (see Zilitinkevich and Esau 2002, 2003; Zilitinkevich et al. 2007a) and becomes comparable with z_1 adopted in operational models. In such cases τ_1 and $F_{\theta 1}$ have nothing in common with τ_* and F_*.

Furthermore, the MO theory, considered for half a century as an ultimate framework for analysing the surface-layer turbulence, is now revised. Zilitinkevich and Esau (2005) have found that, besides L, Eq. 2, which characterise the stabilising effect of local buoyancy forces on turbulence, there are two additional length scales: L_f characterising the effect of the Earth's rotation and L_N characterizing the non-local effect of the static stability in the free atmosphere above the ABL:

$$L_N = \frac{\tau^{1/2}}{N}, \tag{6a}$$

$$L_f = \frac{\tau^{1/2}}{|f|}, \tag{6b}$$

where f is the Coriolis parameter, and $N = (\beta \partial \Theta / \partial z)^{1/2}$ is the Brunt-Väisälä frequency above the ABL. For certainty, we determine N from the temperature profile in the height interval $h < z < 2h$ (see Zilitinkevich and Esau 2005). Its typical atmospheric value is $N \sim 10^{-2}\,\mathrm{s}^{-1}$. Interpolating between the squared reciprocals of the three scales (which gives larger weights to stronger mechanisms that is to smaller scales) a composite turbulent length scale becomes:

$$\frac{1}{L_*} = \left[\left(\frac{1}{L}\right)^2 + \left(\frac{C_N}{L_N}\right)^2 + \left(\frac{C_f}{L_f}\right)^2 \right]^{1/2}, \tag{7}$$

where $C_N = 0.1$ and $C_f = 1$ are empirical dimensionless coefficients.[2] Advantages of this scaling have been demonstrated in the plots of Φ_M and Φ_H versus z/L_* (Figs. 2 and 5 in *op. cit.*) showing an essential collapse of data points compared to the traditional plots of Φ_M and Φ_H vs. z/L.

---

[1] The ABL height is defined as the level at which the turbulent fluxes become an order of magnitude smaller than close to the surface.

[2] In *op. cit.* the coefficient C_N was taken 0.1 for Φ_M and 0.15 for Φ_H. Further analysis has shown that the difference is insignificant, which allows employing one composite length scale given by Eq. 7.

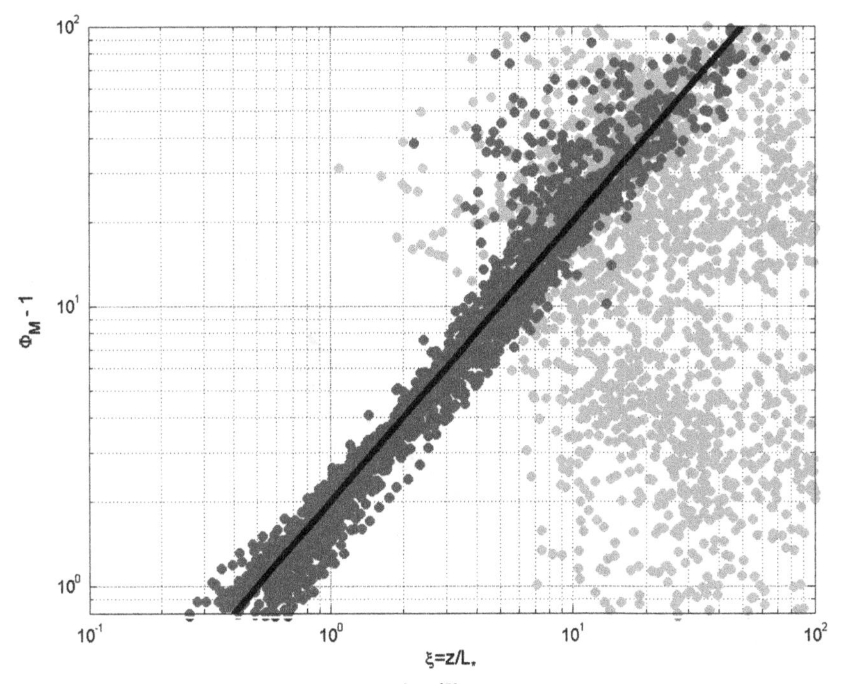

Fig. 1 Dimensionless velocity gradient, $\Phi_M = \frac{kz}{\tau^{1/2}} \frac{dU}{dz}$, in the ABL ($z < h$) and above ($z > h$) versus dimensionless height $\xi = z/L_*$, after the LES DATABASE64. Dark grey points show data for $z < h$; light grey points, for $z > h$; the line shows Eq. 11a with $C_{U1} = 2$

Practical application of this scaling requires information about vertical profiles of turbulent fluxes across the ABL. As demonstrated by Lenshow et al. (1988), Sorbjan (1988), Wittich (1991), Zilitinkevich and Esau (2005) and Esau and Byrkjedal (2007), the ratios τ/τ_* and F_θ/F_* are reasonably accurately approximated by universal functions of z/h, where h is the ABL height (see Eq. 15 below).

As follows from the above discussion, currently used surface-flux calculation schemes need to be improved accounting for

- modern experimental evidence and theoretical developments arguing against the critical Ri concept,
- additional mechanisms and scales, first of all L_N, disregarded in the classical similarity theory for the stable ABL,
- essential difference between the surface fluxes and the fluxes at $z = z_1$.

In the present paper we attempt to develop a new scheme applicable to as wide as possible a range of stable and neutral ABL regimes using recent theoretical developments and new, high quality observations and LES.

2 Mean gradients and the Richardson number

Until recently the ABL was distinguished accounting for only one factor, the potential temperature flux at the surface, F_*: neutral ABL at $F_* = 0$, and stable ABL at $F_* < 0$. Accounting

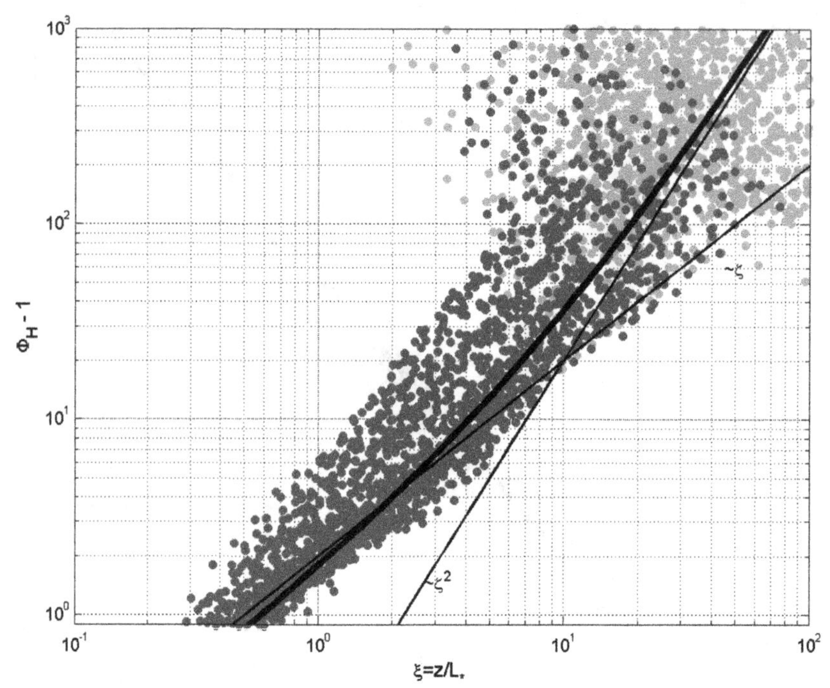

Fig. 2 Same as in Fig. 1 but for the dimensionless potential temperature gradient, $\Phi_H = \frac{k_T \tau^{1/2} z}{-F_\theta} \frac{d\Theta}{dz}$. The bold curve shows Eq. 11b with $C_{\Theta 1} = 1.6$ and $C_{\Theta 2} = 0.2$; the thin lines show its asymptote $\Phi_H = 0.2\xi^2$ and the traditional approximation $\Phi_H = 1 + 2\xi$

for the recently disclosed role of the static stability above the ABL, we now apply a more detailed classification:

- truly neutral (TN) ABL: $F_* = 0$, $N = 0$,
- conventionally neutral (CN) ABL: $F_* = 0$, $N > 0$,
- nocturnal stable (NS) ABL: $F_* < 0$, $N = 0$,
- long-lived stable (LS) ABL: $F_* < 0$, $N > 0$.

Realistic surface-flux calculation schemes should be based on a model applicable to all these types of ABL.

As mentioned in Sect. 1, Eq. 4b gives erroneous asymptotic behaviour at large $\xi = z/L$ and leads to the appearance of the critical Ri. This conclusion is sometimes treated as a failure of the MO theory, but this is not the case. The MO theory states only that Φ_M and Φ_H are universal functions of ξ, whereas the linear forms of the Φ functions, Eq. 4, are derived form the heuristic concept of z-less stratification, which is neither proved theoretically nor confirmed by experimental data.

In fact, this concept is not needed to derive the linear asymptotic formula for the velocity gradient in stationary, homogeneous, sheared flows in very strong static stability. Recall that the flux Richardson number is defined as the ratio of the consumption of turbulent kinetic energy (TKE) caused by the negative buoyancy forces, $-\beta F_\theta$, to the shear generation of the TKE, $\tau dU/dz$:

$$\mathrm{Ri}_f = \frac{-\beta F_\theta}{\tau dU/dz}. \tag{8}$$

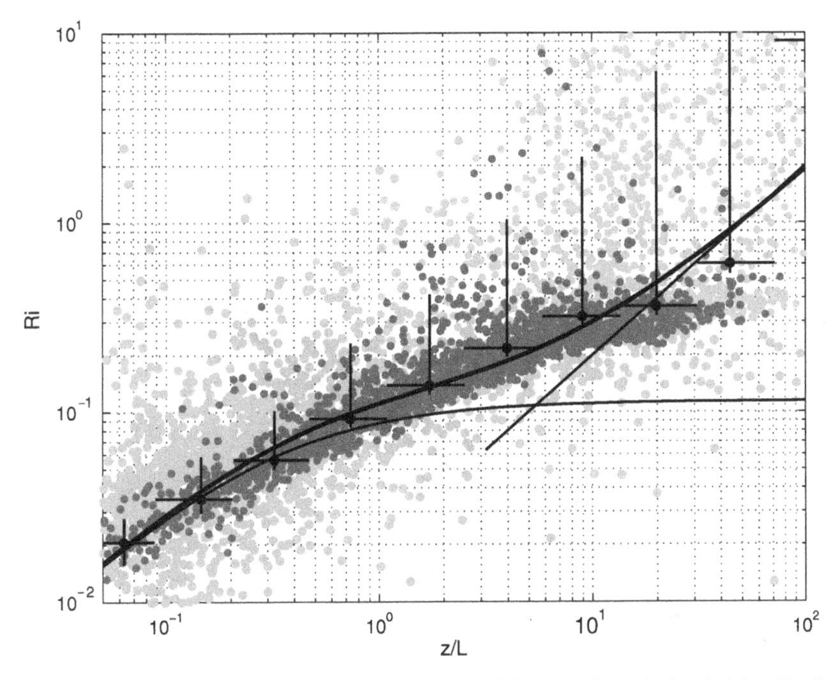

Fig. 3 Gradient Richardson number, Ri, within and above the ABL versus dimensionless height z/L, after the NS ABL data from LES DATABASE64 (dark grey points are for $z < h$ and light grey points, for $z > h$) and observational data from the field campaign SHEBA (green points). Heavy black points with error bars (one standard deviation above and below each point) show the bin-averaged values of Ri after the DATABASE64. The bold curve shows Eq. 10 with Φ_H taken after Eq. 11b, $C_{U1} = 2$, $C_{\Theta 1} = 1.6$ and $C_{\Theta 2} = 0.2$; the steep thin line shows its asymptote: $\mathrm{Ri} \sim z/L$; and the thin curve with a plateau (unrealistic in the upper part of the ABL) shows Eq. 10 with the traditional, linear approximation of $\Phi_H = 1 + 2z/L$

Ri_f (in contrast to the gradient Richardson number, Ri) cannot grow infinitely, otherwise the TKE consumption would exceed its production. Hence Ri_f at very large ξ should tend to a limit, Ri_f^∞ (= 0.2 according to currently available experimental data, see Zilitinkevich et al. 2007b). Then solving Eq. 8 for dU/dz and substituting Ri_f^∞ for Ri_f gives the asymptote

$$\frac{dU}{dz} \to \frac{\tau^{1/2}}{\mathrm{Ri}_f^\infty L}, \tag{9}$$

which in turn gives $\Phi_M \to k(Ri_f^\infty)^{-1}\xi$, and thus rehabilitates Eq. 4 for Φ_M. The gradient Richardson number becomes

$$\mathrm{Ri} \equiv \frac{\beta d\Theta/dz}{(dU/dz)^2} = \left(\frac{k^2}{k_T}\right)\frac{\xi \Phi_H(\xi)}{(1 + C_{U1}\xi)^2}. \tag{10}$$

Therefore to ensure unlimited growth of Ri with increasing ξ (in other words, to guarantee "no critical Ri"), the asymptotic ξ dependence of Φ_H should be stronger than linear. Recalling that the function Φ_H at small ξ is known to be close to linear, a reasonable compromise could be a quadratic polynomial [recall the above quoted conclusion of Esau and Byrkjedal (2007) that $C_{\Theta 1}$ in Eq. 4b increases with increasing z/L).

On these grounds we adopt the approximations $\Phi_M = 1 + C_{U1}\xi$ and $\Phi_H = 1 + C_{\Theta 1}\xi + C_{\Theta 2}\xi^2$ covering both the TN and NS ABL. To extend them to the CN and LS ABL, we

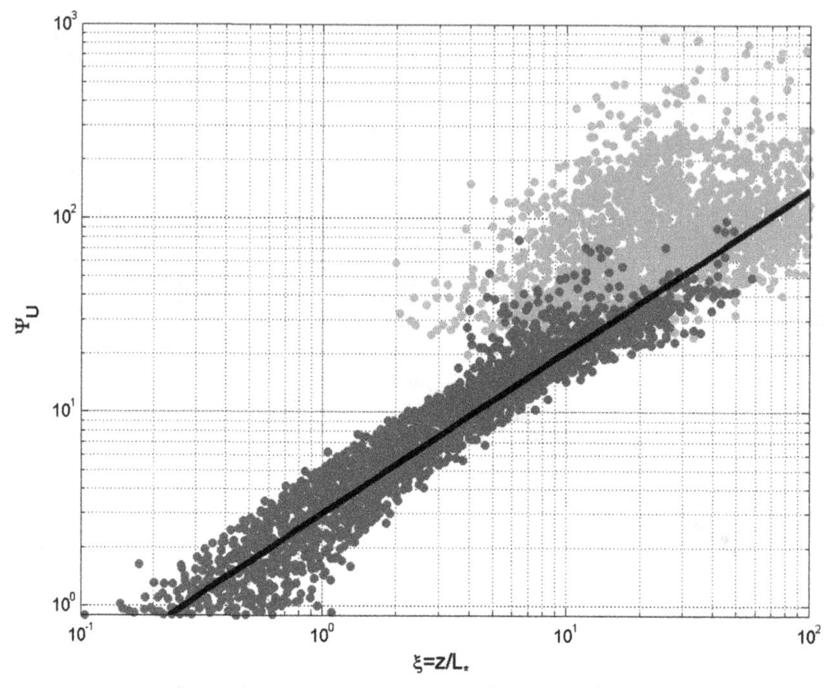

Fig. 4 The wind-speed characteristic function $\Psi_U = k\tau^{-1/2}U - \ln(z/z_{0u})$ versus dimensionless height $\xi = z/L_*$, after the LES DATABASE64. Dark grey points show data for $z < h$; light grey points, for $z > h$. The line shows Eq. 13a with $C_U = 3.0$

employ the generalised scaling, Eqs. 6–7:

$$\Phi_M = 1 + C_{U1}\frac{z}{L_*}, \tag{11a}$$

$$\Phi_H = 1 + C_{\Theta 1}\frac{z}{L_*}\xi + C_{\Theta 2}\left(\frac{z}{L_*}\right)^2. \tag{11b}$$

Comparing Eqs. 9 and 11a gives $\mathrm{Ri}_f^\infty = kC_{U1}^{-1}$. Then taking conventional values of $\mathrm{Ri}_f^\infty = 0.2$ and $k = 0.4$ gives an *a priori* estimate of C_{U1}= 2.

Figures 1 and 2 show Φ_M and Φ_H vs. $\xi = z/L_*$ after the LES DATABASE64 (Beare et al. 2006; Esau and Zilitinkevich, 2006), which includes the TN, CN, NS, and LS ABLs. Figure 2 confirms that the ξ dependence of Φ_H is indeed essentially stronger than linear: With increasing ξ, the best-fit linear dependence $\Phi_H = 1+2\xi$ shown by the thin line diverge from data more and more, and at $\xi \gg 1$ becomes unacceptable. The steeper thin line shows the quadratic asymptote $\Phi_H = 0.2\xi^2$ relevant only for very large ξ. Figure 1 confirms the expected linear dependence. Both figures demonstrate a reasonably good performance of Eq. 11 over the entire ABL depth (data for $z < h$ are indicated by dark grey points) and allows determining the constants $C_{U1} = 2$ (coinciding with the above *a priori* estimate), $C_{\Theta 1} = 1.6$ and $C_{\Theta 2} = 0.2$, with the traditional values of the von Karman constants: $k = 0.4$ and $k_T = 0.47$. For comparison, data for $z > h$ (indicated by light grey points) quite expectedly exhibit wide spread. The composite scale L_* is calculated after Eqs. 6–7 with $C_N = 0.1$ and $C_f = 1$.

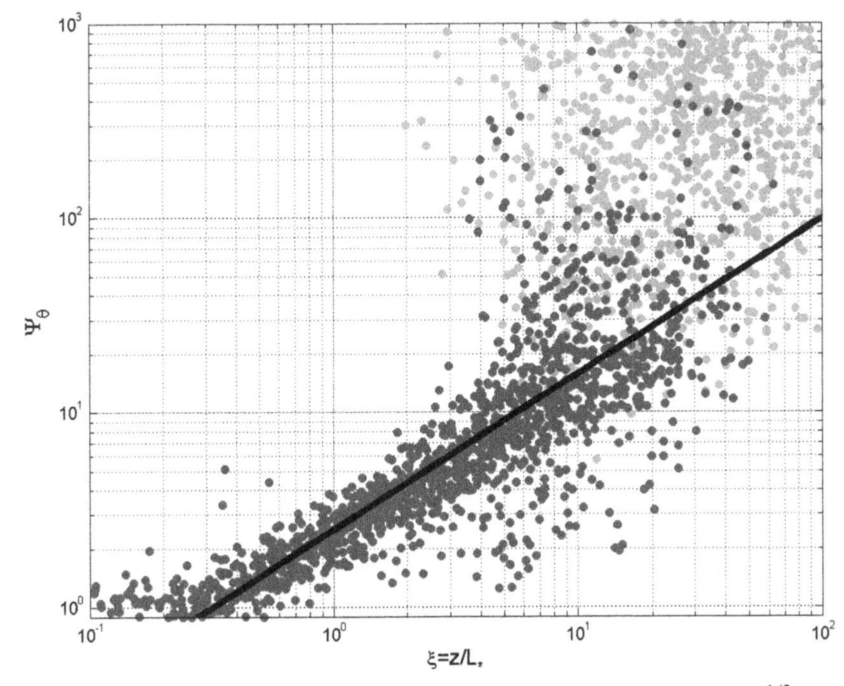

Fig. 5 Same as in Fig. 4 but for the potential-temperature characteristic function $\Psi_\Theta = k_T \tau^{-1/2} (\Theta - \Theta_0) (-F_\theta)^{-1} - \ln(z/z_{0u})$. The line shows Eq. 13b with $C_\Theta = 2.5$

Figure 3 shows the gradient Richardson number, Eq. 10, vs. z/L after the LES data for the TN and NS ABL (indicated by dark and light grey points, as in Figs. 1 and 2) and data from meteorological mast measurements at about 5, 9 and 13 m above the snow surface in the field campaign SHEBA (Uttal et al. 2002) indicated by green points. The bold curve shows our approximation of $\mathrm{Ri} = k^2 k_T^{-1} \xi \Phi_H \Phi_M^{-2}$—taking Φ_M and Φ_H after Eq. 11 with $C_{U1} = 2$, $C_{\Theta 1} = 1.6$ and $C_{\Theta 2} = 0.2$; the thin curve shows the traditional approximation of Ri—taking Φ_M and Φ_H after Eq. 4 with $C_{U1} = 2$ and $C_{\Theta 1} = 2$ (it affords a critical value of Ri≈0.17); and the steep thin line shows the asymptotic behaviour of our approximation, $\mathrm{Ri} \sim z/L$, at large z/L. Heavy points with error bars are the bin-averaged values after LES DATABASE64.

This figure demonstrates consistency between the LES and the field data for such a sensitive parameter as Ri (the ratio of gradients—inevitably determined with pronounced errors). For our analysis this result is critically important. It allows using the LES DATABASE64 on equal grounds with experimental data. Recall that in using LES we have the advantage of fully controlled conditions, which is practically unachievable in field experiments.

We give here one example: dealing with LES data we are able to distinguish between data for the ABL interior, $z < h$ (indicated in our figures by dark grey points) and data for $z > h$ (indicated by light grey points). As seen in Fig. 3, the gradient Richardson number within the ABL practically never exceeds 0.25–0.3, although turbulence is observed at much larger Ri. This observation perfectly correlates with the recent theoretical conclusion that Ri ~ 0.25 is not the critical Ri in the traditional sense (the border between turbulent and laminar regimes) but a threshold separating the two turbulent regimes of essentially different nature: strong, fully developed turbulence at Ri $\ll$0.25; and weak, intermittent turbulence at Ri$\gg$0.25

(Zilitinkevich et al. 2007b). These two are just the regimes typical of the ABL and the free atmosphere, respectively.

3 Surface fluxes

The above analysis clarifies our understanding of the physical nature of the stable ABL but does not immediately give flux–profile relationships suitable for practical applications. To receive analytical approximations of the mean wind and temperature profiles, $U(z)$ and $\Theta(z)$, across the ABL, we apply the generalised similarity theory presented in Sect. 2 to "characteristic functions":

$$\Psi_U = \frac{kU(z)}{\tau^{1/2}} - \ln\frac{z}{z_{0u}}, \tag{12a}$$

$$\Psi_\Theta = \frac{k_T\tau^{1/2}\left[\Theta(z) - \Theta_0\right]}{-F_\theta} - \ln\frac{z}{z_{0u}}, \tag{12b}$$

and employ LES DATABASE64 to determine their dependences on $\xi = z/L_*$.

Results from this analysis presented in Figs. 4 and 5 are quite constructive. Over the entire ABL depth, Ψ_U and Ψ_Θ show practically universal dependences on ξ that can be reasonably accurately approximated by the power laws:

$$\Psi_U = C_U\xi^{5/6}, \tag{13a}$$

$$\Psi_\Theta = C_\Theta\xi^{4/5}, \tag{13b}$$

with C_U= 3.0 and C_Θ= 2.5.

The wind and temperature profiles become

$$\frac{kU}{\tau^{1/2}} = \ln\frac{z}{z_{0u}} + C_U\left(\frac{z}{L}\right)^{5/6}\left[1+\frac{(C_N N)^2+(C_f f)^2}{\tau}L^2\right]^{5/12}, \tag{14a}$$

$$\frac{k_T\tau^{1/2}(\Theta - \Theta_0)}{-F_\theta} = \ln\frac{z}{z_{0u}} + C_\Theta\left(\frac{z}{L}\right)^{4/5}\left[1+\frac{(C_N N)^2+(C_f f)^2}{\tau}L^2\right]^{2/5}, \tag{14b}$$

where $C_N = 0.1$ and $C_f = 1$ (see discussion of Eq. 7). Given $U(z)$, $\Theta(z)$ and N, Eqs. 14a, b allow determination of the turbulent fluxes, τ and F_θ, and the Obukhov length, $L = \tau^{3/2}(-\beta F_\theta)^{-1}$, at the computational level z. Numerical solution to this system is simplified by the fact that the major terms on the right-hand sides are the logarithmic terms, and moreover, the second terms in square brackets are usually small compared to unity. Hence iteration methods should work efficiently. As a first approximation N, unknown until we determine the ABL height, is taken $N = 0$. In the next iterations, it is calculated using Eq. 18.

Given τ and F_θ, the surface fluxes are calculated using dependencies:

$$\frac{\tau}{\tau_*} = \exp\left[-3\left(\frac{z}{h}\right)^2\right], \tag{15a}$$

$$\frac{F_\theta}{F_*} = \exp\left[-2\left(\frac{z}{h}\right)^2\right]. \tag{15b}$$

For details see Zilitinkevich and Esau (2005) and Esau and Byrkjedal (2007).

The ABL height, h, required in Eq. 15 is calculated using the multi-limit h model (Zilitinkevich et al. 2007a, and references therein) consistent with the present analysis. The diagnostic version of this model determines the equilibrium ABL height, h_E:

$$\frac{1}{h_E^2} = \frac{f^2}{C_R^2 \tau_*} + \frac{N|f|}{C_{CN}^2 \tau_*} + \frac{|f\beta F_*|}{C_{NS}^2 \tau_*^2}, \tag{16}$$

where C_R= 0.6, C_{CN}= 1.36 and C_{NS}= 0.51 are empirical dimensionless constants.

More accurately h can be calculated using the prognostic, relaxation equation (Zilitinkevich and Baklanov 2002):

$$\frac{\partial h}{\partial t} + \vec{U} \cdot \nabla h - w_h = K_h \nabla^2 h - C_t \frac{u_*}{h_E}(h - h_E), \tag{17}$$

which therefore should be incorporated in a numerical model employing our scheme. In Eq. 17, h_E is taken after Eq. 16, w_h is the mean vertical velocity at the height $z = h$ (available in numerical models), the combination $C_t u_* h_E^{-1}$ is the inverse ABL relaxation time scale, $C_t \approx 1$ is an empirical dimensionless constant, and K_h is the horizontal turbulent diffusivity (the same as in other prognostic equations of the model under consideration).

Finally, given h, the free-flow Brunt-Väisälä frequency, N, is determined through the root-mean-square value of the potential temperature gradient over the layer $h < z < 2h$:

$$N^4 = \frac{1}{h} \int_h^{2h} \left(\beta \frac{\partial \Theta}{\partial z} \right)^2 dz \tag{18}$$

and substituted into Eq. 14 for the next iteration.

Some problems (first of all, air–sea interaction) require not only the absolute value of the surface momentum flux, $\vec{\tau}_*$, but also its direction. Recalling that our method allows determination of the ABL height, h, and therefore the wind vector at this height, $\vec{U}_h$, the problem reduces to the determination of the angle, α_* between $\vec{U}_h$ and $\vec{\tau}_*$. For this purpose we employ the cross-isobaric angle formulation:

$$\sin \alpha_* = \frac{-fh}{kU_h} \left[-2 + 10 \frac{(-\beta F_* h)^2}{\tau_*^3} + 0.225 \frac{(Nh)^2}{\tau_*} + 10 \frac{(fh)^2}{\tau_*} \right], \tag{19}$$

based on the same generalised similarity theory as the present paper (see Eqs. 7b, 41b, 43 and Fig. 4 in Zilitinkevich and Esau (2005)).

Following the above procedure, Eqs. 14–18 allow calculating the following parameters:

- turbulent fluxes $\tau(z)$ and $F_\theta(z)$ at any computational level z within the ABL,
- surface fluxes, $\vec{\tau}_*$ and F_*,
- ABL height, h,[either diagnostically after Eq. 16 or more accurately, accounting for its evolution after Eqs. 16–17].

Empirical constants that appear in the above formulations are given in Table 1.

The proposed method can be applied, in particular, to the shallow ABL, when the lowest computational level is close to h, and standard approaches completely fail. But it has advantages also in situations when the ABL (the height interval $0 < z < h$) contains several computational levels. In such cases, it provides several independent estimates of h, u_*^2 and F_*, and by this means makes available a kind of data assimilation, namely, more reliable determination of h, u_*^2 and F_* through averaging over all estimates.

Table 1

Constant	In Equation	Comments
$k = 0.4,\ k_T = 0.47$	(1), (3), etc	traditional values
$C_N = 0.1,\ C_f = 1$	(7)	after Zilitinkevich and Esau (2005), slightly corrected
$C_{U1} = 2.0,\ C_{\Theta 1} = 1.6,\ C_{\Theta 2} = 0.2$	(11a,b)	after present paper; $C_{U1} = 2.0$ and $C_{\Theta 1} = 1.6$ correspond to the coefficients $\beta_1 = C_{U1}/k = 5.0$ and $\beta_2 = C_{\Theta 1}/k = 4.0$ in the log-linear laws formulated for $L = u_*^3(-k\beta F_*)^{-1}$
$C_U = 3.0,\ C_\Theta = 2.5$	(13), (14)	after present paper
$C_R = 0.6,\ C_{CN} = 1.36,\ C_{NS} = 0.51$	(16)	after Zilitinkevich et al. (2007a)
$C_t = 1$	(17)	after Zilitinkevich and Baklanov (2002)

4 Concluding remarks

In this paper we employ a generalised similarity theory for the stably stratified sheared flows accounting for non-local features of the atmospheric stable ABL, follow modern views on the turbulent energy transformations rejecting the critical Richardson number concept, and use recent, high quality experimental and LES data to develop analytical formulations for the wind speed and potential temperature profiles across the entire ABL.

Results from our analysis are validated using LES data from DATABASE64 covering the four types of ABL: truly neutral, conventionally neutral, nocturnal stable and long-lived stable. These LES are in turn validated through (shown to be consistent with) observational data from the field campaign SHEBA.

Employing generalised formulae for the dimensionless velocity and potential temperature gradients, Φ_M and Φ_H, Eq. 3, based on the composite turbulent length scale L_*, Eq. 7, and z-dependent turbulent velocity and temperature scales, $\tau^{1/2}$ and $F_\theta \tau^{-1/2}$, we demonstrate that Φ_M and Φ_H are to a reasonable accuracy approximated by universal functions of z/L_* (Φ_M linear, Φ_H stronger than linear) across the entire ABL.

Using the quadratic polynomial approximation for Φ_H, we demonstrate that our formulation leads to the unlimitedly increasing z/L dependence of the gradient Richardson number, Ri, consistent with both LES and field data and arguing against the critical Ri concept.

We employ the above generalised format to the deviations, Ψ_U and Ψ_Θ, Eq. 12, of the dimensionless mean wind and potential temperature profiles from their logarithmic parts [$\sim \ln(z/z_{0u})$] to obtain power-law approximations: $\Psi_U \sim (z/L_*)^{5/6}$ and $\Psi_\Theta \sim (z/L_*)^{4/5}$ that perform quite well across the entire ABL.

On this basis, employing also our prior ABL height model and resistance laws, we propose a new method for calculating the turbulent fluxes at the surface in numerical models.

Acknowledgements This work has been supported by the EU Marie Curie Chair Project MEXC-CT-2003-509742, ARO Project W911NF-05-1-0055, EU Project FUMAPEX EVK4-2001-00281, EU Project TEMPUS 26005, Norwegian Project MACESIZ 155945/700, joint Norway-USA Project ROLARC 151456/720, and NORDPLUS Neighbour 2006-2007 Project 177039/V11.

References

Ayotte KW, Sullivan PP, Andren A, Doney SC, Holtslag AAM, Large WG, McWilliams JC, Moeng C-H, Otte M, Tribbia JJ, Wyngaard J (1996) An evaluation of neutral and convective planetary boundary-layer parameterizations relative to large eddy simulations. Boundary-Layer Meteorol 79:131–175

Beare RJ, MacVean MK, Holtslag AAM, Cuxart J, Esau I, Golaz JC, Jimenez MA, Khairoudinov M, Kosovic B, Lewellen D, Lund TS, Lundquist JK, McCabe A, Moene AF, Noh Y, Raasch S, Sullivan P (2006) An intercomparison of large eddy simulations of the stable boundary layer. Boundary Layer Meteorol 118:247–272

Esau I, Byrkjedal Ø (2007) Application of large eddy simulation database to optimization of first order closures for neutral and stably stratified boundary layers. This issue of Boundary Layer Meteorol

Esau IN, Zilitinkevich SS (2006) Universal dependences between turbulent and mean flow parameters in stably and neutrally stratified planetary boundary layers. Nonlin Processes Geophys 13:135–144

Högström U (1996) Review of some basic characteristics of the atmospheric surface layer. Boundary-Layer Meteorol 78:215–246

Lenschow DH, Li XS, Zhu CJ, Stankov BB (1988) The stably stratified boundary layer over the Great Plains. Part 1: Mean and turbulence structure. Boundary-layer Meteorol 42:95–121

Monin AS, Obukhov AM (1954) Main characteristics of the turbulent mixing in the atmospheric surface layer. Trudy Geophys Inst AN SSSR 24(151):153–187

Nieuwstadt FTM (1984) The turbulent structure of the stable, nocturnal boundary layer. J Atmos Sci 41:2202–2216

Sorbjan Z (1988) Structure of the stably-stratified boundary layer during the SESAME-1979 experiment. Boundary-Layer Meteorol 44:255–266

Tjernstrom M, Zagar M, Svensson G, Cassano JJ, Pfeifer S, Rinke A, Wyser A, Dethloff K, Jones C, Semmler T, Shaw M (2004) Modelling the arctic boundary layer: an evaluation of six ARCMIP regional-scale models using data from the SHEBA project. Boundary-Layer Meteorol 117:337–381

Uttal T, 26 co-authors (2002) Surface heat budget of the arctic Ocean. Bull Amer Meteorol Soc 83:255–275

Wittich KP (1991) The nocturnal boundary layer over Northern Germany: an observational study. Boundary-Layer Meteorol 55:47–66

Yague C, Viana S, Maqueda G, Redondo JM (2006) Influence of stability on the flux-profile relationships for wind speed, phi-m, and temperature, phi-h, for the stable atmospheric boundary layer. Nonlin Processes Geophys 13:185–203

Zilitinkevich SS, Baklanov A (2002) Calculation of the height of stable boundary layers in practical applications. Boundary-Layer Meteorol 105:389–409

Zilitinkevich SS, Esau IN (2002) On integral measures of the neutral, barotropic planetary boundary layers. Boundary-Layer Meteorol 104:371–379

Zilitinkevich SS, Esau IN (2003) The effect of baroclinicity on the depth of neutral and stable planetary boundary layers. Quart J Roy Meteorol Soc 129:3339–3356

Zilitinkevich SS, Esau I (2005) Resistance and heat transfer laws for stable and neutral planetary boundary layers: old theory, advanced and ee-evaluated. Quart J Roy Meteorol Soc 131:1863–1892

Zilitinkevich SS, Grachev AA, Fairall CW (2001) Scaling reasoning and field data on the sea-surface roughness lengths for scalars. J Atmos Sci 58:320–325

Zilitinkevich SS, Perov VL, King JC (2002) near-surface turbulent fluxes in stable stratification: calculation techniques for use in general circulation models. Quart J Roy Meteorol Soc 128:1571–1587

Zilitinkevich SS, Esau I, Baklanov A (2007a) Further comments on the equilibrium height of neutral and stable planetary boundary layers. Quart J Roy Meteorol Soc (In press)

Zilitinkevich SS, Elperin T, Kleeorin N, Rogachevskii I (2007b) A minimal turbulence closure model for stably stratified flows: energy and flux budgets revisited. Submitted to this issue of Boundary Layer Meteorol

Application of a large-eddy simulation database to optimisation of first-order closures for neutral and stably stratified boundary layers

Igor N. Esau · Øyvind Byrkjedal

Abstract Large-eddy simulation (LES) is a well-established numerical technique, resolving the most energetic turbulent fluctuations in the planetary boundary layer. By averaging these fluctuations, high-quality profiles of mean quantities and turbulence statistics can be obtained in experiments with well-defined initial and boundary conditions. Hence, LES data can be beneficial for assessment and optimisation of turbulence closure schemes. A database of 80 LES runs (DATABASE64) for neutral and stably stratified planetary boundary layers (PBLs) is applied in this study to optimize first-order turbulence closure (FOC). Approximations for the mixing length scale and stability correction functions have been made to minimise a relative root-mean-square error over the entire database. New stability functions have correct asymptotes describing regimes of strong and weak mixing found in theoretical approaches, atmospheric observations and LES. The correct asymptotes exclude the need for a critical Richardson number in the FOC formulation. Further, we analysed the FOC quality as functions of the integral PBL stability and the vertical model resolution. We show that the FOC is never perfect because the turbulence in the upper half of the PBL is not generated by the local vertical gradients. Accordingly, the parameterised and LES-based fluxes decorrelate in the upper PBL. With this imperfection in mind, we show that there is no systematic quality deterioration of the FOC in the strongly stable PBL provided that the vertical model

---

The submission to a special issue of the "*Boundary-Layer Meteorology*" devoted to the NATO advanced research workshop "*Atmospheric Boundary Layers: Modelling and Applications for Environmental Security*".

I. N. Esau (✉)
G.C. Rieber Climate Institute, Nansen Environmental and Remote Sensing Centre, Thormohlensgt, 47, Bergen 5006, Norway
e-mail: igore@nersc.no

I. N. Esau · Ø. Byrkjedal
Bjerknes Centre for Climate Research, Bergen, Norway

Ø. Byrkjedal
Geophysical Institute, University of Bergen, Bergen, Norway

Atmospheric Boundary Layers. A. Baklanov & B. Grisogono (eds.),
doi: 10.1007/978-0-387-74321-9_5,

resolution is better than 10 levels within the PBL. In agreement with previous studies, we found that the quality improves slowly with the vertical resolution refinement, though it is generally wise not to overstretch the mesh in the lowest 500 m of the atmosphere where the observed, simulated and theoretically predicted stably stratified PBL is mostly located.

Keywords Atmospheric turbulence · First-order turbulence closure · Large-eddy simulation · Parameterization accuracy · Stable boundary layer

1 Introduction

Modelling of the planetary boundary layer (PBL) is a challenge for meteorological numerical simulations. Models need simplified turbulence-closure schemes to account for complex physical processes and to approximate strongly curved, non-linear mean vertical profiles of meteorological variables in this layer. As in any simplified model of physics, a turbulence closure scheme requires validation and the constraint of involved empirical constants and universal functions. This has been done in a very large number of studies involving atmospheric observations (e.g., Businger et al. 1971; Louis 1979; Businger 1988; Högström 1988) and more recently involving numerical data from large-eddy simulations (LES). A comprehensive validation of the closure schemes requires a dataset covering the entire parameter space in which models are intended to operate. Although recent observational efforts such as SHEBA (Uttal et al. 2002) and CASES-99 (Poulos et al. 2002) provided relatively accurate data throughout the surface layer and sometimes the entire PBL, any direct intercomparisons are difficult to interpret. The difficulties lurk in our inability to control the PBL evolution and its governing parameters. Therefore the performance of parameterizations in the reference tests considerably depends on modellers' skills and understanding of suitable numerical initial and boundary conditions.

In these circumstances, the intercomparisons with LES data could be a useful exercise as shown in pioneering works of Deardorff (1972) and Moeng and Wyngaard (1989). The LES data are advantageous as they provide the whole range of turbulence statistics throughout the entire PBL under controlled conditions. Moreover, the LES is conducted in idealized conditions, which are consistent with the background assumptions behind the closure schemes, thus excluding the potential influence of unaccounted physical processes on the data scatter and biases in intercomparisons. It is not surprising then that LES data are attractive. Many studies have exploited the possibilities to use LES data for turbulence research (e.g., Holtslag and Moeng 1991; Andren and Moeng 1993; Galmarini et al. 1998; Xu and Taylor 1997; Nakanishi 2001; Noh et al. 2003), including most recently the GEWEX Atmospheric Boundary Layer Study (GABLS). GABLS involved 11 different LES codes and 20 different turbulence closures as described in Beare et al. (2006) and Cuxart et al. (2006, hereafter C06) [also available on www.gabls.org].

GABLS is a case study, however, with little emphasis on the sensitivity of closures to variations in external parameters. Ayotte et al. (1996), whose work was based on intercomparisons between 10 LES runs and corresponding runs of seven single-column models with different turbulence closures, have shown that the closure performance can change significantly for neutral and convective PBLs. To our knowledge, a similar multi-case study has not been done yet for the stably stratified PBL, perhaps due to the considerable computational cost of simulating the stably stratified PBL. Based on our experience, about two orders of magnitude more computer time is needed to obtain accurate LES data for a typical nocturnal clear-sky PBL as compared to a truly neutral PBL. This is due to the strong reduction of characteristic turbulence scales both internal and integral followed by the need of adequate mesh refinement and a much smaller model timestep.

Although higher-order turbulence-closure schemes have gradually become more computationally affordable in meteorological simulations, especially in meso-meteorological ones (e.g., Tjernström et al. 2005), the majority of global-scale models (e.g., five of seven operational models in C06) still rely on simpler, first-order closures (FOC) similar to those described in Holt and Raman (1988).

To study the FOC performance, a single-column model is usually initiated with prescribed vertical profiles of velocity and temperature, which is thought to characterize the chosen case, and iterations continue until a quasi-stationary steady-state solution is achieved. In this model state, both the mean variables and the turbulent fluxes differ from the data used for comparisons, as C06 have clearly shown. It is therefore difficult to trace the reasons for differences and to identify the specific component of the FOC responsible for the deviations. Here, we exploit a different approach. We do not study an equilibrium solution, but instead we test different components of the FOC against data from a LES database of 80 neutral and stably stratified runs, hereafter DATABASE64. It is reasonable to expect that, given the stationary, steady-state mean profiles for the wind speed and temperature, an ideal parameterization would recover fluxes and tendencies similar to those from DATABASE64. Any discrepancy would indicate an internal inconsistency in the parameterization.

The performance of the FOC is also expected to change with vertical mesh resolution. In C06, the performance was studied at a vertical resolution of 6.25 m. Realistic resolutions are much coarser, and there are typically 2–4 model levels within the stably stratified PBL in actual simulations. Such a coarse resolution could significantly alter the FOC performance, because the closure relies on vertical gradients computed on a finite mesh. For instance, Lane et al. (2000) showed considerable convergence of the single-column model results on consequently refined meshes. More recently, Roeckner et al. (2006) reported significant improvement in the representation of the near-surface high-latitude climate with resolution refinement in the ECHAM5 model. In contrast, C06 quoted results fairly insensitive to the vertical resolution for a mesh spacing less than 50 m. Ayotte et al. (1996) suggest a way to resolve this contradiction, showing a considerable improvement of accuracy for a resolution of up to 10 levels in the PBL. This gives a spacing of about 50 m for a typical conventionally neutral PBL, but the spacing should be much finer to resolve a long-lived stably stratified PBL (Zilitinkevich and Esau 2003).

Our study aims to assess and to optimise key elements of first-order turbulence closure for the stably stratified PBL using LES data in DATABASE64. Attention is paid to stability functions, which describe the turbulence mixing efficiency as a function of fluid stratification. We also aim to clarify whether improvements in model resolution could significantly affect the vertical mixing or whether the FOC should be modified at the physical level.

The structure of this paper is as follows. Section 2 describes the large-eddy simulation code and the DATABASE64. Section 3 describes a general first-order closure and provides details of the optimized FOC fitted to DATABASE64. Section 4 gives the FOC quality assessment over the range of stably stratified PBL and resolutions, while Sect. 5 discusses possible effects of non-local fluxes on the FOC performance. Finally, Sect. 6 outlines conclusions.

2 Large-eddy simulation data

Turbulence-resolving simulations have been conducted with the Nansen Environmental and Remote Sensing Center large-eddy simulation code LESNIC. The code solves three-dimensional momentum, temperature and continuity equations for an incompressible Boussinesq fluid, and employs a fully conservative second-order central difference scheme for the skew-symmetric form of the advection term, a fourth-order Runge-Kutta scheme for time stepping, and a direct fractional-step pressure correction scheme for the continuity preservation in the horizontally periodic domain. The computational mesh is the staggered C-type mesh, which demands only fluxes as boundary conditions at the top and bottom of the domain. LESNIC employs a dynamic mixed closure, which recalculates the Smagorinsky constant in the eddy diffusivity part of closure at every timestep. A detailed description of LESNIC can be found in Esau (2004) and intercomparisons can be found in Fedorovich et al. (2004) and Beare et al. (2006).

LESNIC has been used to compute a set of cases, referred to as DATABASE64 (Esau and Zilitinkevich 2006), and from this 80 LES runs are suitable for the present study. All runs have been computed using the equidistant mesh with 64^3 grid nodes. The aspect ratio between the vertical and horizontal grid spacings varies from 1:1 to 1:4, with the majority of data computed at the 1:2 mesh. The physical resolution varies from run to run, maintaining about 25–45 vertical levels within the fully developed PBL.

The turbulent boundary layer comprised only 1/2 to 2/3 of the depth of the computational domain, an arrangement that ensures that the largest eddies, which may occupy the entire PBL, were not affected by the limited horizontal size of the LESNIC domain. The lateral boundary conditions were periodic in all runs. At the surface, the turbulent flux of potential temperature was prescribed and therefore is considered in the study as an external parameter, whereas the turbulent flux of momentum was computed instantly and pointwise using the logarithmic wall-law. With these boundary conditions, turbulent surface flux parameterizations cannot be evaluated accurately, as the surface temperature has not been prescribed in the LESNIC but retrieved from the data by means of a parameterization.

The vertical profiles of the mean variables and the turbulence statistics have been computed from instant resolved-scale fluctuations. For instance, a vertical flux of the potential temperature, τ_θ, is computed as

$$\begin{aligned}\tau_\theta(z) &= \overline{< w'\theta' >} + \tau_{\mathrm{SGS}} \\ &= \overline{< (w(x,y,z,t) - < w(x,y,z,t) >)(\theta(x,y,z,t) - < \theta(x,y,z,t) >) >} + \tau_{\mathrm{SGS}},\end{aligned}$$

where $w(x,y,z,t)$, $\theta(x,y,z,t)$ are instant values of the vertical component of velocity and the potential temperature at every grid node. The angular brackets denote horizontal averaging over the computational domain, and the overbar denotes time averaging. All runs in DATABASE64 have been calculated for 16 hours of model time, and only the last hour was used for the time averaging. A subgrid-scale flux, τ_{SGS}, is provided from the dynamic mixed closure employed in the LESNIC. Comparing DATABASE64 with the LESNIC runs at 128^3 mesh and at 256^3mesh (for one truly neutral case), we estimated that about 80% of the fluctuation kinetic energy and 65% of the vertical velocity variations at the first computational level were resolved. Thus, τ_{SGS} is about 20–40% of τ_θ at the first level and reduces with height.

3 Optimization of the first-order turbulence closure

3.1 Description

The prognostic tendency due to the turbulent flux divergence (e.g., Holt and Raman 1988; C06) can be expressed as

$$\frac{\partial \psi}{\partial t} = -\nabla_z \tau_\psi, \tag{1}$$

where ψ is a prognostic variable, and τ_ψ is its vertical turbulent flux. To avoid model specific details, we consider the closure in Eq. 1 as applied to a horizontally homogeneous boundary layer of incompressible, barotropic flow with a constant turbulent temperature flux at the surface. Such conditions describe a realistically shallow and stably stratified PBL. The vertical flux τ_ψ is parameterized following a flux-gradient assumption (Businger et al. 1971;Deardorff 1972) through an eddy diffusivity, K_ψ, and the vertical gradient of a mean variable as

$$\tau_\psi = K_\psi \nabla_z \psi. \tag{2}$$

The eddy-diffusivity parameterization generally follows Louis (1979); there is, however, a rich variety of individual variations in different models, and a general form reads

$$K_\psi = l^2 \left| \nabla_z \vec{U} \right| f_\psi. \tag{3a}$$

Another non-local formulation (e.g., Troen and Mahrt 1986) based on the predefined shape of the K_ψ profile is also popular, and for cubic profiles reads

$$K_\psi = \kappa \tau_0^{1/2} \phi_\psi^{-1} z (1 - z/H)^2, \tag{3b}$$

where $\phi_\psi = 1 + C_\psi z/L$; f_ψ is a stability correction function; l is a mixing length scale; C_ψ is an empirical constant; z is the height above the surface; and H is the boundary-layer depth. Both f_ψ and l are universal functions of the gradient Richardson number, $Ri = g/\theta_0 \nabla_z \theta \left(\nabla_z \vec{U}\right)^{-2}$. The mean variables are: $\vec{U}$— the horizontal wind velocity, θ is the potential temperature, and $L = -\tau_0^{3/2}/g\theta_0^{-1}\tau_{\theta 0}$ is the Obukhov length scale. The index 0 denotes surface values of the momentum flux, $\tau(z) = \sqrt{< u'w' >^2 + < v'w' >^2}$, and temperature flux, $\tau_\theta(z) =< \theta' w' >$. The constants κ, g, θ_0 are the von Karman constant, 0.41, the gravity acceleration, 9.8 m s^{-2}, and the reference potential temperature, 300 K respectively. Theory and observational data suggest that the stability functions f_m, f_θ for momentum and temperature should differ asymptotically, as the ratio of the corresponding turbulent fluxes, known as the Prandtl number Pr, changes with Ri. However, there are no obvious reasons for l to be different for momentum and temperature, as is often assumed in C06 parameterizations.

3.2 Fitting of empirical functions

There is a great variety of parameterizations for f_ψ and l, as these have been thought to be the best variables for modification; surveys of the proposed functions can be found in Holt and Raman (1988), Derbyshire (1999), Mahrt and Vickers (2003) and C06. It is interesting to determine these functions and to assess their universality using data from DATABASE64. An original concept behind the distinction between f_ψ and l was to use $l = l(z)$ to account for the eddy-damping effect of an impenetrable wall (Nikuradze 1933; Blackader 1962) and to use $f_\psi = f_\psi(Ri)$ to account for the effect of the static stability (Djolov 1973; Louis 1979). Here, we will refer only to the original concept, excluding from the discussion the existing multitude of l modifications that involve stability corrections (e.g., Mahrt and Vickers 2003). Figure 1 shows the mixing length scale obtained in truly neutral runs as

$$l = \tau_\psi^{1/2} \left(\left| \nabla_z \vec{U} \right| \nabla_z \vec{U} \right)^{-1/2}. \tag{4}$$

The mixing length scale is indeed a universal function in the lower half of the PBL, and can be approximated with a simple relationship (Blackader 1962) as

$$l = (1/\kappa z + 1/l_0)^{-1}, \tag{5}$$

where $\kappa = 0.41$ is the von Karman constant. It is still used in many models. The best asymptotic fit of l_0/H to the DATABASE64 (Fig. 2) is $l_0/H = 0.3$, close to the empirical fit in meteorological models (e.g., Ballard et al. 1991). The l_0/H ratio gives $100\,\mathrm{m} < l_0 < 200\,\mathrm{m}$ for the near-neutral PBL, which includes the frequently quoted value $l_0 = 150\,\mathrm{m}$ (Viterbo et al. 1999). The mixing length scale exhibits irregular fluctuations in the middle of the PBL, fluctuations that are thought to be due to the inflection point on the velocity profile where the mean velocity gradient is close to zero. This feature may be related to roll structures (Brown 1972) or low-level jet development (Thorpe and Guymer 1977). Thus, the large values of l may be due to an inconsistency between the local flux-profile assumption and the more complex non-local nature of PBL turbulence. To summarise the literature on l_0 (e.g., Weng and Taylor 2003), we compute the ratio

$$l_0 = C_l q_n = C_l \frac{\int_0^\infty z E^{1/2} \mathrm{d}z}{\int_0^\infty E^{1/2} \mathrm{d}z}. \tag{6}$$

where E is the turbulent kinetic energy and C_l is an empirical constant. Weng and Taylor cited the range $0.1 < C_l < 0.25$ but argued that C_l should be as small as 0.055. Using the proportionality $q_n = 0.625H$ (see Fig. 3), we determine $C_l = (1/0.625)\,l_0/H = 0.48$ from DATABASE64, so our C_l is an order of magnitude larger than those previously cited and would require retuning of other turbulent constants.

Using the fitted mixing length scale from Eq. 5 in combination with Eq. 2 and Eq. 3a, one can compute the stability correction functions from DATABASE64 as

$$f_\psi = \frac{\tau_\psi}{\left| \nabla_z \vec{U} \right| \nabla_z \psi (1/\kappa z + 1/l_0)^{-2}}. \tag{7}$$

These functions for momentum and temperature are shown in Fig. 4 where it should be noted that the presented LES data behave similarly to the SABLES98 data shown in Yague et al. (2006). It is clearly seen that popular approximations in C06 are in rather poor agreement with both DATABASE64, a high-resolution LESNIC run for the GABLS test case (Beare et al. 2006) and theoretical considerations presented by Zilitinkevich and Esau (2007). An empirical fit based on the bin-averaged DATABASE64 data reads

$$f_m = (1 + a_m Ri)^{-2} + b_m Ri^{1/2}, \tag{8a}$$

$$f_\theta = (1 + a_\theta Ri)^{-3} + b_\theta, \tag{8b}$$

where $a_m = 21$, $b_m = 0.005$, $a_\theta = 10$, $b_\theta = 0.0012$. It is interesting that similar functions with exponents -2 and -3 correspondingly have appeared in Derbyshire (1999). These

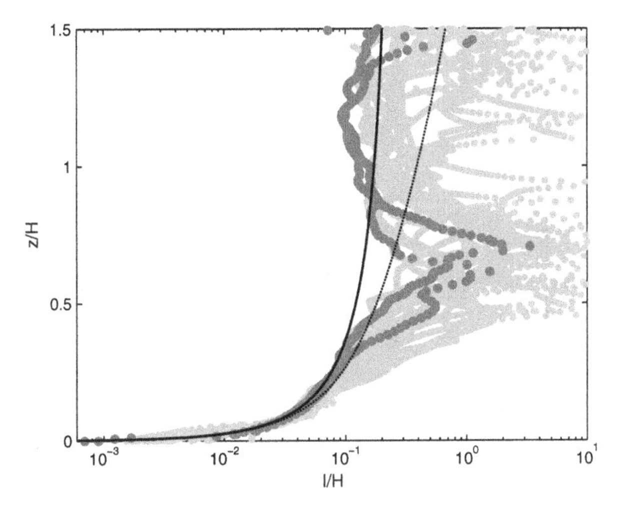

Fig. 1 Profiles of the normalized mixing length scale, l/H, where H is the PBL thickness, computed for the truly neutral LES runs from DATABASE64 (light dots) and for higher resolution 128^3 and 256^3 LES runs (dark dots). The solid curve represents the Blackader (1962) approximation after Eq. 5 and $l_0/H = 0.3$. The dashed line represents Nikuradze (1933) formulation given by $l = -0.06z^4 + 0.24z^3 - 0.2z^2 + \kappa z$

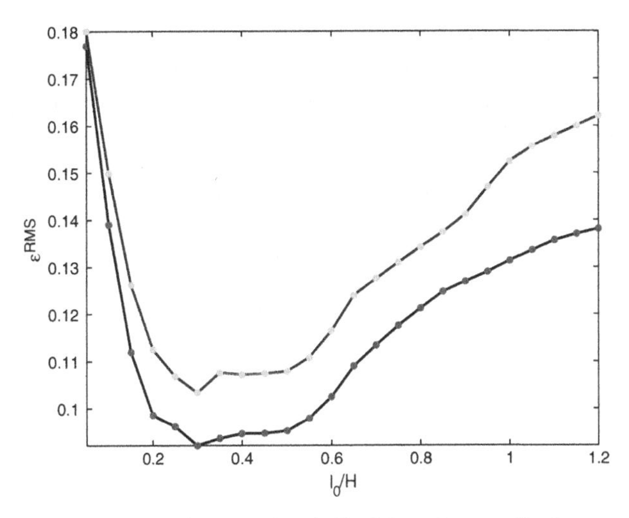

Fig. 2 Dependence of the mean relative error given in Eq. 9 from the normalized asymptotic length scale. Dark dots and solid curve—errors in momentum flux; light dots and dotted curve—errors in temperature flux

functions demonstrated physically consistent behaviour in equilibrium runs of a single-column model for the stably stratified PBL, where no decoupling between the surface and the atmosphere has been observed at even the strongest stabilities.

Figure 4 reveals substantial inconsistency between DATABASE64 and the formulations for the stability correction functions in the meteorological models. Small revisions of the Louis (1979) formulations do not improve the approximations, which have wrong asymptotes at large Ri and therefore the incorrect Prandtl number.

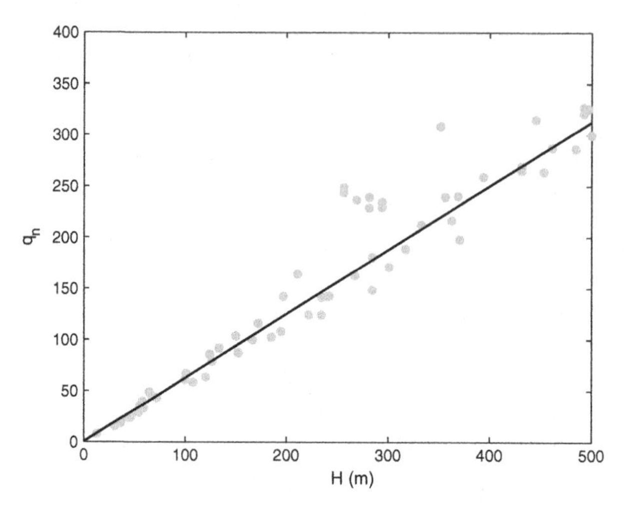

Fig. 3 Dependence of the integral normalized TKE on the PBL thickness: light dots—DATABASE64 data; solid line—the best linear fit to the data

Figure 5 shows values of the constants C_ψ in Eq. 3b determined for the non-dimensional gradients ϕ_ψ, where commonly cited values are between 4 and 7 (e.g., Högström 1988). DATABASE64 gives a rather certain estimation for momentum with $C_u = 5.9$ (= 2.5 in terms of the Zilitinkevich and Esau (2005) formulation). As expected from the Prandtl number behaviour, the constant C_θ= 7.3, matched to our LES data, does not fit the proposed linear dependence. According to Zilitinkevich and Esau (2005) both constants should appear in Eq. 3b where a generalised turbulence length scale is used instead of the surface Obukhov length scale. It is clearly seen in Eq. 3b that the linear approximation for the temperature non-dimensional gradient is not suitable, since it leads to a limitation $\mathrm{Pr} < C_\theta / C_u < 2$ on the Prandtl number. Advanced theory discussed by Zilitinkevich et al. (2007b) predicts a quadratic dependence on the non-dimensional height.

4 Quality assessment

4.1 Quality dependence on turbulence structure

Figure 6 shows vertical profiles of correlations between the FOC fluxes and the fluxes directly calculated from DATABASE64; correlations at the lowest few layers are affected by the prescribed surface fluxes. Above this layer, there is generally good agreement between DATABASE64 and the FOC in the lower half of the PBL, with the mean correlations as large as 0.9 for all cases. In the upper PBL, the correlations deteriorate, with a minimum of about 0.2.

A detailed analysis reveals that the temperature flux decorrelates only for the conventionally neutral PBL, and the long-lived PBL with a significant temperature inversion at its top. In the inversion layer, even small eddies are able to produce temperature fluctuations in the presence of large mean temperature gradients. This flux is not accounted for in the FOC. C06 showed that the majority of turbulence closures failed to develop even a relatively weak inversion in the GABLS case after 9 hours of simulation. A similar conclusion has been

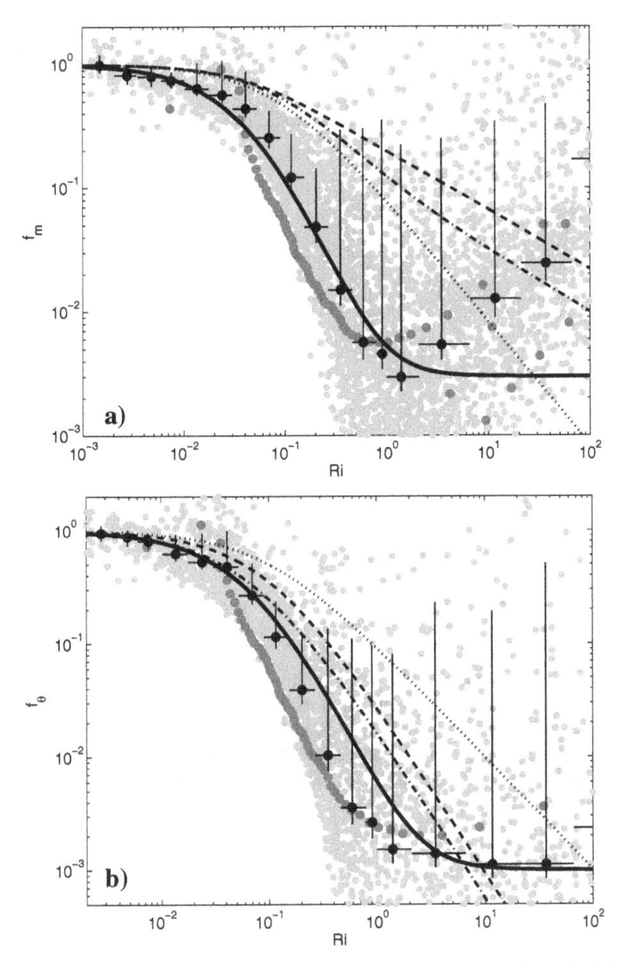

Fig. 4 Stability correction functions as functions of the gradient Richardson number: $f_m(Ri)$—for momentum (**a**); $f_h(Ri)$—for temperature (**b**). Symbols are: light dots — DATABASE64 data; dark dots – LES simulations for the GABLS case (Beare et al. 2006); black dots with error bars—bin-averaged values of equally weighted DATABASE64 data, the horizontal bars denote the bin width, the vertical bars denote one standard deviation within each bin for the data found above and below the bin-average value. Curves are: solid curve—the best fit to the bin-averaged DATABASE64 data; dashed curve—the approximation used in the ECMWF, the ARPEGE (MeteoFrance) and other models (Louis 1982; C06); dash-dotted curve—an improved MeteoFrance approximation discussed in C06; dotted curve—the approximation used in the UK MetOffice model

drawn from intercomparisons between Moderate Resolution Imaging Spectrometer (MODIS) satellite data and the National Center for Enviromental Prediction (NCEP) reanalysis data (Liu and Key 2003), which are the most consistent data that large-scale numerical models can produce. The correlations remain considerably higher in the upper nocturnal PBL where the entrainment flux is insignificant, with the average correlation about 0.8 for the temperature flux but 0.4 for the momentum flux. These results are in general agreement with Galmarini et al. (1998), who demonstrated nearly perfect reproduction of a weakly stratified nocturnal PBL using a second-order turbulence closure.

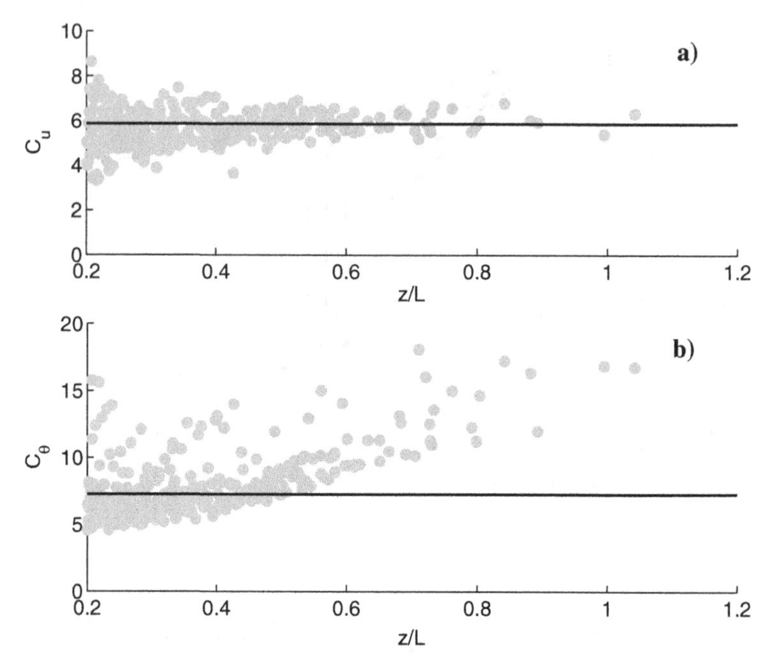

Fig. 5 Constants in flux-profile relationships after Eq. 3b: C_u— for momentum (**a**) and C_θ— for the temperature (**b**) non-dimensional gradients as functions of the height normalized by the surface Obukhov length scale. Light dots denote DATABASE64 data for the lowest 1/3 of the PBL with exception of the surface layer (1–2 computational levels in the LES). The bold lines are the mean values of the constants C_u= 5.9 and C_θ= 7.3. It is clearly seen that the linear approximation for the non-dimensional temperature gradient is not suitable

4.2 Quality dependence on stability

Even the optimized FOC is of course never perfect. A relative root-mean-square error can be calculated as

$$\varepsilon_{\psi}^{RMS} = \left(\sum_{n=1}^{N_z} \left(\tau_{\psi}^{FOC}(z_n) - \tau_{\psi}^{LES}(z_n) \right)^2 \right)^{1/2} \left(\sum_{n=1}^{N_z} \left(\tau_{\psi}^{LES}(z_n) \right)^2 \right)^{-1/2}, \tag{9}$$

where superscripts "FOC" and "LES" denote respectively the fluxes from the FOC initiated with the mean profiles from DATABASE64 and the fluxes computed directly in the LES. N_z is the number of LES levels within the PBL and z_n is the height of the nth level. After 15 h of integration, the runs in DATABASE64 provide quasi-stationary, steady-state mean variables and fluxes, and it is reasonable to expect consistency between the FOC and the LES fluxes computed from the equilibrium temperature and wind profiles, i. e., $\varepsilon_{\psi}^{RMS} \to 0$. However, Fig. 7 shows this is not the case. The error does not depend systematically on the PBL thickness, which is an integral measure of the PBL stability (Zilitinkevich and Esau 2003). The error is larger only for those cases that are absolutely dominated by the entrainment fluxes from the capping inversions, and such cases are particularly difficult for the FOC to reproduce. The minimum asymptotic error for DATABASE64 is $\varepsilon^{LES} > 0.05$ both for momentum and temperature fluxes. The existence of a rather large minimum asymptotic error has also been found by Ayotte et al. (1996).

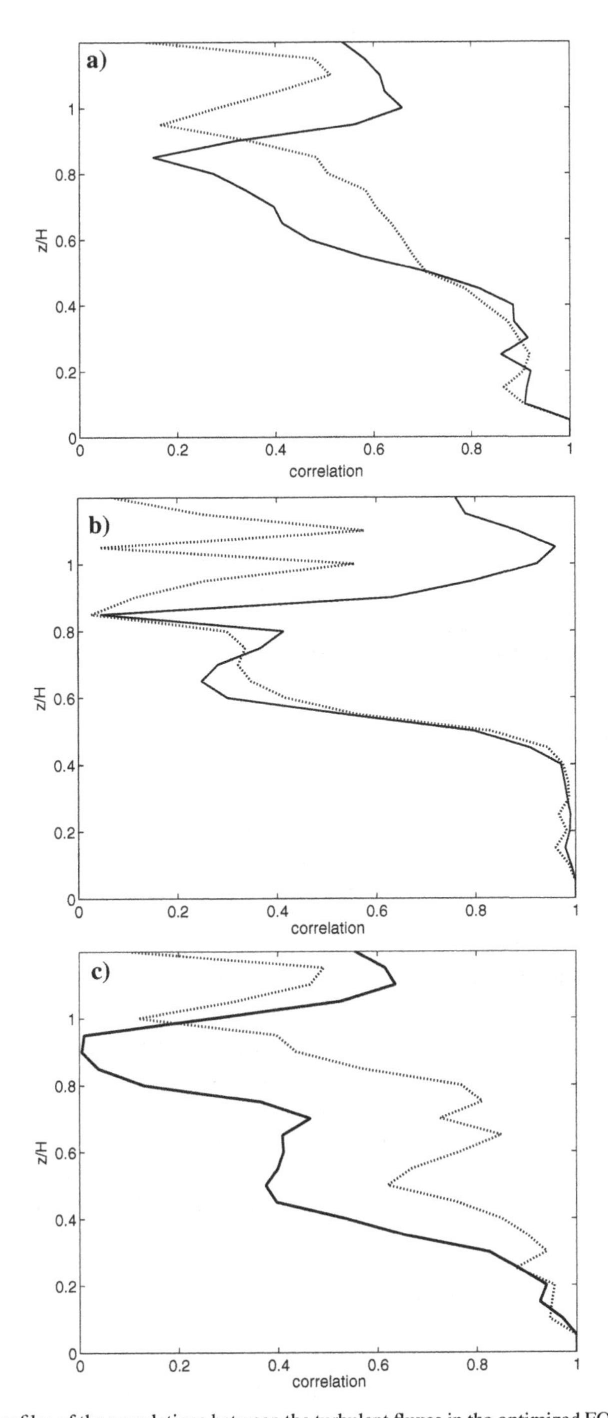

Fig. 6 Vertical profiles of the correlations between the turbulent fluxes in the optimized FOC and the directly computed LES fluxes from DATABASE64. Solid curve is for the momentum and the dotted curve for the temperature flux correlations: (**a**) averaged over all cases; (**b**) averaged over conventionally neutral cases; (**c**) averaged over nocturnal cases. High correlations near the surface, $z/H < 0.05$, are an artificial product of the numerical analysis

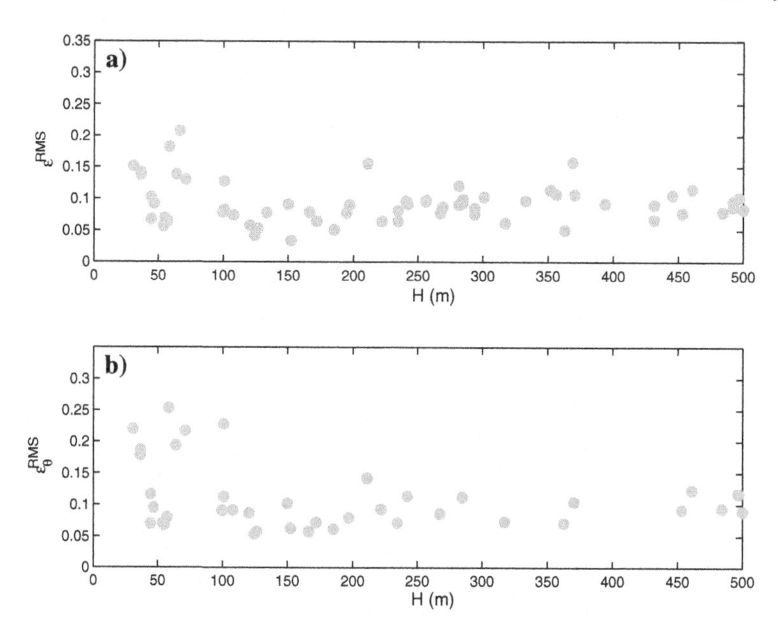

Fig. 7 Relative root-mean square error over the boundary layer after Eq. 9 for the momentum (**a**) and temperature (**b**) fluxes computed for the optimized FOC from DATABASE64 wind and temperature profiles

4.3 Quality dependence on vertical resolution

The fine vertical resolution in DATABASE64 is unattainable in most meteorological models, where model resolution varies, but is usually less than seven levels within a relatively deep, weakly stratified PBL (Tjernström et al. 2005). The resolution deteriorates to a single level for a long-lived and strongly stratified polar PBL (Cassano 2001). Ayotte et al. (1996) found that ε^{LES} is reached in neutral and convective PBLs at resolutions $N_z > 10$, and Figure 8 partially supports their conclusion. The error slowly decays with resolution refinement, though the largest quality improvement is however seen in the interval $10 < N_z < 15$. To obtain Fig. 8, we have computed ε_ψ^{RMS} for a number of regular meshes with inter-level spacing 5, 10, 15, 20, 30, 50 and 70 m for every run from DATABASE64. Then the average ε_ψ^{RMS} for every N_z was computed, subtracting ε^{LES} for every LES run.

Another important aspect of the vertical resolution is a numerical approximation of strongly-curved vertical profiles on coarse meshes using finite-difference numerical schemes. In the majority of models, calculations of the FOC require calculating the vertical gradients with the second-order central differences as, e.g., for the vertical flux divergence below

$$\nabla_z^\delta \tau_\psi = \delta^{-1}\left(\tau_\psi(z_n) - \tau_\psi(z_{n-1})\right), \tag{10}$$

where $\delta = z_n - z_{n-1}$ is the distance between nth and nth−1 model levels at which the variable ψ is found.

We can quantify distortions introduced by the finite-difference operator ∇_z^δ in the non-linear universal non-dimensional profiles of the turbulent fluxes (Zilitinkevich and Esau 2005). The proposed analytical approximations of these profiles are

$$\tau(z)/\tau_0 = \exp(-8/3(z/H)^2), \tag{11a}$$

$$\tau_\theta(z)/\tau_{\theta 0} = \exp(-2(z/H)^2), \tag{11b}$$

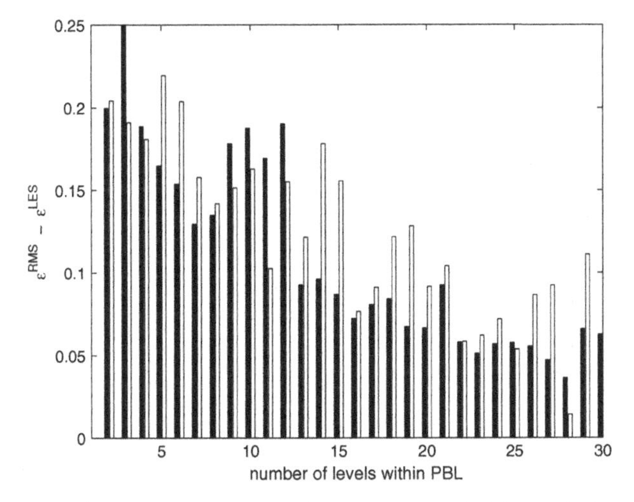

Fig. 8 The FOC quality dependence on model resolution within the PBL. The quality is computed after Eq. 9 with subtraction of the minimum error for every LES run. Dark bars are for the momentum flux and light bars for the temperature flux computed by the optimized FOC from the DATABASE64 wind and temperature profiles

where τ_0, $\tau_{\theta 0}$ are surface values of the momentum and temperature fluxes respectively. The vertical profiles of the relative distortions in velocity and temperature tendencies can be computed as

$$\varepsilon_\psi(z) = \frac{\left(\frac{\partial \psi}{\partial t}\,|_\delta - \frac{\partial \psi}{\partial t}\right)}{\frac{\partial \psi}{\partial t}} = \frac{(\nabla_z^\delta \tau_\psi - \nabla_z \tau_\psi)}{\nabla_z \tau_\psi}, \tag{12}$$

and Figure 9 shows the profiles of $\varepsilon_\psi(z)$ for different regular meshes with N_z stated accordingly. Note that the operator ∇_z^δ introduces a negative relative error in the model tendencies within the PBL and a positive error immediately above the PBL, and the same must be true for the Richardson number and the fluxes. To maintain the differentiation errors reasonably within 10% of accuracy, the model should have $N_z > 6$.

5 Discussion

The data and results presented here point to an important role of non-local turbulent mixing within the stably stratified PBL. The definition "non-local" is used here in a broad sense to indicate inconsistency between the FOC assumption of a linear flux-gradient correlation and the observed de-correlations, disregarding the physical mechanisms behind these inconsistencies. The observed features seemingly contradict the commonly accepted hypothesis that the turbulent eddies in the stably stratified PBL are small and therefore their generation/dissipation is governed by the local gradients. In fact this local mixing hypothesis is supported only by data from the lower nocturnal PBL (Nieuwstadt 1984) where effects of the capping inversion are relatively weak. In this case, the turbulence scale is limited by the local Obukhov length scale $\Lambda = \tau^{3/2}/\kappa g \theta_0^{-1} \tau_\theta$ but the PBL thickness is larger than the mean eddy size, i.e., $\Lambda < H$. Zilitinkevich (2002) suggested that in many cases the stability of the free atmosphere imposes stronger limitations on the turbulence scale than Λ.

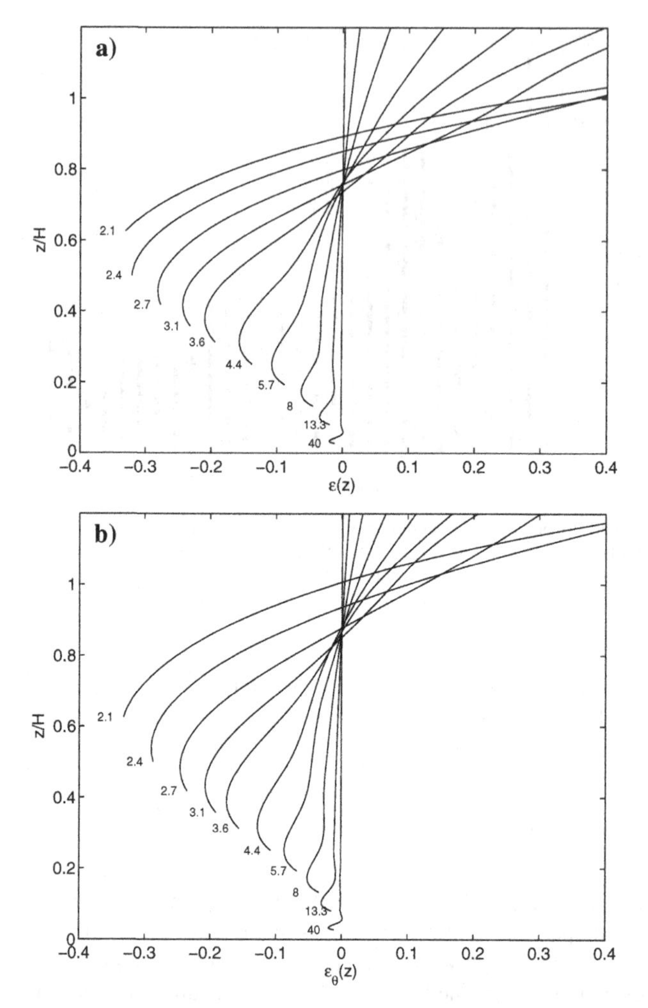

Fig. 9 Profiles of relative errors in momentum (**a**) and temperature tendencies (**b**). The errors are computed using Eq. 12 for different equidistant meshes with $N_z = H/\delta$ indicated at the bottom of the corresponding curve

Zilitinkevich and Esau (2002, 2003) and Zilitinkevich et al. (2007a) showed these limitations with the LES data. In the atmosphere, the cases with $\Lambda > H$ are frequently observed in mid- to high latitudes during wintertime where the strong negative radiation imbalance and large-scale subsidence in anticyclones act to create and strengthen temperature inversions (Overland and Guest 1991). A typical long-lived stably stratified PBL, as has been observed at the Surface Heat Budget of the Arctic (SHEBA) site, is between 100 and 200 m thick and capped by a relatively strong potential temperature inversion with $\nabla_z^\theta \approx 6\,\mathrm{K\,km^{-1}}$. It is rather unusual to observe, and very difficult to simulate, situations with $\Lambda << H$ under conditions with a significant wind speed.

Following Deardorff (1972) and Holtslag and Moeng (1991), one can estimate the relative importance of the non-local, countergradient temperature flux in the entire stably stratified PBL. DATABASE64 analysis reveals that the countergradient temperature flux is larger in the

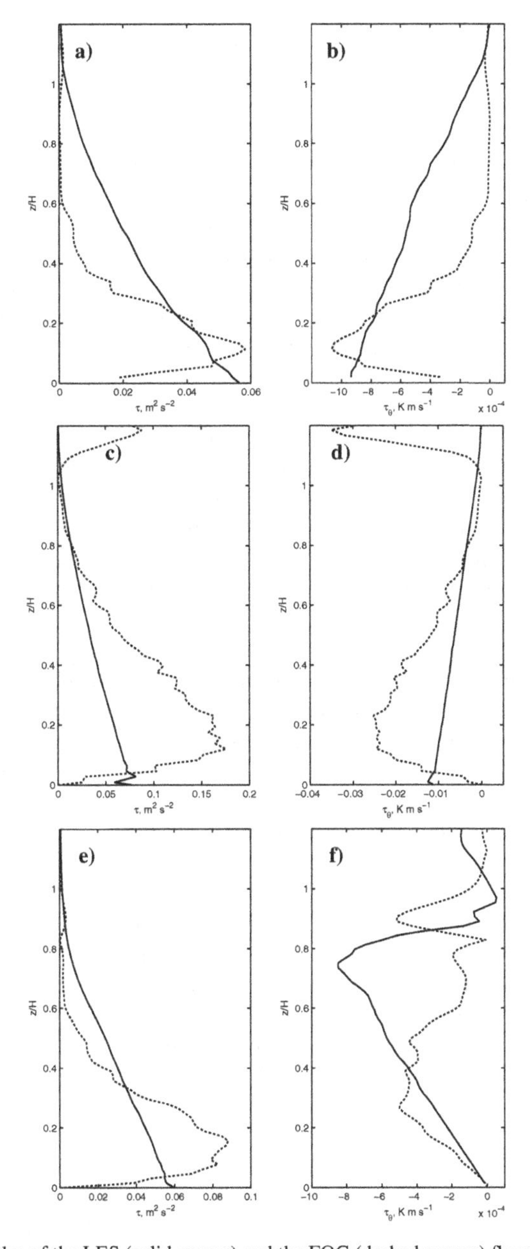

Fig. 10 Vertical profiles of the LES (solid curves) and the FOC (dashed curves) fluxes for momentum (**a, c, e**) and temperature (**b, d, f**) for three distinct cases: (**a, b**) the nocturnal PBL with zonal geostrophic wind, $U_g = 5\,\mathrm{m\,s^{-1}}$, the surface temperature flux $\tau_{\theta 0} = 0.001\,\mathrm{K\,m\,s^{-1}}$, at latitude 45 °N and surface roughness 0.1 m; (**c, d**) the long-lived stably stratified PBL in the GABLS case, with zonal geostrophic wind, $U_g = 8\,\mathrm{m}$ $\mathrm{s^{-1}}$, the variable surface temperature flux with the mean value of $\tau_{\theta 0} \approx 0.07\,\mathrm{K\,m\,s^{-1}}$, at latitude 73 °N and surface roughness 0.1 m; (**e, f**) the conventionally neutral PBL with zonal geostrophic wind, $U_g = 5\,\mathrm{m\ s^{-1}}$, the surface temperature flux $\tau_{\theta 0} = 0\,\mathrm{K\,m\,s^{-1}}$, at latitude 45 °N and surface roughness 0.1 m. The FOC fluxes in $z/H < 0.05$ are a numerical artefact since as no surface flux parameterization has been applied

LES runs with $\Lambda < H$. Moreover, this flux constitutes a large fraction of the total flux in the upper PBL in all LES runs with a developed capping inversion. Hence DATABASE64 data only partially support the Holtslag and Boville (1993) assumption that the countergradient flux is small and can be neglected in stable conditions.

Our perspective is that the FOC fails due to two different physical reasons. Consider the flux in the FOC as a multiplication of the three-dimensional eddy diffusivity characterizing the turbulent mixing and the three-dimensional mean gradient of a meteorological variable. In the asymptotic regime of the nocturnal PBL, the vertical gradients in the PBL are created only through turbulent mixing, which compensates for the lower surface temperature and the surface friction. Thus, the gradients depend on the mixing. The mixing is however suppressed as the strongest gradients are observed at the surface. Pressure variations and eddy interactions with the mean flow result in the development of eddy activity in the layers with small or no gradients. In such a layer at the top of the PBL, one can observe mixing (as DATABASE64 shows) that is considerably stronger than would be the case if imported from the lower layers. In fact the transport terms in the TKE budget are fairly small ($\sim 10^{-5}\,\mathrm{m}^2\,\mathrm{s}^{-3}$ and $\sim 10^{-6}\,\mathrm{K}^2\,\mathrm{s}^{-1}$) in the upper nocturnal PBL. In the asymptotic regime of the conventionally neutral PBL, the gradients are imposed through free atmospheric stability. Thus, the mixing imported to the top of the PBL from lower and weakly stratified layers is stronger than locally generated mixing. The TKE transport terms increase by a factor 10 or more in comparison with the nocturnal PBL.

As we have seen the top of the PBL and the capping inversion are the most difficult layers to parameterize in the framework of the first-order closure. The inversion layer is important, especially at high latitudes where the fluxes (especially the temperature and moisture fluxes) formed in that layer counteract radiative surface cooling, low-level clouds and fog formation. The FOC intrinsically underestimate fluxes, which, if not corrected, would allow the surface temperature to decrease to much lower values than have been observed. The flux enhancement, disclosed in C06, is in part to prevent such a PBL decoupling from the surface (Derbyshire 1999). Figure 10 provides intercomparisons between the LES and FOC fluxes for several high-resolution runs. Generally, the FOC overestimates fluxes in the lower PBL and underestimates them in the upper PBL. Taking into account the fact that single-column models based on the FOC in C06 predict a deeper PBL than the LES, it could be argued that in such a simulation, the entire observed PBL would be situated within the layer of overestimated fluxes. We note that the small FOC fluxes at $z/H < 0.05$ are a numerical artefact since no surface flux parameterization has been employed here.

6 Conclusions

In this study, we have used a large-eddy simulation database DATABASE64 of medium vertical resolution (25–45 levels within the PBL) to optimize the simplest but still popular first-order closure and to assess its quality over a range of governing parameters in the stably stratified PBL. The analysis has been also supported by three higher resolution LES runs, one of which was from a GABLS simulation (Beare et al. 2006). Our LES numerically resolves the most energetic fluctuations of velocity and potential temperature, and allows the use of high-quality quasi-stationary, steady state mean profiles as input for flux calculations in the FOC. The FOC fluxes were compared with directly computed turbulent fluxes from DATABASE64.

To benefit from this approach, we optimized elements of the FOC, such as the mixing length scale and the stability correction functions, to achieve the best agreement with 80 LES experiments for neutral and stably stratified PBLs in DATABASE64. We have demonstrated a general inconsistency of the traditionally used constants and universal functions with the LES data. New stability functions have been proposed that correct asymptotes in the regimes both of strong and weak mixing in accordance with DATABASE64 and the recent development in turbulence theory and observations from the SABLES98 campaign. The correction excludes the need for a critical Richardson number.

We have analysed the FOC quality as a function of the vertical model resolution and the PBL stability. The FOC is, of course never perfect, because in the upper PBL the mixing and temperature gradients are essentially non-local. Accordingly, the parameterized and LES fluxes decorrelate in the upper PBL. As in previous studies, we have shown that the simulation quality improves slowly with the vertical resolution refinement, and it is generally sufficient to have at least 10 levels within the PBL to achieve the FOC accuracy close to the asymptotically feasible values. As even 10 levels might be too costly for most meteorological models, it is wise not to overstretch the mesh in the lowest 500 m of the atmosphere where the majority of observed, simulated and theoretically predicted stably stratified PBLs are located.

Although other reasons might exists, we speculate that the common flux enhancement needed in numerical models using FOC originates from internal model inconsistencies: the coarse vertical mesh, which smoothes vertical gradients, overmixing in the lower PBL, which biases and smoothes the mean profiles, and undermixing in the upper PBL, which reduces warm air entrainment. Thus, to satisfy the surface energy budget in meteorological models, where the turbulent flux is but one of the budget components, the FOC should provide larger fluxes than is generally obtained from the flux-gradient assumption.

Acknowledgements This work has been supported by the Norwegian project POCAHONTAS 178345/S30, joint Norwegian-USA project PAACSIZ 178908/S30. Cooperation in frameworks of the NORDPLUS Neighbour 2005–2007 network FI-51 was essential for this study. The authors thank Prof. N.-G. Kvamstø (Geophysical Institute of Bergen University) and Prof. S. S. Zilitinkevich (Helsinki University) for supporting discussions.

References

Andren A, Moeng C-H (1993) Single-point Closures in a neutrally stratified boundary Layer. J. Atmos Sci 50:3366–3379

Ayotte KW, Sullivan PP, Andrén A, Doney SC, Holtslag AAM, Large WG, McWilliams JC, Moeng C-H, Otte MJ, Tribbia JJ, Wyngaard JC (1996) An evaluation of neutral and convective planetary boundary-layer parameterizations relative to large eddy simulations. Boundary-Layer Meteorol 79:131–175

Ballard SP, Golding BW, Smith RNB (1991) Mesoscale model experimental forecasts of the Haar of Northeast Scotland. Mon Wea Rev 119(9):2107–2123

Beare RJ, MacVean MK, Holtslag AAM, Cuxart J, Esau I, Golaz JC, Jimenez MA, Khairoudinov M, Kosovic B, Lewellen D, Lund TS, Lundquist JK, McCabe A, Moene AF, Noh Y, Raasch S, Sullivan P (2006) An intercomparison of large eddy simulations of the stable boundary layer. Boundary-Layer Meteorol 118(2):247–272

Blackader AK (1962) The vertical distributions of wind and turbulent exchange in neutral atmosphere. J Geophys Res 67:3095–3120

Brown RA (1972) On the inflection point instability of a stratified ekman boundary layer. J Atmos Sci 29(5):850–859

Businger JA, Wyngaard JC, Izumi Y, Bradley EF (1971) Flux-profile relationships in the atmospheric surface layer. J Atmos Sci 28:181–189

Businger JA (1988) A note on the businger-dyer profiles. Boundary-Layer Meteorol 42:145–151

Cassano JJ, Parish TR, King JC (2001) Evaluation of turbulent surface flux parameterizations for the stable surface layer over Halley, Antarctica. Mon Wea Rev 129:26–46

Cuxart J, Holtslag AAM, Beare RJ, Bazile E, Beljaars A, Cheng A, Conangla L, Freedman MEkF, Hamdi R, Kerstein A, Kitagawa H, Lenderink G, Lewellen D, Mailhot J, Mauritsen T, Perov V, Schayes G, Steeneveld GJ, Svensson G, Taylor P, Weng W, Wunsch S, Xu K-M (2006) Single Column model intercomparison for the stably stratified atmospheric boundary layer. Boundary-Layer Meteorol 118(2):273–303

Deardorff JW (1972) Parameterization of the planetary boundary layer for use in general circulation models. Mon Wea Rev 100:93–106

Derbyshire SH (1999) Boundary layer decoupling over cold surfaces as a physical boundary instability. Boundary-Layer Meteorol 90:297–325

Djolov GD (1973) Modeling of the interdependent diurnal variation of meteorological elements in the boundary layer. Ph.D. Thesis, Univ. of Waterloo, Waterloo, Ont., Canada, 127 pp

Esau I (2004) Simulation of ekman boundary layers by large eddy model with dynamic mixed subfilter closure. J Environ Fluid Mech 4:273–303

Esau I, Zilitinkevich SS (2006) Universal dependences between turbulent and mean flow parameters in stably and neutrally stratified planetary boundary layers. Nonlin Proces Geophys 13:122–144

Fedorovich E, Conzemius R, Esau I, Chow FK, Lewellen DC, Moeng, C-H, Pino D, Sullivan PP, Vila J (2004) Entrainment into sheared convective boundary layers as predicted by different large eddy simulation codes. Paper P4.7., 16th Symposium on Boundary Layers and Turbulence, Amer Meteorol Soc, January 2004, pp 1–14

Galmarini S, Beets C, Duynkerke PG, Vilà-Guerau de Arellano J (1998) Stable nocturnal boundary layers: a comparison of one-dimensional and large-eddy simulation models. Boundary-Layer Meteorol 88:181–210

Holt T, Raman S (1988) A review and comparative evaluation of multi-level boundary layer parameterizations for first order and turbulent kinetic energy closure schemes. Rev Geophys 26:761–780

Holtslag AAM, Moeng C-H (1991) Eddy diffusivity and countergradient transport in the convective atmospheric boundary layer. J Atmos Sci 48:1690–1698

Holtslag AAM, Boville B (1993) Local versus nonlocal boundary-layer diffusion in a global climate model. J Climate 6:1825–1842

Högström U (1988) Non-dimensional wind and temperature profiles in the atmospheric surface layer: a re-evaluation. Boundary-Layer Meteorol 42:55–78

Lane DE, Somerville R, Iacobellis SF (2000) Sensitivity of cloud and radiation parameterizations to changes in vertical resolution. J Climate 13:915–922

Liu Y, Key J (2003) Detection and analysis of clear-sky, low-level atmospheric temperature inversions with MODIS. J Atmos Ocean Tech 20:1727–1737

Louis J-F (1979) A parametric model of vertical eddy fluxes in the atmosphere. Boundary-Layer Meteorol 17(2):187–202

Mahrt L, Vickers D (2003) Formulation of turbulent fluxes in the stable boundary Layer. J Atmos Sci 60:2538–2548

Moeng C-H, Wyngaard JC (1989) Evaluation of turbulent transport and dissipation closures in second-order modeling. J Atmos Sci 46:2311–2330

Nakanishi M (2001) Improvement of the mellor–Yamada turbulence closure model based on large-eddy simulation data. Boundary-Layer Meteorol 99:349–378

Nieuwstadt FTM (1984) The turbulent structure of the stable nocturnal boundary layer. J Atmos Sci 41:2202–2216

Nikuradze J (1933) Strömungsgesetze in Rauhen Rohren. Forsch Arb Ing-Wes 361, Ausgabe B, Band 4, 2–22

Noh Y, Cheon WG, Hong SY, Raasch S (2003) The improvement of the K-profile model for the PBL using LES. Boundary-Layer Meteorol. 107:401–427

Overland JE, Guest PS (1991) The arctic snow and air temperature budget over sea ice during winter. J Geophys Res 96:4651–4662

Poulos GS, Blumen W, Fritts DC, Lundquist JK, Sun J, Burns SP, Nappo C, Banta R, Newsom R, Cuxart J, Terradellas E, Balsley B, Jensen M (2002) CASES-99: a comprehensive investigation of the stable nocturnal boundary layer. Bull Amer Meteorol Soc 83:555–581

Roeckner E, Brokopf R, Esch M, Giorgetta M, Hagemann S, Kornblueh L, Manzini E, Schlese U, Schulzweida U (2006) Sensitivity of simulated climate to horizontal and vertical resolution in the ECHAM5 Atmosphere Model. J Climate 19(16):3771–3791

Tjernstrom M, Zagar M, Svensson G, Cassano JJ, Pfeifer S, Rinke A, Wyser K, Dethloff K, Jones C, Semmler T, Shaw M (2005) Modelling the arctic boundary layer: an evaluation of six ARCMIP regional-scale models with data from the SHEBA project. Boundary-Layer Meteorol 117:337–381

Thorpe AJ, Guymer TH (1977) The nocturnal jet. Quart J Roy Meteorol Soc 103:633–653

Troen I, Mahrt L (1986) A simple model of the atmospheric boundary layer: sensitivity to surface evaporation. Boundary-Layer Meteorol 37:129–148

Uttal T, Curry JA, Mcphee MG, Perovich DK, Moritz RE, Maslanik JA, Guest PS, Stern H, Moore JA, Turenne R, Heiberg A, Serreze MC, Wylie DP, Persson OG, Paulson CA, Halle C, Morison JH, Wheeler PA, Makshtas A, Welch H, Shupe MD, Intrieri JM, Stamnes K, Lindsey RW, Pinkel R, Pegau WS, Stanton TS, Grenfeld TC (2002) Surface heat budget of the arctic ocean. Bull Amer Meteorol Soc 83:255–275

Viterbo P, Beljaars ACM, Mahouf J-F, Teixeira J (1999) The representation of soil moisture freezing and its impact on the stable boundary layer. Quart J Roy Meteorol Soc 125:2401–2426

Weng W, Taylor PA (2003) On modelling the one-dimensional atmospheric boundary layer. Boundary-Layer Meteorol 107(2):371–400

Xu D, Taylor PA (1997) On turbulence closure constants for atmospheric boundary-layer modelling: neutral stratification. Boundary-Layer Meteorol 84:267–287

Yague C, Viana S, Maqueda G, Redondo JM (2006) Influence of stability on the flux-profile relationships for wind speed and temperature for the stable atmospheric boundary Layer. Nonlin Proces Geophys 13:185–203

Zilitinkevich SS (2002) Third-order transport due to internal waves and non-local turbulence in the stably stratified surface layer. Quart J Roy Meteorol Soc 128:913–925

Zilitinkevich SS, Esau I (2002) On integral measures of the neutral barotropic planetary boundary Layers. Boundary-Layer Meteorol 104(3):371–379

Zilitinkevich SS, Esau I (2003) The effect of baroclinicity on the depth of neutral and stable planetary boundary Layers. Quart J Roy Meteorol Soc 129:3339–3356

Zilitinkevich SS, Esau I (2005) Resistance and heat transfer laws for stable and neutral planetary boundary layers: old theory, advanced and re-evaluated. Quart J Royal Meteorol Soc 131:1863–1892

Zilitinkevich SS, Esau I (2007) Similarity theory and calculation of turbulent fluxes at the surface for the stably stratified atmospheric boundary layer. Boundary-Layer Meteorol (this issue)

Zilitinkevich SS, Esau I, Baklanov A (2007a) Further comments on the equilibrium height of neutral and stable planetary boundary layers. Quart J Roy Meteorol Soc 133: 265–271

Zilitinkevich SS, Elperin T, Kleeorin N, Rogachevskii I (2007b) Energy- and Flux-budget (EFB) turbulence closure model for stably stratified flows. Part I: steady-state, homogeneous regimes. Boundary-Layer Meteorol (this issue)

The effect of mountainous topography on moisture exchange between the "surface" and the free atmosphere

Andreas P. Weigel · Fotini K. Chow · Mathias W. Rotach

Abstract Typical numerical weather and climate prediction models apply parameterizations to describe the subgrid-scale exchange of moisture, heat and momentum between the surface and the free atmosphere. To a large degree, the underlying assumptions are based on empirical knowledge obtained from measurements in the atmospheric boundary layer over flat and homogeneous topography. It is, however, still unclear what happens if the topography is complex and steep. Not only is the applicability of classical turbulence schemes questionable in principle over such terrain, but mountains additionally induce vertical fluxes on the meso-γ scale. Examples are thermally or mechanically driven valley winds, which are neither resolved nor parameterized by climate models but nevertheless contribute to vertical exchange. Attempts to quantify these processes and to evaluate their impact on climate simulations have so far been scarce. Here, results from a case study in the Riviera Valley in southern Switzerland are presented. In previous work, measurements from the MAP-Riviera field campaign have been used to evaluate and configure a high-resolution large-eddy simulation code (ARPS). This model is here applied with a horizontal grid spacing of 350 m to detect and quantify the relevant exchange processes between the valley atmosphere (i.e. the ground "surface" in a coarse model) and the free atmosphere aloft. As an example, vertical export of moisture is evaluated for three fair-weather summer days. The simulations show that moisture exchange with the free atmosphere is indeed no longer governed by turbulent motions alone. Other mechanisms become important, such as mass export due to topographic narrowing or the interaction of thermally driven cross-valley circulations. Under certain atmospheric

---

A. P. Weigel (✉) · M. W. Rotach
Federal Office of Meteorology and Climatology, MeteoSwiss,
Zurich, Switzerland
e-mail: andreas.weigel@meteoswiss.ch

F. K. Chow
Department of Civil and Environmental Engineering,
University of California,
Berkeley, USA

Atmospheric Boundary Layers. A. Baklanov & B. Grisogono (eds.),
doi: 10.1007/978-0-387-74321-9_6, © Springer Science+Business Media B.V. 2007

conditions, these topographical-related mechanisms exceed the "classical" turbulent contributions a coarse model would see by several times. The study shows that conventional subgrid-scale parameterizations can indeed be far off from reality if applied over complex topography, and that large-eddy simulations could provide a helpful tool for their improvement.

Keywords Large-eddy simulations · Moisture fluxes · Mountain meteorology · Surface exchange

1 Introduction

Numerical weather and climate prediction models are known to be very sensitive to surface moisture fluxes (e.g. Beljaars et al. 1996; Viterbo and Betts 1999). As these underlying surface exchange processes occur on turbulent scales, they cannot be directly resolved but need to be parameterized. In practice, surface moisture fluxes are estimated on the basis of land-surface models (see, e.g., Pitman 2003). The transfer of moisture is thereby usually described by some form of semi-empirical aerodynamic transfer coefficient that, at best, is based on similarity functions as presented by Businger et al. (1971) and Deardorff (1972), for example. Given, however, that these similarity functions have typically been obtained from measurements over flat and homogeneous topography, and given that the major part of the planetary land surface is hilly or mountainous, the general validity of such similarity functions over complex topography is questionable in principle (Rotach 1995). The same applies for the vertical transport of other quantities such as heat, momentum and pollutants.

This is not the only problem, since microscale turbulence is not the only process responsible for the vertical exchange of moisture. Transport mechanisms on the meso-γ scale, such as thermally and mechanically driven mountain winds (reviewed, e.g., by Barry 1992; Whiteman 1990, 2000), appear to be an additional component to be considered, in particular because their characteristic length scales are closely linked to the scales of the underlying topography. Because the full resolution of mountainous topography may require a horizontal grid spacing of 100 m (Young and Pielke 1983), it is clear that the effect of, say, mountain winds is not considered at all by typical weather and climate models and needs to be incorporated into their subgrid-scale (SGS) parameterizations. Indeed, Noppel and Fiedler (2001) identified mountain venting, i.e. overshooting slope winds as described by Kossmann et al. (1999), as an important exchange mechanism with the free atmosphere. With a conceptual model they showed that the effect of this mechanism has a magnitude that is comparable to the impact of microscale turbulent fluxes alone. Clearly, coarse numerical climate models, which do not resolve the topography of narrow valleys, are not able to capture this process.

Another recent example of the strong impact of mountain winds on vertical exchange is given by the aircraft measurements of Henne et al. (2004) in the Leventina and Mesolcina Valleys in southern Switzerland, where considerable net export of valley air (and thus of moisture and pollutants) into the free atmosphere is observed. According to their estimates, valley air is transported into the free atmosphere at surprisingly high rates of up to 33% of the entire valley air volume per hour. The observation of elevated moisture layers in the lower troposphere on the leeward side of the Alpine arc can perhaps only be explained by such non-turbulent processes of moisture export (Henne et al. 2005), emphasizing the importance of these transport mechanisms also for climatological considerations. Thus, a thorough evaluation of the discrepancies between the "real" fluxes of moisture (as well as heat, momentum, scalars, etc.) over complex topography and those a coarse model would parameterize appears to be required. This has so far not been possible, mainly due to a lack of experimental evidence.

In this study we focus on high-resolution numerical modelling as a new approach that may eventually serve to fill this gap in knowledge. Indeed, increases in computational power have meanwhile made it possible to simulate entire valley domains with sub-kilometre resolution (e.g., Gohm et al. 2004; Zängl et al. 2004; Chow et al. 2006; Weigel et al. 2006a) and thus to explicitly consider those topography-related motions that are neither resolved by coarse models nor accounted for in their SGS parameterizations. We present a case study in which such high-resolution simulations are applied to the Riviera Valley in southern Switzerland (Fig. 1), a typical medium-sized Alpine valley. Model output data are used to directly calculate the vertical exchange of moisture between the valley and the free atmosphere above, and to identify the dominant transport mechanisms. In particular, we seek to quantify the extent to which the "real" moisture fluxes deviate from those a coarse model would see over the same terrain. To exclude the complicating effect of microphysical processes, only fair-weather days are considered. For the Riviera Valley, a comprehensive dataset exists from the Mesoscale Alpine Program (MAP)-Riviera field campaign (Rotach et al. 2004), allowing a careful evaluation and calibration of the model performance. Thus, the benefits of high-resolution numerical modelling are combined with the "truth" provided by measurements. As a modelling tool, we use the Advanced Regional Prediction System (ARPS, Xue et al. 2000, 2001), a non-hydrostatic, compressible large-eddy simulation (LES) code.

The paper is organized as follows: Section 2 provides information on the measurement campaign as well as the numerical set-up applied. Section 3 characterizes the dynamical and thermal characteristics of the valley atmosphere on the days considered in this study. In Section 4, the "real" vertical fluxes of air mass and moisture between the Riviera Valley and the free atmosphere are calculated and compared to the exchange a coarse atmospheric model would parameterize. Concluding remarks are given in Section 5.

2 Experimental and numerical set-up

2.1 The MAP-Riviera field campaign

The MAP-Riviera measurement campaign was carried out from summer through autumn in 1999 with the aim to investigate in detail atmospheric boundary-layer processes occurring in a typical U-shaped Alpine valley (Rotach et al. 2004). The Riviera Valley has a length of approximately 15 km, a base width of about 1.5 km and an average depth of 2–2.5 km. The valley side walls have slopes of 30–35°. The dataset obtained includes sonic anemometer and profile measurements at various surface stations, radiosoundings, as well as aircraft measurements. A detailed account of the experimental set-up is provided in Rotach et al. (2004) and Weigel and Rotach (2004), and the key findings are summarized in Rotach and Zardi (2006), where all MAP related boundary-layer projects are reviewed.

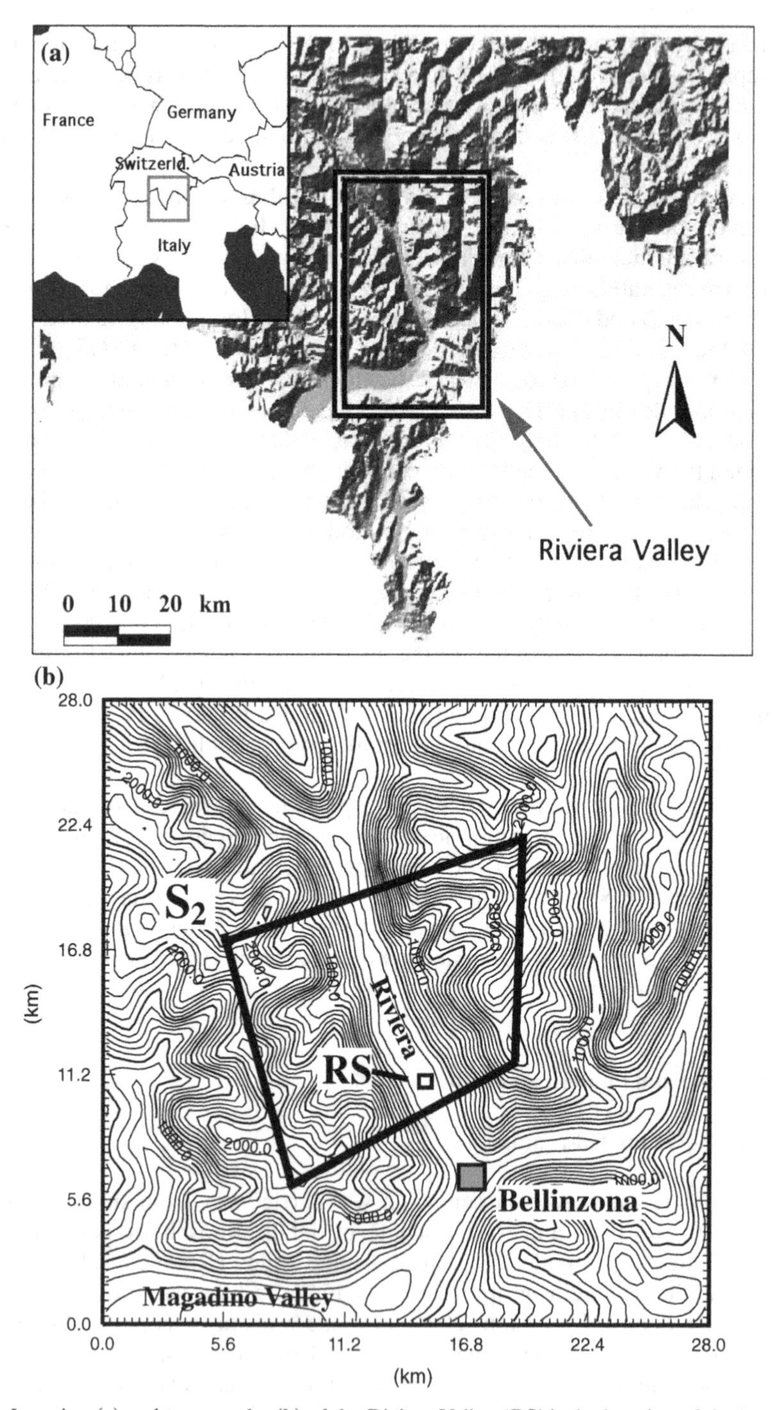

Fig. 1 Location (**a**) and topography (**b**) of the Riviera Valley. 'RS' is the location of the launch site for the radiosonde measurements. S_2 denotes the interface through which vertical moisture fluxes are calculated; (**b**) also represents the simulation domain of the 350-m grid

In this study we investigate three fair weather days of the measurement campaign, namely 21, 22 and 25 August 1999.

2.2 Numerical set-up

The model ARPS is applied in a one-way nesting mode to simulate the aforementioned three days of the measurement campaign. A grid of 9 km horizontal spacing is initialized from European Centre for Medium-Range Weather Forecasts (ECMWF) analysis data and is then successively nested down to grids of finer resolution (3 km, 1 km, 350 m). The finest domain (350-m grid spacing) is displayed in Fig. 1b. Due to computational limitations, simulations with 150-m horizontal grid spacing could only be carried out on sub-domains of the Riviera valley (Weigel et al. 2006a) and will therefore not be further considered in this study. The grid is vertically stretched using a hyperbolic tangent function, while the minimum vertical resolution is 30 m at the lowest level. For all three days, the simulations start at 1800 UTC of the previous day. More details on initialization, simulation domains and boundary conditions can be found in Chow et al. (2006) and Weigel et al. (2006a); the model set-up applied here is equivalent to the so-called "LU-SM"-set-up described in the above two references. Therein, extensive verification studies with MAP-Riviera measurements are contained, ensuring that on the three days considered here ARPS indeed accurately reproduces the complex thermal and dynamic structure observed in the valley. Here, we only provide a summary of these model evaluations.

Table 1 shows mean errors (bias) and root-mean-square errors (rmse) of simulated potential temperature, wind speed and wind direction in comparison with measurements from a surface station at the location "RS" (as shown in Fig. 1b); 46 half-hourly values (beginning at 0015 UTC) have been used to calculate the scores. For potential temperature and wind speed, the errors between measurements and simulations can be considered small, especially when compared to other typical modelling studies of this kind (e.g. Zhong and Fast 2003; Zängl et al. 2004). The simulation of 21 August thereby performs slightly worse, probably due to rainfall over the Swiss Alps on 20 August, the day of initialization for the simulation of 21 August, which makes the model initialization more error prone. The high values in rmse of wind direction are primarily due to light nighttime winds, leading to large directional fluctuations. Table 2 shows analogous intercomparisons for radiosondes up to 6 km launched from site "RS", where values are averages over 6–7 daily radiosoundings (launched between 0000 UTC and 2100 UTC). Again, the errors can be considered small. Note that no radiosonde wind comparison can be done for 21 and 22 August due to measurement failures. Comparison plots between simulation results and radiosonde, surface as well as airborne measurements are shown in Figs. 2, 3 and 4 (described in more detail below). Finally, it has been shown that ARPS is even able to capture the turbulence characteristics of the valley atmosphere. Time series of turbulent kinematic surface heat flux at various locations in the valley are accurately reproduced (Chow et al. 2006, their Fig. 9), and profiles of simulated turbulent kinetic energy reveal the same scaling characteristics as observed from the airborne measurements (Weigel 2005).

Table 1 Mean errors (bias) and root-mean-square errors (rmse) for simulations compared to surface measurements at location 'RS' in Fig. 1b (summarized from Weigel et al. 2006a; Chow et al. 2006)

Date	θ bias (K)	θ rmse (K)	U bias (m s^{-1})	U rmse (m s^{-1})	Φ bias (°)	Φ rmse (°)
21 August	−1.76	2.18	−1.67	2.41	3.01	54.82
22 August	−0.19	0.88	−1.03	1.80	−7.93	90.44
25 August	−0.41	0.69	0.57	1.28	−11.05	63.21

θ is potential temperature, U wind speed and Φ wind direction. 46 half-hourly values (beginning at 0015 UTC) have been used to calculate bias and rmse. The high values in Φ rmse are primarily due to light night-time winds, leading to large directional fluctuations

Table 2 Mean errors (bias) and root-mean-square errors (rmse) for simulations of potential temperature (θ), wind speed (U) and wind direction (Φ) in comparison with radiosoundings up to 6 km from location 'RS' (summarized from Weigel et al. 2006a; Chow et al. 2006)

Date	θ bias (K)	θ rmse (K)	U bias (m s^{-1})	U rmse (m s^{-1})	Φ bias (°)	Φ rmse (°)
21 August	0.28	1.23	NA	NA	NA	NA
22 August	0.32	0.85	NA	NA	NA	NA
25 August	−0.22	0.94	−0.12	2.04	−5.99	45.73

The values are averages over 6–7 daily radiosoundings (launched between 0000 UTC and 2100 UTC). Radiosonde wind observations for 21 and 22 August are not available (NA)

3 Thermal and dynamical characteristics

All three days considered in this study are characterized by fair weather conditions. 25 August was totally cloud-free, while 21 and 22 August showed some transient cumulus cloud formation along the mountain ridges, resulting in temporary shading and short-term decreases in turbulent surface fluxes. On all three days, synoptic-scale flow (determined from radiosoundings at 0600 UTC in 4000 m) had a direction of 270–300° with wind speeds of about 10 m s^{-1} (Weigel and Rotach 2004).

In Fig. 2a–c profiles of potential temperature as obtained from radiosonde measurements (launched from site 'RS' in Fig. 1b) are shown for the three days. A detailed description of their temporal evolution and of their "fine structure" is provided in Weigel and Rotach (2004). Here we only classify them by their macroscopic appearance: 21 and 22 August are comparable in that their afternoon temperature profiles are characterized by a more or less uniform slightly stable stratification up to about 3500–4000 m (all altitudes are above mean sea level), which is significantly higher than the surrounding ridges and peaks. The lapse rate is roughly 0.001–0.002 K m^{-1}. On 25 August, on the other hand, a very pronounced and persistent inversion is observed between about 800 and 2000 m with a lapse rate on the order of 0.005 K m^{-1}. Despite sunny conditions, the development of a well-mixed layer is suppressed on all three days. This has been shown to be due to the combined effect of cold-air advection in the along-valley direction and of a cross-valley circulation that leads to subsidence over the valley centre (Weigel et al. 2006a). Note that the temperature profiles observed on 21 and 22 August exhibit a mid-level inversion seen between 3 km and 4 km altitude. Its existence is consistent with the notion of a double-layered boundary-layer structure over complex topography, as has been proposed by Henne et al. (2004) and De Wekker et al. (2004). Figures 2d–f show the corresponding profiles of potential

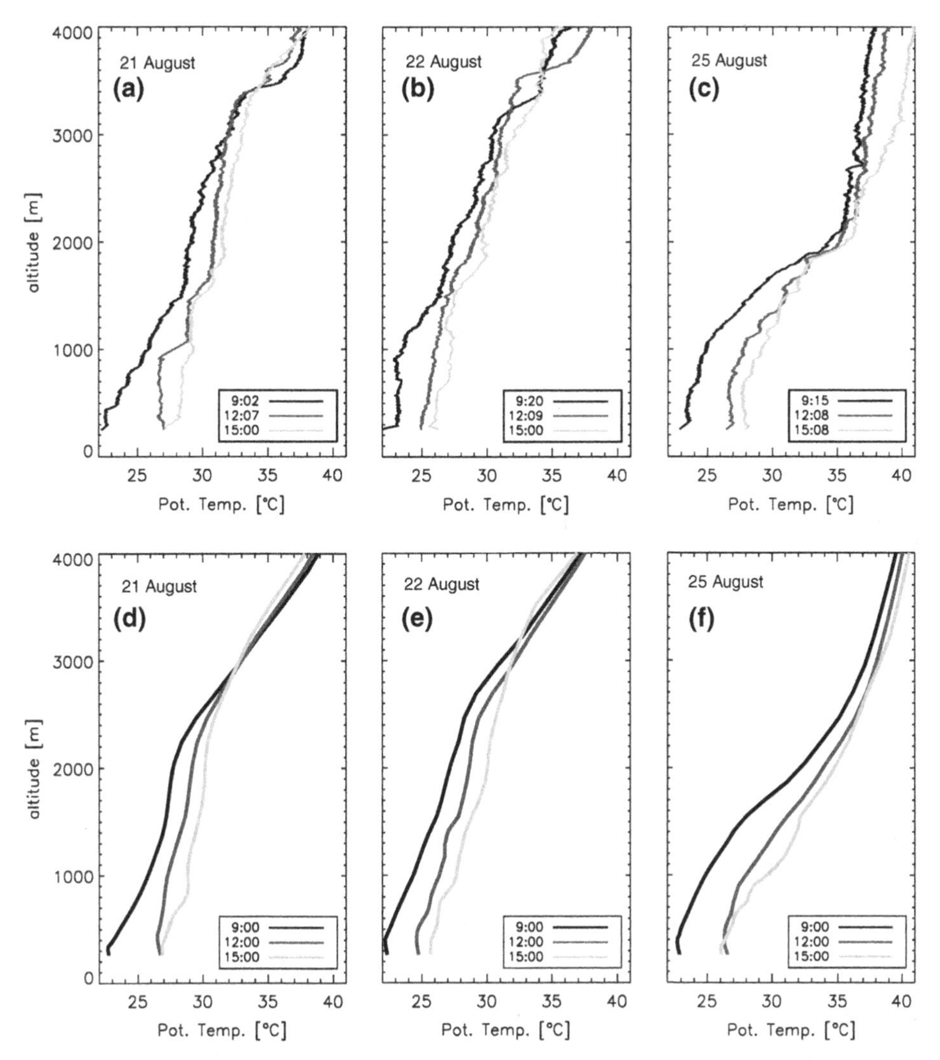

Fig. 2 Profiles of potential temperature as obtained from radiosoundings (upper row) and ARPS (lower row) on the three days considered in this study. The radiosondes were launched and the simulation data extracted, respectively, at the location 'RS' as shown in Fig. 1b at the indicated times UTC

temperature obtained from the simulations. Consistent with the scores of Table 2, the simulations agree well with the observations, and the temporal evolution of the profiles is well reproduced. However, the model fails to reproduce the mid-level inversion observed on 21 and 22 August, probably due to reduced vertical resolution in this altitude.

On the three days considered in this study, a complicated and pronounced thermally driven flow pattern develops. As is typical for mountainous topography under sunny conditions, this flow pattern consists of a superposition of local slope winds (directed normal to the valley axis and along the slopes), channelled and thermally induced valley winds (parallel to the valley axis) as well as secondary circulations in

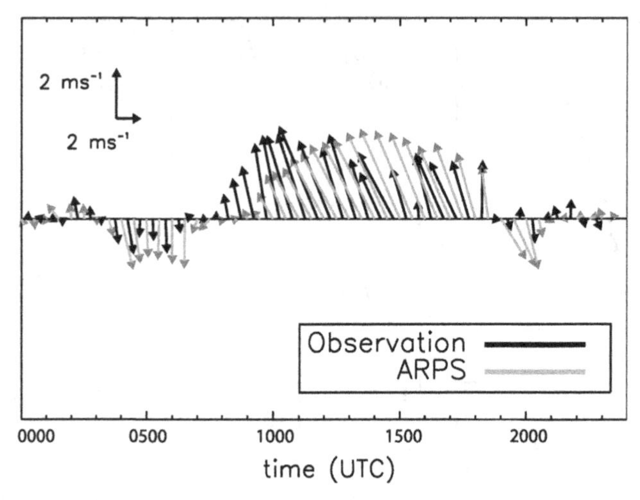

Fig. 3 Measured (black) and simulated (grey) time series of surface winds at location 'RS' on 25 August

the cross-valley direction. The *slope winds* develop along the insolated valley side walls, and are also observed to overshoot into the free atmosphere, i.e. sometimes mountain venting occurs (Weigel et al. 2006a). The *up-valley winds* in the Riviera Valley typically start at about 1000 UTC (local daylight saving time = UTC + 2). As an example, Fig. 3 shows the time series of measured and simulated surface winds at the location 'RS' (see Fig. 1b) for 25 August. The diurnal cycle of weak down-valley flow at night and strong thermally driven up-valley flow during the day is clearly visible and well reproduced by ARPS, although the modelled morning transition lags the measurements by 1–2 h, possibly due to uncertainties in soil moisture initialization (see Chow et al. 2006). 21 and 22 August reveal a very similar pattern (not shown here) that is equally well captured by the model (Weigel et al. 2006a). Once the up-valley flow regime is established, a distinct three-dimensional flow pattern can be observed. The air, which has to flow around a sharp corner at the southern valley entrance, meanders into the Riviera Valley similarly to a channelled water flow. Due to centrifugal forces, in the valley entrance region the up-valley flow has its maximum (on the order of $10\,\mathrm{m\,s^{-1}}$) next to the eastern valley side wall in the southern valley entrance region, while farther north it is uniformly spread over the entire valley diameter. This can be seen in Fig. 4a where simulated wind vectors from a bird's-eye view at an altitude of 900 m on 21 August at 1500 UTC are shown, and in Fig. 4b, which displays the up-valley wind speed in a slice through the asymmetric flow as indicated by the black bar in Fig. 4a. The corresponding aircraft measurements shown in Fig. 4c demonstrate that, again, the model simulations are consistent with the observations. Simulations and observations for 22 and 25 August reveal the same flow pattern (not shown here). Finally, from the aircraft measurements it has been shown (Weigel and Rotach 2004) that the flow curvature in the southern valley half induces complicated secondary circulation structures, which can also be seen in the simulations (Weigel et al. 2006a). However, since the focus of our study is the interaction between the valley atmosphere and the free atmosphere aloft rather than intra-valley processes, this will not be further considered here.

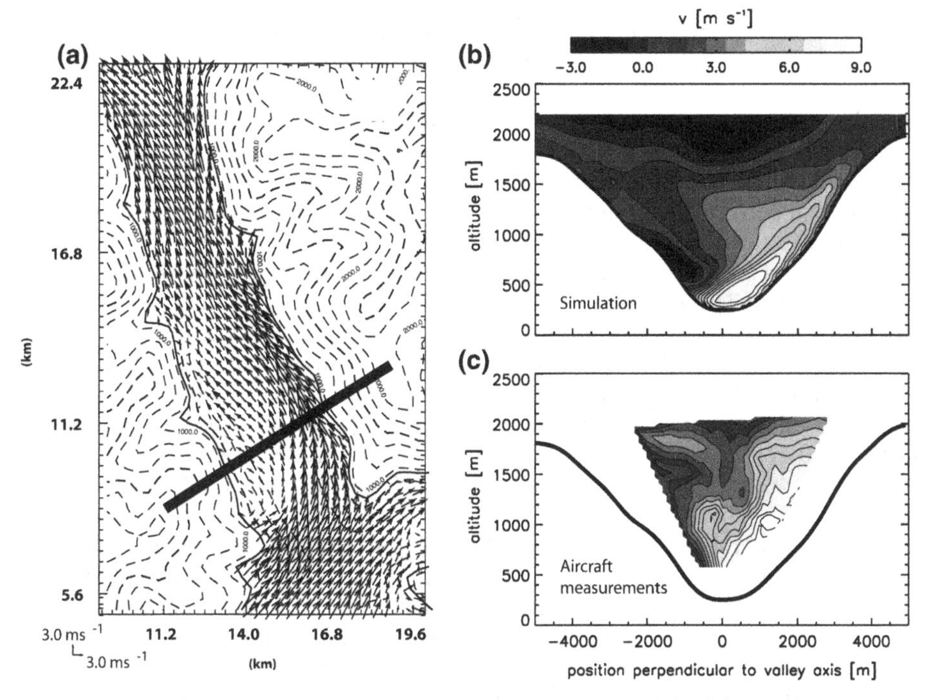

Fig. 4 Up-valley flow in the Riviera Valley on 21 August at around 1500 UTC (adapted from Weigel et al. 2006a). (**a**) shows simulated wind vectors from a bird's-eye view at an altitude of 900 m. The up-valley wind speed in a slice through the asymmetric flow as indicated by the black line in is displayed in (**b**). (**c**) shows the corresponding airborne observations

Given the ability of the model to accurately reproduce the Riviera atmosphere, ARPS is now used to detect and quantify the relevant exchange processes between the valley and the free atmosphere aloft—a task that has not been possible on the basis of MAP-Riviera measurements alone due to their limitations in temporal and spatial resolution.

4 Vertical fluxes of mass and moisture

4.1 Mass fluxes

We begin by calculating the vertical exchange of air mass between the valley atmosphere and the free atmosphere aloft. Quantitative knowledge of this process is important for determining the moisture exchange, as not only air mass is transported but also the water vapour contained therein. The calculations are carried out on the basis of the ARPS model simulations as described above. The fluxes are evaluated over a horizontal interface S_2 intersecting the valley side walls at an altitude of $z_0 = 2.0\,\mathrm{km}$ (see Fig. 1b). This altitude of 2 km also corresponds approximately to the level at which daytime winds go through transition from the up-valley direction to the direction of the synoptic-scale flow. The northern and the southern boundaries of S_2 are set locally perpendicular to the valley axis. The net vertical flux of air mass through this interface from the valley into the free atmosphere, M_{net}, is given by

$$M_{\text{net}}(z_0) = \frac{1}{|S_2|} \iint\limits_{S_2} w(x, y, z_0)\rho(x, y, z_0)\, \mathrm{d}x\, \mathrm{d}y. \tag{1}$$

w is the vertical velocity component, ρ the density of air and $|S_2|$ the area of interface S_2. Positive fluxes are directed upwards. Figure 5 displays the time series of M_{net} between 0600 and 2000 UTC (solid lines); the magnitude of this flux reveals both a distinct diurnal pattern and pronounced day-to-day differences. On all three days, M_{net} is negative in the morning (on 21 August until about 0930 UTC, on 22 August until 1130 UTC and on 25 August until 0830 UTC), implying that air is transported downward into the valley. The afternoon fluxes differ significantly on the three days. On 21 August, M_{net} continuously increases up to a peak value of about $0.22\,\text{kg}\,\text{s}^{-1}\,\text{m}^{-2}$; on 22 August, the afternoon mass flux is significantly lower and relatively constant (on the order of $0.03\,\text{kg}\,\text{s}^{-1}\,\text{m}^{-2}$), and on 25 August it is close to zero, partially even slightly negative. Averaged over the typical period of thermally driven up-valley flow (1000–1900 UTC), net vertical export of air into the free atmosphere amounts to 186% of the entire valley air mass on 21 August. This is of comparable magnitude to the estimates reported by Henne et al. (2004) for a similar valley. On 22 August, net export is considerably lower at 84%, and it is almost negligible on 25 August at 7%.

Two processes can lead to vertical mass flux through S_2 out of the valley: (i) As in most Alpine valleys, the cross-sectional area of the Riviera Valley decreases in the up-valley direction, i.e. the valley becomes narrower. Mass conservation then requires either the along-valley flow to be accelerated or valley air to escape vertically from the valley. (ii) Spatial variations in the along-valley acceleration can lead to local zones of flow convergence or divergence. This could for example be due to abrupt changes in surface roughness (as investigated for water-land transitions by Samuelsson and Tjernström, 2001), due to upslope winds that are not fully compensated by subsidence (Henne et al. 2004), or due to variations in the local horizontal pressure gradients. For example, if the local along-valley pressure gradients north of the Riviera valley are significantly smaller than those south of the Riviera, up-valley winds would be decelerated in the Riviera Valley and vertical mass fluxes would be the consequence.

An upper boundary for the vertical mass flux due to the aforementioned "narrowing effect", M_{narrow}, is estimated in the following way. Let M_{along} be the air flux in the along-valley direction (velocity v) at the southern valley mouth (at $y_0 = 0$), and let C_0 be the corresponding valley cross-sectional area, i.e. the vertical plane through the southern boundary of S_2. This set-up is schematically illustrated in Fig. 6. M_{along} can then be expressed by

$$M_{\text{along}} = \frac{1}{|C_0|} \iint\limits_{C_0} v(x, y_0, z)\, \rho(x, y_0, z)\, \mathrm{d}x\, \mathrm{d}z. \tag{2}$$

The cross-section of the Riviera Valley decreases by about 18% from the southern to the northern valley mouth, and if we assume that the flow does not accelerate, mass conservation requires 18% of the along-valley flow to be vertically exported through S_2. Then, a rough estimate of M_{narrow} is given by

$$M_{\text{narrow}} = 0.18 \frac{|C_0|}{|S_2|} M_{\text{along}}. \tag{3}$$

Time series of M_{narrow} are shown in Fig. 5 (dashed lines). On 25 August, M_{narrow} is much larger than M_{net}, thus not providing a good estimate of the net vertical air flux.

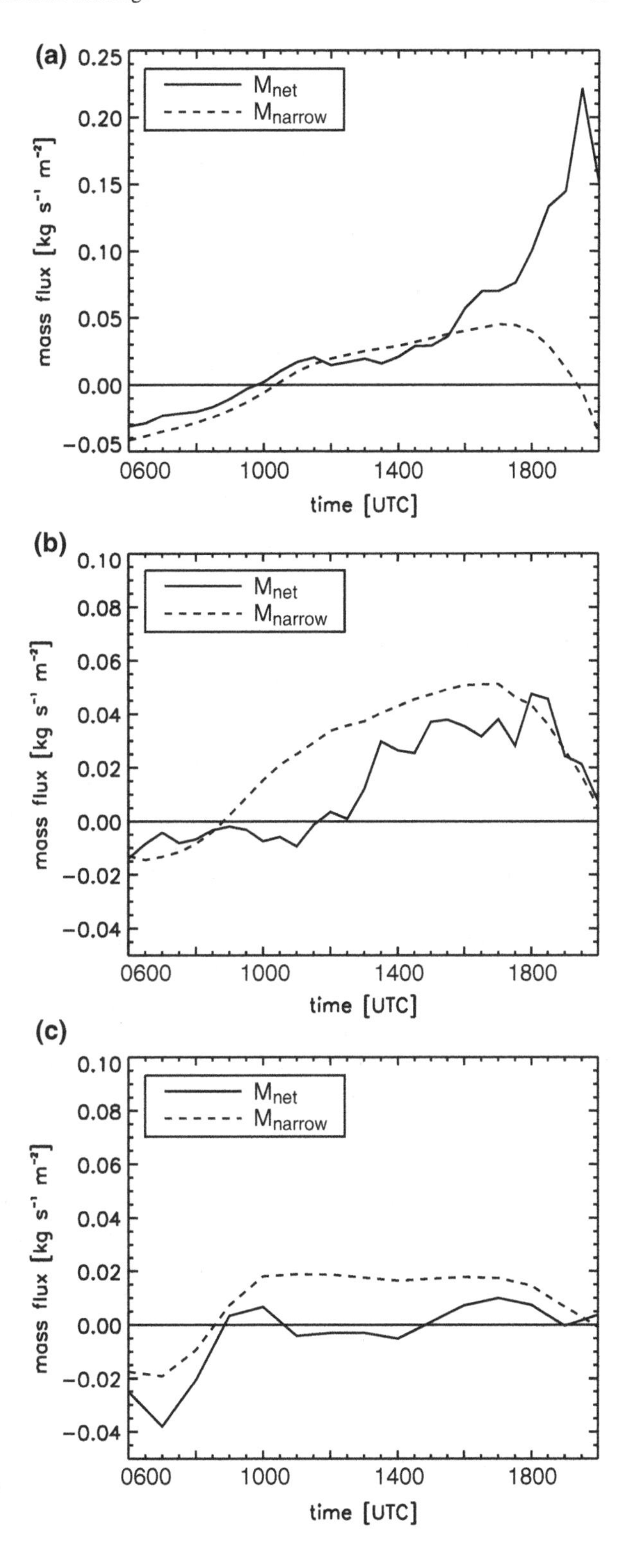

Fig. 5 Vertical mass fluxes through S_2 (Fig. 1b). The plots show the time series of M_{net} (net vertical mass flux) and M_{narrow} (estimate of the vertical mass flux due to the narrowing effect) on (**a**) 21 August, (**b**) 22 August and (**c**) 25 August. Note the different scale in (**a**)

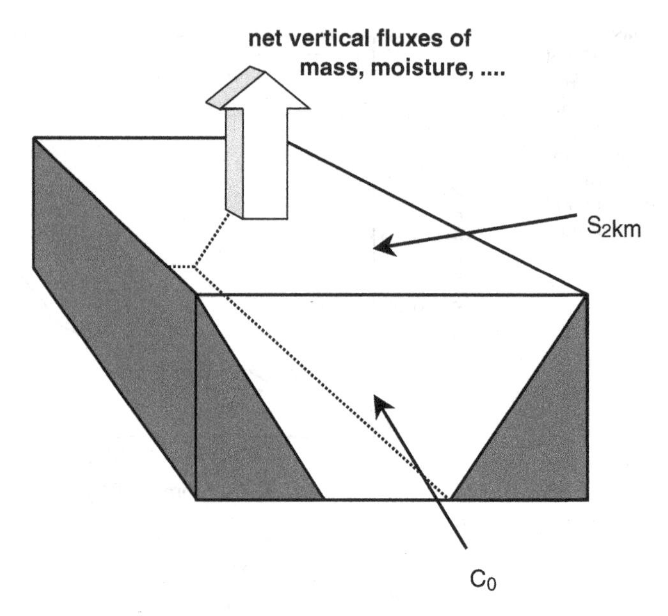

Fig. 6 Schematic representation of the interfaces C_0 and S_2, which are used for the flux calculations

This is probably due to the relatively stable stratification observed in the valley on that day (see Fig. 2c). Under such conditions, the narrowing effect appears to favour a flow acceleration in the along-valley direction rather than vertical export of air. On 21 August and 22 August, on the other hand, the valley atmosphere is less stably stratified (Fig. 2a, b, respectively), and by and large, M_{narrow} is of a magnitude comparable to M_{net}. In the evening of 21 August (after 1600 UTC), however, M_{net} sharply exceeds M_{narrow}. This is due to a local convergence of horizontal flow in the Riviera Valley, as the valley winds north of the Riviera Valley turn to the down-valley direction earlier than do those south of the Riviera (not shown). The reasons for such an asymmetric evening transition of the valley winds have not been investigated. Outside this transition period, however, the "narrowing effect" and the associated mass flux, M_{narrow}, appear to give a reasonable approximation of the vertical export of valley air, as long as the stratification is not too stable.

4.2 Moisture fluxes

In the following we determine the vertical flux of moisture from the valley atmosphere into the free atmosphere aloft. As indicated above, the direct export of valley air represents an important mechanism in this context, because that way not only air mass is exchanged between the "surface" (the valley/ridge system) and the free atmosphere, but also water vapour. The moisture flux associated with this mean vertical flow, L_{mean}, is

$$L_{\text{mean}} = \rho \langle w \rangle \langle q \rangle, \tag{4}$$

where q is specific humidity. The brackets denote spatial averaging over S_2. There is also a second mechanism of moisture exchange to be considered, because superimposed on the mean vertical mass flow are thermally or mechanically driven cross-valley circulations. They typically lead to an export of moist air along the slopes while dryer air is simultaneously imported via subsidence over the valley centre (Kuwagata and Kimura 1995, 1997). These circulations can therefore be associated with an additional vertical moisture flux. Finally, a third mechanism of moisture exchange is given by resolved-scale (RS) and subgrid-scale (SGS) turbulence. These three processes (mean vertical air flow, local circulations and turbulence) are schematically illustrated in Fig. 7a–c for an idealized valley topography.

For practical reasons, moisture transport due to RS turbulence and local circulations will henceforth be treated jointly as one exchange mechanism, because they occur on overlapping length scales. This contribution can be estimated in the following way: let w' and q' be the circulation-related and RS-turbulent fluctuations in vertical velocity and specific humidity, such that $\langle w' \rangle = 0$ and $\langle q' \rangle = 0$. Again, the angular brackets refer to the area of averaging, S_2. The associated moisture flux due to these fluctuations, L_{fluc}, is then given by

$$L_{\mathrm{fluc}} = \rho \langle w' q' \rangle. \tag{5}$$

This leaves the moisture flux due to SGS turbulence, L_{SGS}, as a last component to be considered. However, in the average over S_2, L_{SGS} turns out to be negligibly small (not shown) and will therefore henceforth be omitted.

Using this decomposition, the total vertical moisture flux L_{tot} can be expressed by

$$\begin{aligned} L_{\mathrm{tot}} &= L_{\mathrm{mean}} + L_{\mathrm{fluc}} + L_{\mathrm{SGS}} \\ &\approx L_{\mathrm{mean}} + L_{\mathrm{fluc}} \\ &= \rho \langle wq \rangle. \end{aligned} \tag{6}$$

Time series of L_{tot}, L_{mean} and L_{fluc} are shown in Fig. 8. L_{tot} and L_{mean} are calculated from model output data, using Eqs. 6 and 4, while L_{fluc} is obtained as a residual from Eq. 6. On all three days, L_{fluc} reveals a similar diurnal pattern, reaching a maximum of about $0.5 \times 10^{-4} - 1 \times 10^{-4}\,\mathrm{kg\,s^{-1}\,m^{-2}}$ at around 1300 UTC. While L_{fluc} is the dominating moisture transport term on 25 August, it appears to be of negligible

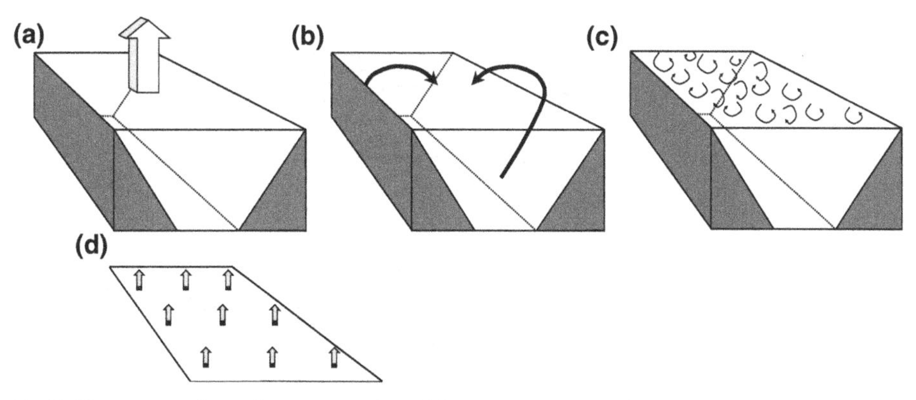

Fig. 7 Upper row: schematic representation of the processes responsible for vertical exchange over a steep valley: (**a**) is the moisture flux associated with net vertical air flow, for example as a consequence of valley geometry or local horizontal flow convergence. (**b**) is the moisture flux due to cross-valley circulations and mountain venting, and (**c**) illustrates turbulent transport. Lower row (**d**): Moisture exchange as seen by a coarse numerical model not resolving the valley

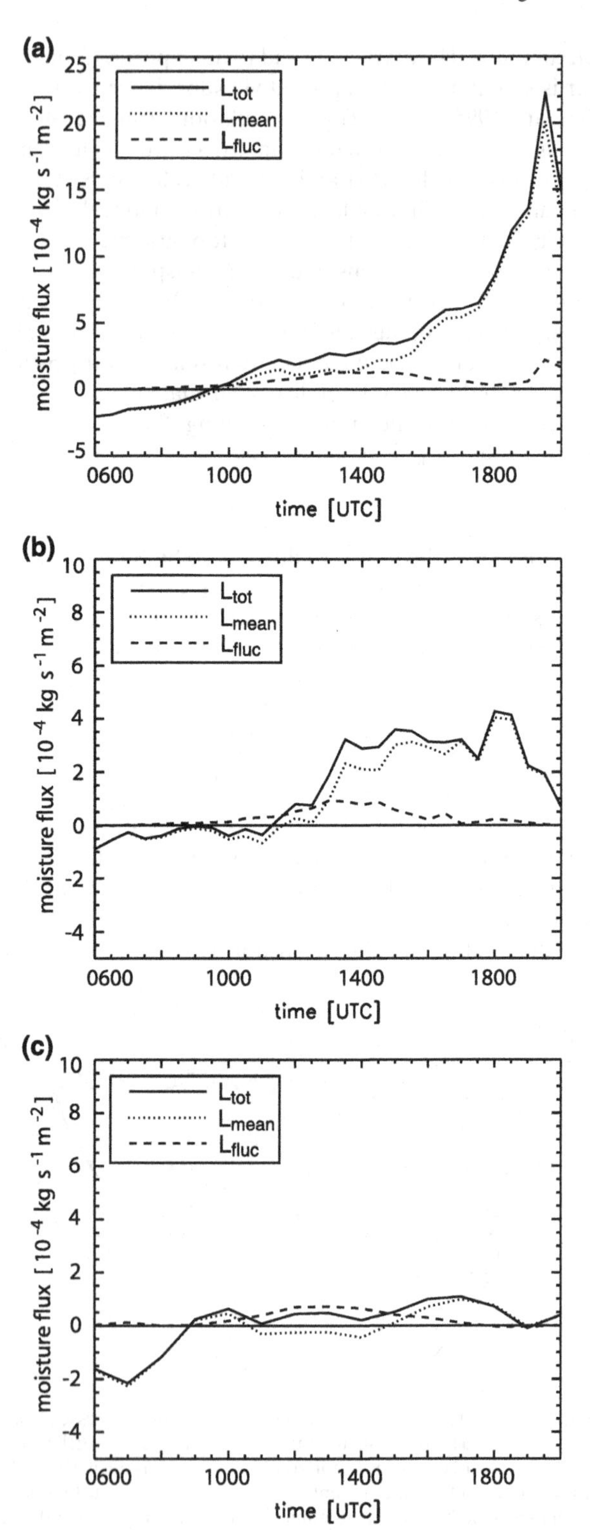

Fig. 8 Vertical moisture fluxes through S_2. The plots show the time series of L_{tot} (total vertical moisture flux), L_{mean} (moisture flux due to vertical export of air) and L_{fluc} (moisture flux due to valley circulations and resolved-scale turbulence) on (**a**) 21 August, (**b**) 22 August and (**c**) 25 August. Note the different scale in (**a**)

magnitude compared to L_{mean} when strong vertical mass exchange is present. On 22 August, a relatively constant moisture flux of $3 \times 10^{-4} - 4 \times 10^{-4}\,\mathrm{kg\,s^{-1}\,m^{-2}}$ is observed in the afternoon, and on 21 August L_{mean} peaks in a sharp maximum of $20 \times 10^{-4}\,\mathrm{kg\,s^{-1}\,m^{-2}}$, which can be associated to the corresponding peak in the time series of M_{net} (Fig. 5).

4.3 What would a coarse model see?

The examples evaluated show that moisture exchange reveals a strong day-to-day variability that is mainly governed by the net vertical export of valley air. Since this export seems to be determined by aspects such as local stratification and valley wind direction it is closely linked to the scales of the underlying topography. Typical numerical weather and climate prediction models, however, do not resolve valleys of the Riviera scale. Their smoothed topography would show a gentle dip instead, or even just be flat. Consequently, a coarse model would not "see" L_{mean} and L_{fluc}, that is the moisture fluxes induced by valley topography. Rather, moisture exchange would be estimated entirely on the basis of the parameterized turbulent surface fluxes E (illustrated in Fig. 7d). A question arises as to how big the discrepancies are between "real" and parameterized vertical moisture exchange.

To obtain a quantitative estimate, we calculate the time series of E as obtained from the ARPS land-surface model on the 350-m grid, averaged over the entire valley surface below S_2. Note that E obtained from the coarser nesting levels is of comparable magnitude (not shown). Time series of E are plotted, together with L_{tot}, the "real" moisture flux into the free atmosphere, in Fig. 9. The results can be summarized as follows: firstly, in contrast to L_{tot}, E does not show any significant day-to-day variability on the three days considered. Secondly, the diurnal cycle of E is highly uncorrelated to that of L_{tot}. While E has its maximum at around 1200 UTC, the maximum of "real" moisture flux is in the late afternoon. Finally, at least on 21 and 22 August (where stratification is only slightly stable), the magnitudes of E and L_{tot} differ significantly. Averaged over the entire period of positive surface moisture fluxes (0600–1800 UTC), the net moisture transport into the free atmosphere exceeds surface moisture fluxes (predicted by the land-surface model) by a factor of 3.8 on 21 August and a factor of 2.7 on 22 August. Advection by the up-valley flows supplies this extra moisture flux.

Thus, the moisture exchange obtained from conventional SGS parameterizations applied over complex topography can deviate significantly from reality, both with respect to magnitude and temporal resolution. Indeed, the net vertical export of valley air into the free atmosphere appears to be the key aspect to be considered for realistic parameterizations.

5 Summary and conclusions

In this paper we presented an evaluation of the daytime vertical exchange of mass and moisture between the Riviera Valley, a typical medium-sized Alpine valley, and the free atmosphere aloft under fair-weather conditions. This has been done with high-resolution large-eddy simulations (LES) using ARPS as a modelling tool. The model itself has been evaluated and calibrated beforehand on the basis of measurements from the MAP-Riviera field campaign (Chow et al. 2006; Weigel et al. 2006a). The results can be summarized as follows:

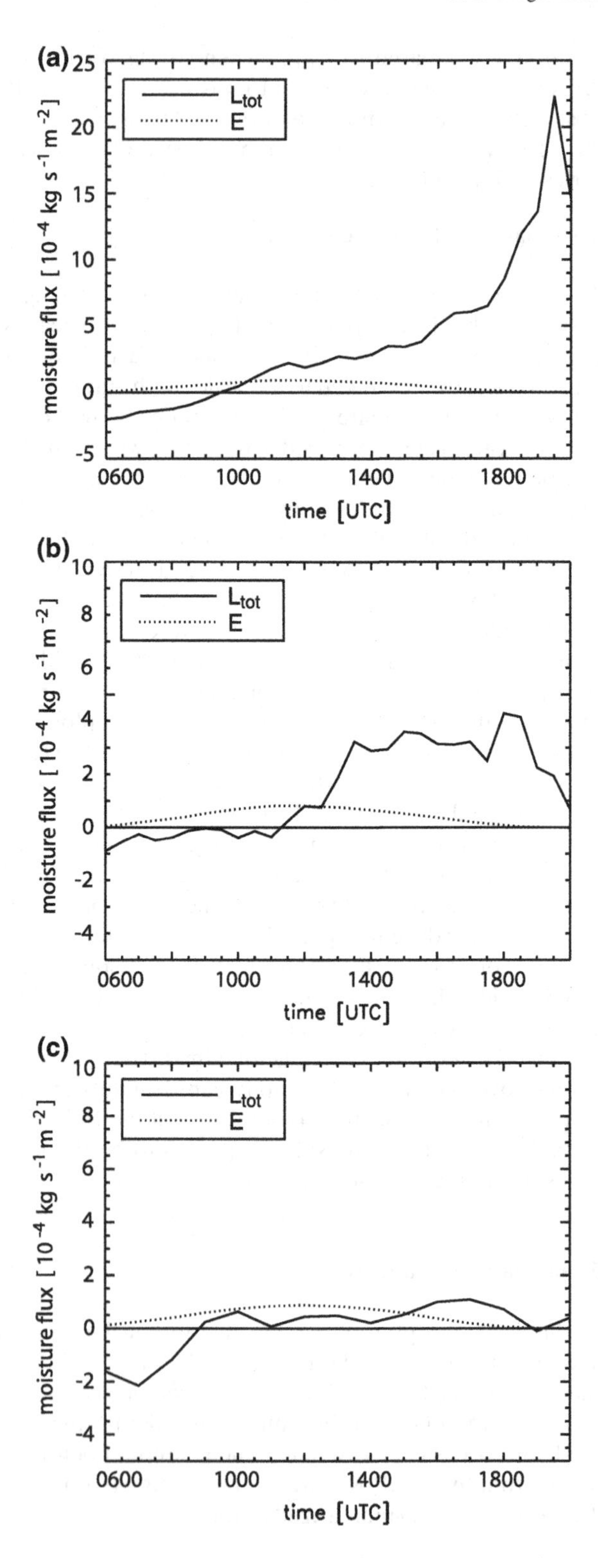

Fig. 9 Total vertical moisture flux L_{tot} through S_2 and turbulent surface moisture flux E on the valley surface underneath S_2 on (**a**) 21 August, (**b**) 22 August and (**c**) 25 August. Note the different scale in (**a**)

(i) Depending on the stability of the valley atmosphere, considerable vertical moisture fluxes out of the valley with magnitudes of up to $4 \times 10^{-4} - 20 \times 10^{-4}\,\mathrm{kg\,s^{-1}\,m^{-2}}$ can be observed. These appear to be mainly due to the narrowing of the valley cross-section and, depending on the valley flow structure, due to horizontal flux convergence in the valley.

(ii) In the daytime average, the amount of moisture carried out of the valley that way can be about 3–4 times larger than the amount of moisture evaporated from the surface underneath (as long as the stratification is not too stable).

(iii) Under stable conditions in the valley atmosphere, vertical mass exchange is suppressed. The export of moisture then seems to be mainly due to the effect of cross-valley circulations and resolved-scale turbulence. The associated moisture fluxes are of comparable magnitude to the turbulent surface fluxes predicted by the land-surface model.

Our results demonstrate that classical subgrid-scale parameterizations in coarse atmospheric models, which only consider surface turbulent exchange, can deviate significantly from realistic fluxes if applied over highly mountainous terrain, at least when the stratification is only slightly stable or even neutral. The key to realistic parameterizations seems to lie in an appropriate representation of the mean vertical exchange of air between a valley atmosphere and the free atmosphere aloft. Of course, this study is insufficient to provide such parameterizations, given that only three days in a single valley are examined. Yet, our simulations suggest that a simple valley-narrowing argument may be helpful to provide a first estimate of the air carried out of the valley. From this net vertical mass flux, the corresponding exchange of moisture (as well as pollutants, aerosols, and other scalars) can be easily obtained, if the specific humidity (pollutant concentration, aerosol concentration, etc.) of the air is known.

Overall, LES appears to be a useful tool to obtain more realistic estimates of the exchange processes that occur over steep topography and are very difficult to measure. What has been shown here for the example of moisture could analogously be carried out for heat and momentum. Indeed, LES could and should be applied systematically for a wider range of meteorological situations and valley topographies with the ultimate aim of improving the SGS parameterizations of coarse models for flow over complex terrain.

Acknowledgments This work has been supported by the Swiss National Science Foundation (Grants 20-68320.01 and 20-100013) (APW) and by National Science Foundation Grants ATM-0073395 and ATM-0453595 (FKC). We gratefully acknowledge the National Center for Atmospheric Research (sponsored by the U.S. National Science Foundation) for providing the computing time used in this research.

References

Barry RG (1992) Mountain weather and climate, 2nd edn. Routledge, London and New York, 402 pp

Beljaars ACM, Viterbo P, Miller MJ, Betts AK (1996) The anolamous rainfall over the United States during July 1993: sensitivity to land surface parameterization and soil moisture. Mon Wea Rev 124:362–383

Businger JA, Wyngaard JC, Izumi Y, Bradley EF (1971) Flux–profile relationships in the atmospheric surface layer. J Atmos Sci 28:181–189

Chow FK, Weigel AP, Street RL, Rotach MW, Xue M (2006) High-resolution large-eddy simulations of flow in a steep alpine valley. Part I: Methodology, verification and sensitivity studies. J Appl Meteorol Climatol 45:63–86

Deardorff JW (1972) Numerical investigation of neutral and unstable planetary boundary layers. J Atmos Sci 29:91–115

De Wekker SFJ, Steyn DG, Nyeki S (2004) A comparison of aerosol-layer and convective boundary-layer structure over a mountain range during STAAARTE '97'. Boundary-Layer Meteorol 113:249–271

Gohm A, Zängl G, Mayr GJ (2004) South Foehn in the Wipp Valley on 24 October 1999 (MAP IOP 10): verification of high-resolution numerical simulations with observations. Mon Wea Rev 132:78–102

Henne S, Furger M, Nyeki S, Steinbacher M, Neininger B, de Wekker SFJ, Dommen J, Spichtinger N, Stohl A, Prévôt ASH (2004) Quantification of topographic venting of boundary layer air to the free troposphere. Atmos Chem Phys 4:497–509

Henne S, Furger M, Prévôt ASH (2005) Climatology of mountain venting induced elevated moisture layers in the lee of the Alps. J Appl Meteorol 44:620–633

Kossmann M, Corsmeier U, de Wekker SFJ, Fiedler F, Kalthoff N, Güsten H, Neininger B (1999) Observation of handover processes between the atmospheric boundary layer and the free troposphere over mountainous terrain. Contrib Atmos Phys 72:329–350

Kuwagata T, Kimura F (1995) Daytime boundary layer evolution in a deep valley. Part I: Observations in the Ina Valley. J Appl Meteorol 34:1082–1091

Kuwagata T, Kimura F (1997) Daytime boundary layer evolution in a deep valley. Part II: Numerical simulation of the cross-valley circulation. J Appl Meteorol 36:883–895

Noppel H, Fiedler F (2001) Mesoscale heat transport over complex terrain by slope winds – a conceptual model and numerical simulations. Boundary-Layer Meteorol 104:73–97

Pitman AJ (2003) The evolution of, and revolution in, land surface schemes designed for climate models. Int J Climatol (23), 479–510

Rotach MW (1995) On the boundary layer over mountainous terrain – a frog's perspective. MAP-Newsletter 3:31–32

Rotach MW, Calanca P, Graziani G, Gurtz J, Steyn DG, Vogt R, Andretta M, Christen A, Cieslik S, Connolly R, De Wekker SFJ, Galmarini S, Kadygrov EN, Kadygrov V, Miller E, Neininger B, Rucker M, van Gorsel E, Weber H, Weiss A, Zappa M (2004) The turbulence structure and exchange processes in an Alpine valley: the Riviera project. Bull Amer Meteorol Soc 85:1367–1385

Rotach MW, Zardi D (2006) On the boundary layer structure over highly complex terrain: Key findings from MAP. Quart J Roy Meteorol Soc, accepted

Samuelsson, P, Tjernström, M (2001) Mesoscale flow modification induced by land-lake surface temperature and roughness differences. J Geophys Res 106(D12):12419–12435

Viterbo P, Betts AK (1999) The impact of the ECMWF reanalysis soil water on forecasts of the July 1993 Mississippi flood. J Geophys Res 104(D16):19361–19366

Weigel AP, Chow FK, Rotach MW, Street RL, Xue M (2006a) High-resolution large-eddy simulations of flow in a steep alpine valley. Part II: Flow structure and heat budgets. J Appl Meteorol Climatol 45:87–107

Weigel AP, Rotach MW (2004) Flow structure and turbulence characteristics of the daytime atmosphere in a steep and narrow Alpine valley. Quart J Roy Meteorol Soc 130: 2605–2627

Weigel AP (2005) On the atmospheric boundary layer over highly complex topography. Ph.D. thesis, Swiss Federal Institute of Technology (ETH). http://e-collection.ethbib.ethz.ch/cgi-bin/show.pl?type=diss&nr=15972

Whiteman CD (1990) Observations of thermally developed wind systems in mountainous terrain. In: Blumen W (ed) Atmospheric processes over complex terrain. American Meteorological Society, Boston, pp 5–42

Whiteman CD (2000) Mountain meteorology. Fundamentals and applications. Oxford University Press, New York, Oxford, 355 pp

Xue M, Droegemeier KK, Wong V, Shapiro A, Brewster K, Carr F, Weber D, Liu Y, Wang D (2001) The Advanced Regional Prediction System (ARPS) – A multi-scale nonhydrostatic atmospheric simulation and prediction tool. Part II: Model physics and applications. Meteorol Atmos Phys 76:143–165

Xue M, Droegemeier KK, Wong V (2000) The Advanced Regional Predicition System (ARPS) – a multi-scale nonhydrostatic atmospheric simulation and prediction model. Part I: Model dynamics and verification. Meteorol Atmos Phys 75:161–193

Young GS, Pielke RA (1983) Application of terrain height variance spectra to mesoscale modeling. J Atmos Sci 40:2555–2560

Zängl G, Chimani B, Häberli C (2004) Numerical simulations of the foehn in the Rhine Valley on 24 October 1999 (MAP IOP 10). Mon Wea Rev 132:368–389

Zhong S, Fast J (2003) An evaluation of the MM5, RAMS, and Meso-Eta models at subkilometer resolution using VTMX field campaign data in the Salt Lake Valley. Mon Wea Rev 131:1301–1322

The influence of nonstationarity on the turbulent flux–gradient relationship for stable stratification

L. Mahrt

Abstract Extensive eddy-correlation datasets are analyzed to examine the influence of nonstationarity of the mean flow on the flux–gradient relationship near the surface. This nonstationarity is due to wavelike motions, meandering of the wind vector, and numerous unidentified small-scale mesoscale motions. While the data do not reveal an obvious critical gradient Richardson number, the maximum downward heat flux increases approximately linearly with increasing friction velocity for significant stability.

The largest of our datasets is chosen to more closely examine the influence of stability, nonstationarity, distortion of the mean wind profile and self-correlation on the flux-gradient relationship. Stability is expressed in terms of z/L, the gradient Richardson number or the bulk Richardson number over the tower layer. The efficiency of the momentum transport systematically increases with increasing nonstationarity and attendant distortion of the mean wind profile. Enhancement of the turbulent momentum flux associated with nonstationarity is examined in terms of the nondimensional shear, Prandtl number and the eddy diffusivity.

Keywords Intermittency · Nocturnal boundary layer · Nonstationarity · Prandtl number · Stable boundary layer

1 Introduction

The stable boundary layer is often nonstationary on small time scales of less than an hour due to a variety of mesoscale motions, including gravity waves (e.g., Finnigan et al. 1984; Finnigan 1999; Chimonas 2002, 2003; Nappo 2002; Cooper et al. 2006), horizontal meandering-like motions (Kristensen et al. 1981; Lilly 1983; Herring and Métais 1989;

L. Mahrt (✉)
College of Oceanic and Atmospheric Sciences, Oregon State University,
104, Ocean Adm Building, Corvallis, OR 97331, USA
e-mail: mahrt@nwra.com

Atmospheric Boundary Layers. A. Baklanov & B. Grisogono (eds.),
doi: 10.1007/978-0-387-74321-9_7,

Etling 1990; Riley and Lelong 2000; McWilliams 2004; Anfossi et al. 2005), density currents and solitons (Sun et al. 2004) and numerous motions more difficult to categorize. Such motions constantly change the "mean" vertical gradients. We will refer to the changing mean flow as "nonstationarity of the mean flow" in contrast to the normal application of the term nonstationarity to turbulence statistics (e.g., Panofsky and Dutton 1984).

As a result of nonstationarity of the mean flow, the turbulence is continuously modified, often viewed as intermittency of the turbulence. The definition of intermittency varies substantially between studies (e.g., Howell and Sun 1999; Coulter and Doran 2002; Moraes et al. 2004; Salmond 2005; Acevedo et al. 2006). Grachev et al. (2005) and Mahrt and Vickers (2006) never find completely vanishing turbulence near the surface, even though the turbulence may become extremely weak. Cases of well-defined intermittency with on-off behaviour of turbulence are relatively rare (Nakamura and Mahrt 2005). As a result, quantitative measures of intermittency depend on method, scale and specified thresholds.

Even for stationary large-scale flows, the turbulence is internally intermittent. For stable flows, this internal intermittency may be induced by the interplay between the turbulence and the mean shear, such that the gradient Richardson number varies about a critical or equilibrium value (Atlas et al. 1970; Kim and Mahrt 1992; Ohya 2001; Pardyjak et al. 2002; Fernando 2003). Other recent studies detailing internal intermittency include the modelling studies of Derbyshire (1995, 1999) and van de Wiel et al. (2002) and the wind-tunnel study of Ohya and Uchida (2003). Perhaps the turbulence can be considered as stationary on time scales that are large compared to the scale of internal intermittency, although the background atmospheric flow will seldom be stationary on such longer time scales.

Businger (2005) proposes that the time dependence of mean profiles on time scales larger than the turbulence, but smaller than the record length, might enhance the eddy diffusivity for momentum, but reduce the diffusivity for heat. This relative enhancement of the momentum flux would be due to the shear generation of turbulence during periods of enhanced momentum gradient and buoyancy suppression of turbulence during periods of enhanced potential temperature gradient. In addition, the turbulent flux–gradient relationship is probably affected by the failure of the turbulence to maintain equilibrium with the changing mean flow, as evident for well-defined gravity waves studied by Finnigan and Einaudi (1981). Separation between internal intermittency and nonstationarity of the turbulence due to mesoscale motions from atmospheric data is not completely possible. In this study, we will focus on response of the turbulence to nonstationarity of the mean flow associated primarily with mesoscale motions.

Examination of the influence of stability on the turbulence must recognize that nonstationarity due to nocturnal mesoscale motions emerges mainly for weak winds (Anfossi et al. 2005), and that the nocturnal boundary layer is most stable for clear sky weak wind conditions. Therefore the influences of stability and nonstationarity on the flux–gradient relationship are also difficult to separate using atmospheric observations. Kondo et al. (1978) and others found that the Prandtl number increases to values that are significantly larger than unity for large gradient Richardson number, and intermittent turbulence. The increase of the eddy Prandtl number with strong stability has also been attributed to the influence of pressure fluctuations, including momentum transport by nonlinear gravity waves. However, direct atmospheric evidence is not available. Hicks (1976) found that the Prandtl number estimated in terms of the ratio of nondimensional gradients, ϕ_h/ϕ_m, should be increased when only a fraction of the record is turbulent. Kondo et al. (1978), Ueda et al. (1981), Kim and Mahrt (1992), Ohya (2001), Strang and Fernando (2001), Monti et al. (2002) and Mahrt and Vickers (2006) also found increasing Prandtl number with increasing stability. Beljaars and Holtslag (1991) accounted for less efficient transfer of heat compared to momentum for intermittent turbulence in very stable conditions by modifying the ψ-stability functions for the integrated Monin–Obukhov similarity theory while Lee et al. (2006) modify the parameterized eddy diffusivities. In contrast to the above studies, Howell and Sun (1999) found decreasing Prandtl number with increasing stability for strong stability. It is not clear if these differences arise from different analyses techniques or additional influences not represented by the gradient Richardson number. Our data indicates that large self-correlation between the Prandtl number and the gradient Richardson number, not previously investigated, prevents examination of the physical significance of the relationship.

Existing observational and theoretical studies have emphasized stationary homogeneous stable boundary layers or stationary flows over well-defined changes of surface conditions without explicit consideration of nonstationarity of the mean flow. Excluding the most stable conditions where the boundary-layer depth is difficult to define, a successful theoretical framework for stationary flow is emerging (Zilitinkevich and Calanca 2000; Zilitinkevich et al. 2002; Sukoriansky et al. 2005) and some scaling relationships have been extended to stronger stability (Basu et al. 2006). It is not known how scaling relationships degrade with nonstationarity, one of the main subjects of this investigation. In the next section, we describe the data and detail the computation of the gradients and fluxes, which can become problematic in stable conditions.

2 Data

2.1 Sites

As the primary dataset, we analyze four months of nocturnal eddy-correlation data from the Fluxes over Snow-covered Surfaces II (FLOSSII) carried out from 1 December 2002 to 31 March 2003 in the North Park Basin of north-west Colorado, U.S.A. For instrumentation details, see Mahrt and Vickers (2005). The 30-m tower provided seven levels of eddy-correlation data over a grass surface, sometimes partially or completely snow covered. Scattered brush and tall grass beginning about 100 m upwind from the tower site may exert some influence on the turbulence, particularly in the upper part of the tower layer.

We also analyze one month of eddy-correlation data from a 60-m tower with seven levels of eddy-correlation data over a grassland in south central Kansas, U.S.A in CASES-99 (Poulos et al. 2002; Mahrt and Vickers 2002; Sun et al. 2002). All four of the datasets are quality controlled following Vickers and Mahrt (1997). We include only records between sunset and sunrise where the heat flux is downward.

Data from the Air–Sea Interaction Tower (ASIT) collected during the CBLAST experiment in late summer of 2003 are analyzed (Edson et al. 2004). The offshore tower is located 3 km south of Martha's Vineyard in 15 m of water. The 20-Hz turbulence measurements collected by a CSAT3 sonic anemometer (Campbell Scientific, Inc.) and a colocated LI-7500 open path gas analyzer (LI-COR, Inc.) at approximately 5 m above the sea surface are used to calculate eddy-correlation fluxes of momentum, sensible heat and latent heat. Slow response measurements include mean air temperature and humidity at multiple levels and sea-surface radiative temperature. Additionally, we analyze eddy-correlation data collected from Point Barrow, Alaska from March through May 2005 with a CSAT3 sonic anemometer at 2 m over short tundra, consisting mainly of sedge shoots with tussocks 0.1 to 0.2 m height, often partially or completely covered by snow. FLOSSII will be used as the primary dataset since it includes the most data and also includes the most cases of large nonstationarity of the mean flow.

2.2 Calculation of vertical gradients

Vertical gradients are computed from 1-hr averaged variables for estimation of the nondimensional gradients and eddy diffusivities. This study excludes cases with $z/L > 1$ where vertical gradients in the presence of significant nonstationarity of the mean flow were sensitive to the method of calculation and directional shear was sometimes significant (here, z is height, L is the Obukhov length). A subsequent study examines this problem in detail. Computation of the vertical gradient of potential temperature is vulnerable to offset errors in near-neutral cases. The sonic temperatures were corrected for the influence of moisture content. After compositing over all of the stable cases for FLOSSII, the vertical profiles of the aspirated temperatures were relatively smooth while temperatures from the sonic anemometer indicated different offsets at each level. On the other hand, for winds weaker than about $1\,\mathrm{m\,s^{-1}}$, the aspirated thermistors sometimes showed erratic behaviour. Consequently, we have "calibrated" the sonic anemometers by comparing the composited sonic temperatures with the composited aspirated temperature measurements for wind speeds greater than $2\,\mathrm{m\,s^{-1}}$ and then applied these calibrations to all of the sonic temperatures. Performing this calibration procedure for smaller one-month subsets of data indicates that the calibration was stable and not time dependent.

For $z/L < 1$, directional shear was generally weak except for a few very weak wind cases. Here we fit the speed profile. The vertical gradients of potential temperature and wind are computed for each hour by first fitting the calibrated sonic temperatures and wind to a log-linear form. The log-linear fit is partly motivated by surface-layer similarity theory, but the coefficients are determined by a statistical fit, without information on the surface roughness length or Obukhov length to allow expected deviations from surface-layer similarity.

As a measure of the badness of the fit of the wind and potential temperature profiles, we have computed the root-mean-square (rms) error of the profile fit, normalized by the standard deviation of the deviation of the wind or potential temperature from their vertical averages. This normalized error is generally small and exceeds 50% for only 0.1% of the data, which lead to outliers in the analysis. We do not exclude any records based on errors in the profile fit.

2.3 Calculation of fluxes

The separation between turbulence and mesoscale motions becomes critical with weak turbulence, since failure to remove even a small fraction of the mesoscale motion from the weak turbulence signal can lead to serious contamination of the estimated turbulent fluxes and large random flux errors. Mesoscale fluxes are generally contaminated by large sampling problems (e.g., Chimonas 1984). In order to avoid serious contamination of the computed fluxes by inadvertently captured mesoscale motions, the averaging time for defining the perturbations, τ, is specified separately for each record (Vickers and Mahrt 2006; Acevedo et al. 2006; van den Kroonenberg and Bange 2007). This method relies on the systematic behaviour of the "cospectra" for the turbulent heat flux, when expressed in terms of the multi-resolution flux decomposition. The multi-resolution decomposition is the simplest possible orthogonal wavelet basis where the bases functions are simple unweighted averages over a dyadic scale (see references above). Once the perturbations are computed and multiplied, the covariances are averaged over one hour to reduce random errors. Even for weak turbulence, the characteristics of the properly extracted turbulence signal sharply contrast to those of the mesoscale (Vickers and Mahrt 2006).

After discarding records with air flow through the tower ($\pm 60°$) from the boom angle of 270°, 1010 FLOSSII records are retained between sunset and sunrise with downward heat flux ($< -0.001\,\mathrm{K\,m\,s^{-1}}$) at 2 m. To eliminate cases where the vertical gradient of potential temperature is too small to estimate, we require that the vertical gradient of potential temperature is positive and greater than $0.01\,\mathrm{K\,m^{-1}}$, which eliminates three additional records. Elimination of upward momentum flux at 2 m, due to either a wind maximum below 2 m or countergradient momentum flux, reduces the number of records to 978. The countergradient momentum fluxes are sometimes associated with reversal of the sign of the shear with height. If we had used 5-min averages to define the turbulent fluctuations, instead of the variable averaging width, then the number of records with countergradient momentum flux would have increased substantially. We also require that the magnitude of the cross-wind momentum flux is less than the magnitude of the along-wind momentum flux further reducing the number of records to 962. The requirements on the mean shear and momentum flux eliminate almost exclusively cases of very small friction velocity u_* ($< 0.05\,\mathrm{m\,s^{-1}}$); nonetheless, 40 of 74 cases with $u_* < 0.05\,\mathrm{m\,s^{-1}}$ remain. We will return to these difficult, but important, cases in a subsequent study.

2.4 Self-correlation

Examination of the dependence of the observed nondimensional gradients on stability must recognize serious contamination by self-correlation since the nondimensional gradients include scaling by u_*, which also appears in z/L (Hicks 1981; Andreas 2002; Mahrt et al. 2003). Such self-correlation between ϕ_m, and z/L can exceed the physical correlation for stable conditions (Klipp and Mahrt 2004). With pure self-correlation, the nondimensional shear becomes proportional to $(z/L)^{1/3}$. Grachev et al. (2005) find this behaviour for very stable conditions, and interpret the result as de-correlation between u_* and the mean shear. Combining this interpretation with traditional self-correlation thinking indicates that the ϕ_m–z/L relation degenerates into self-correlation when the correlation between the mean shear and u_* is less than the self-correlation. This problem of serious self-correlation can be reduced, but not eliminated, by replacing z/L with the traditional gradient Richardson number (Klipp and Mahrt 2004; Sorbjan 2006).

To test the possibility of significant self-correlation, we construct random datasets from the original pool of data, as in Klipp and Mahrt (2004). Using the randomized data, new values for ϕ_m and z/L are computed and the linear correlation coefficient between them is calculated. This process is repeated 100 times and the resulting 100 correlation coefficients are averaged. Since the random data no longer retain any physical connection between the fundamental variables, the average correlation for the randomized datasets is a measure of self-correlation due to common variables.

2.5 Nonstationarity of the mean flow

To quantify the influence of nonstationarity of the mean flow on the flux-gradient relationship, we define the nonstationarity for each record as the standard deviation of the log of the subrecord gradient Richardson number $\sigma_{\ln(Ri)}$. Here, Ri is the gradient Richardson number evaluated from the fitted profiles of wind and temperature based on the 10-min averages and the standard deviation is computed over 1-hr records. Use of the log of the gradient Richardson number reduces the role of outliers. A single simple definition of nonstationarity cannot properly represent the variety of complex processes leading to nonstationarity, but we find this simple measure to be a useful tool for exploring the role of nonstationarity.

The qualitative dependence of the flux–gradient relationship on the nonstationarity on the mean flow is not sensitive to the choice of the subrecord averaging width. The standard deviation based on 10-min averages includes the influence of mesoscale motions on scales greater than 10 min as well as nonstationarity of the large-scale flow. This standard deviation is generally dominated by temporary increases of the gradient Richardson number due to mesoscale reduction of the shear and occasional rapid reduction of the gradient Richardson number due to enhanced shear. The influence of trend associated with the larger-scale flow is less important.

Admittedly, the standard deviation based on only six 10-min averages within the 1-hr records do not contain enough samples to accurately estimate the true standard deviation, but we are seeking only an index for dividing the data into broad classes of nonstationarity. We define classes of weak, intermediate and strong nonstationarity in terms of the intervals of the log of the gradient Richardson number equal to 0–0.5, 0.5–2.0 and >2.0, respectively. For the restriction $z/L < 1$, these nonstationarity classes contain 501, 207 and 51 records, respectively. The frequency of strong nonstationarity of the mean flow increases with increasing stability.

2.6 Bin averaging

In this study, we avoid explicit exclusion of outliers since such procedures can lead to bias and results can be sensitive to criteria for excluding outliers. On the other hand, averages of various quantities are not always well posed with outliers because of their disproportional influence on the average. For the present data, outliers manifest themselves mainly in ratios associated with extremely small values of the denominator, as occurs with the nondimensional gradients, eddy diffusivities and Prandtl number.

To avoid averaging problems associated with such outliers, we first average fluxes and gradients over all records within a given bin, such as an interval of one of the stability parameters, and then compute ratios. This approach is not without problems in that the records within a stability interval do not represent identical conditions and the usual estimates of standard errors are not applicable with averaging prior to division operations. Bins with less than 10 points are excluded from the bin analysis.

3 Relationship between the surface stress and heat flux

Before examining the flux–gradient relationship, we explore the distribution of observations in u_*-heat flux space (Fig. 1). For a broad range of values of u_* up to about $0.4\,\mathrm{m\,s^{-1}}$, the heat flux is approximately bounded by a linear function of the friction velocity (Fig. 1), such that

$$H_{\max}(u_*) = C_s u_*. \tag{1}$$

This maximum value of the downward heat flux, $H_{\max}(u_*)$, can be defined without additional information on the mean shear and stratification, apparently because the fluxes and vertical gradients mutually adjust. With stronger turbulence and near-neutral stability, the heat flux is more limited by the weakness of the stratification and temperature fluctuations, and the functional relationship between the heat flux and the friction velocity breaks down. This less stable regime is most evident for the

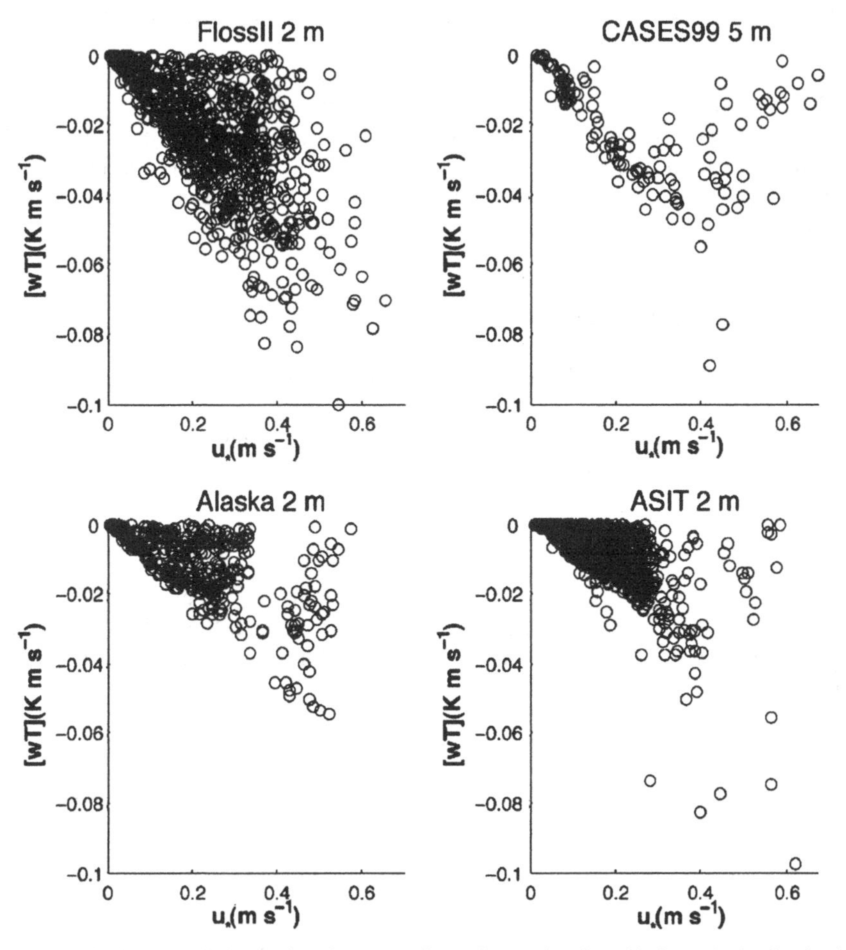

Fig. 1 The distribution of observations in terms of heat flux and surface friction velocity for the four datasets

CASES-99 data. For the FLOSSII and CASES-99 sites, the slope of the change of maximum downward heat flux with increasing friction velocity (C_s, Eq. 1) is approximately 0.17 K (Fig. 1). For the smoother Alaska and ASIT sites, the slope C_s is closer to 0.1 K. C_s is dimensional and not expected to be universal.

One is tempted to relate the maximum value of the downward heat flux to a critical flux Richardson number criterion although a critical flux Richardson number is not evident from the data, perhaps due to large scatter for very stable conditions. The local scaling arguments of Sorbjan (2006) lead to a prediction similar to Eq. 1 and imply that C_s cannot be strictly constant.

The approximately linear relationship between the maximum downward heat flux and friction velocity can be expressed in terms of the observed small variation of the temperature fluctuation scale

$$\theta_* \equiv \frac{\overline{w'\theta'}}{u_*}. \tag{2}$$

For our datasets, θ_* increases with u_* for very small u_* and then becomes more independent of u_*, similar to the clear sky relationship in Van Ulden and Holtslag (1985) and the curves in Holtslag and De Bruin (1988).

Cases of nearly vanishing heat flux and small, but significant, u_* occur for all of the datasets (not easily visualized in Fig. 1) and increase in frequency with height above ground. These records correspond to near-neutral stability in terms of z/L, but can correspond to strong stratification, weak shear and, therefore, large values of the gradient Richardson number. In these cases, z/L is misleading as a stability parameter.

4 Stability and Nonstationarity

For the remainder of this study, we focus on the FLOSSII dataset, which is larger than the other three datasets and contains more records with strong nonstationarity of the mean flow. We will pose the flux–gradient relationship in terms of the nondimensional gradient for an arbitrary variable F

$$\phi_F \equiv \frac{\kappa z \partial[F]/\partial z}{F_*}. \tag{3}$$

where [ ] designates an average over the record length. For the along-wind momentum, $[F] = [u]$ and $F_* = u_*$. For heat, $F = [\theta]$ and $F_* = \theta_*$. If the stability parameter is defined to be z/L, then the nondimensional gradient for $z/L < 1$ is often approximated as

$$\phi_F = 1 + \beta_F \frac{z}{L}. \tag{4}$$

For future use, the eddy diffusivity for momentum within surface-layer similarity theory, is approximated as

$$K_m = \frac{\kappa z u_*}{\phi_m}. \tag{5}$$

The eddy diffusivity for heat is often related to K_m by specifying the eddy Prandtl number.

We now examine the linear correlation between the nondimensional gradients and the different stability parameters for $z/L < 1$ or $Ri < 0.25$, where a linear dependence of nondimensional gradients on the stability is most applicable. Nonetheless, the possible nonlinear dependence on stability and the occurrence of some outliers require that we interpret the linear correlation coefficients with caution.

The correlation coefficients between the nondimensional gradients and the different stability parameters tend to decrease with increasing nonstationarity, depending on the particular variables. The correlation between ϕ_m and z/L averages about 0.8, but the same correlation for the randomized data is a little more than 0.6. The strong self-correlation does not necessarily invalidate the similarity theory, but rather implies that the similarity theory for ϕ_m (z/L) cannot be evaluated by existing analysis.

On the other hand, the correlation between ϕ_h and z/L for stationary conditions is much larger than that for the randomized data (0.89 compared to −0.25), but decreases to 0.5 for the most nonstationarity class (−0.25 for the randomized data). The common variable between the nondimensional temperature gradient and z/L is θ_* (Sect. 3), which varies much less than u_*. The self-correlation for the randomized data is negative so that the positive correlation between ϕ_h and z/L cannot be due to self-correlation and, therefore, the relation between ϕ_h and z/L is assumed to be physical.

The self-correlation between ϕ_m and the gradient Richardson number is due to the occurrence of mean shear in both variables. This self-correlation is small and negative for the randomized data and does not contribute to the positive correlation between ϕ_m and Ri of about 0.6. The positive self-correlation due to the occurrence of the vertical temperature gradient in the numerators of both Ri and ϕ_h is more significant and prevents confident establishment of the significance of the relationship between ϕ_h and Ri. For stationary conditions, the correlation between ϕ_h and Ri is about 0.65 for the original data and 0.4 for the randomized data. In summary, the above results support a physical relationship between ϕ_m and Ri and between ϕ_h and z/L but not between the other two combinations of variables.

Using traditional 5-min averages to define perturbations instead of the variable averaging width would have substantially increased the range of u_* through much larger random flux error, resulting in even larger self-correlation. That is, using an inappropriate large averaging length to compute the perturbations actually increases the correlation between ϕ_m and z/L through increased random variation of u_*. This effect is greatest with larger values of the nonstationarity and stability.

As an aside, the nondimensional gradients are well related to the bulk Richardson number, Ri_B, evaluated from the potential temperature gradient between the 2- and 30-m levels and the wind speed at 30 m. In spite of no self correlation, the correlation between the nondimensional gradients and Ri_B is > 0.5 for all classes of the nonstationarity, and averages about 0.65 for both nondimensional gardients. Randomizing the values used in the nondimensional gradients and the bulk Richardson number leads to near zero correlation, as expected from the absence of shared variables.

This finding seems to be consistent with the observations of Banta et al. (2007) that near-surface turbulence is highly correlated to winds at higher levels, in their case, the magnitude of the low-level jet. The computation of the bulk Richardson number is less vulnerable to errors associated with small differences between large numbers. In addition, surface winds for significant stability can be perturbed by modest time dependence of the mean flow (Sect. 6.2) and even weak surface heterogeneity.

5 Systematic influence of nonstationarity of the mean flow

Even though self-correlation strongly influences the relationship between nondimensional shear and z/L, nonstationarity of the mean flow also significantly influences this relationship. We therefore examine the influence of the nonstationarity on the nonlinear dependence of the nondimensional gradients on stability for $z/L < 1$. The nondimensional shear for intermediate stability ($0.1 < z/L < 1$) (Figs. 2, 3) generally decreases with increasing nonstationarity of the mean flow. That is, the record-averaged momentum flux is larger with respect to the record-averaged mean shear when the within-record standard deviation of the subrecord gradient Richardson number is greater. In other words, the impact of stability on the momentum flux–gradient relationship is less with large nonstationarity. The main impact of nonstationarity increases dramatically between the weak and intermediate nonstationary classes of nonstationarity and seems to saturate when the nonstationarity parameter exceeds roughly unity, as will be seen in subsequent sections. The scatter is large partly because of difficulties estimating vertical gradients with profiles disturbed by nonstationarity. The solid line in Fig. 2 represents the linear prediction with $\beta = 7$, which reasonably fits the class of most stationary records for weak stability. The increase of ϕ_m is well defined for the stationary class (Fig. 3) but the increase is small and poorly defined for the most nonstationary class. This "leveling off" of ϕ_m has been found in a number of studies (see Yagüe et al. 2006 and references therein.)

The decrease of ϕ_m with increasing nonstationarity of the mean flow, for a given value of stability, is also supported by expressing stability in terms of the gradient Richardson number, where self-correlation between ϕ_m and Ri is not important for the allowed range of stability. The nondimensional shear in the intermediate stability interval of Ri = 0.1–0.25, averages about half for the more nonstationarity classes compared to that for the more stationary class. However the interpretation

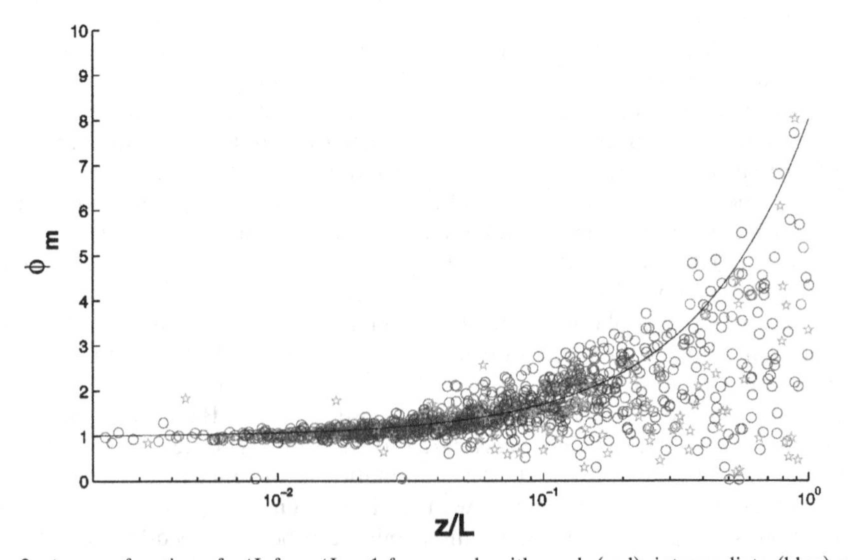

Fig. 2 ϕ_m as a function of z/L for $z/L < 1$ for records with weak (red), intermediate (blue) and strong (green) nonstationarity of the mean flow (Sect. 2.5)

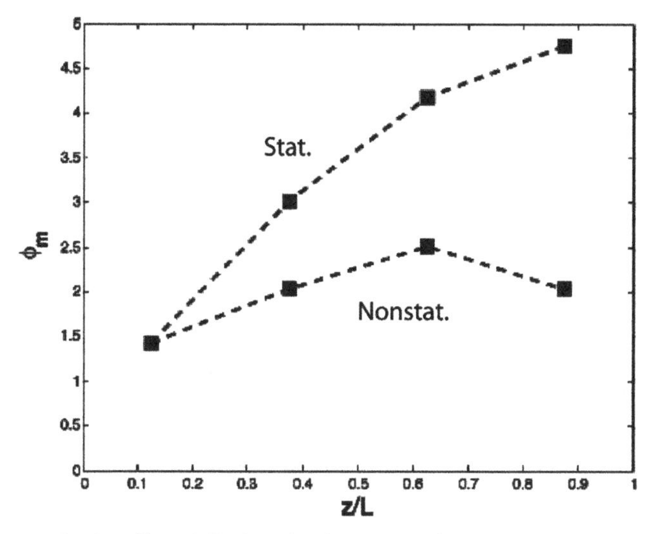

Fig. 3 Bin-averaged values (Sect. 2.6) of ϕ_m for the more stationary class and the combined nonstationary classes

is complicated by a general increase of such nonstationarity of the mean flow with increasing gradient Richardson number.

The influence of nonstationarity on the nondimensional temperature gradient is less definable (not shown). As a result, the eddy Prandtl number increases with nonstationarity (Sect. 7). For the smaller CASES-99, the impact of nonstationarity on the flux-gradient relationship for both heat and momentum is more difficult to define because of fewer cases of significant nonstationarity.

6 Nonstationarity mechanisms

6.1 Businger terms

The nonlinear dependence of the transfer coefficients on stability leads to changes in the time-averaged flux–gradient relationship and effective diffusivity for nonstationary flow, analogous to that found for spatial averages in heterogeneous flows Mahrt (1987). Businger (2005) formalized the influence of nonstationarity in terms of a simple two-state flow. We write Businger's development in a more general form to include arbitrary time dependence of an arbitrary variable, F. The arbitrary variable is decomposed into the record-averaged flow, $[F]$, the deviation of the subrecord average from the record-averaged flow, $\tilde{F}$, and the deviation due to turbulent fluctuations, F', such that

$$F = [F] + \tilde{F} + F' \tag{6}$$

where the total subrecord average can be written as

$$\overline{F} = [F] + \tilde{F}. \tag{7}$$

The subrecord-averaged turbulent flux ($\overline{w'F'}$) can be expressed in terms of an eddy diffusivity as

$$\overline{w'F'} = -K_F \frac{\partial \overline{F}}{\partial z} \quad (8)$$

where the subrecord eddy diffusivity, K_F, relates the subrecord-averaged flux to the subrecord-averaged vertical gradient. The subrecord eddy diffusivity varies within the record due to the influence of the mesoscale motions on the turbulence. This subrecord eddy diffusivity can be partitioned as

$$K_F = [K_F] + \tilde{K_F} \quad (9)$$

where $[K_F]$ is the average of the subrecord diffusivities over the record.

Averaging the subrecord flux over the record, we can write

$$[\overline{w'F'}] = -[([K_F] + \tilde{K_F})(\frac{\partial [F]}{\partial z} + \frac{\partial \tilde{F}}{\partial z})], \quad (10)$$

which becomes

$$[\overline{w'F'}] = -[K_F]\frac{\partial [F]}{\partial z} - [\tilde{K_F}\frac{\partial \tilde{F}}{\partial z}]. \quad (11)$$

The product of tilde terms represents the contribution of subrecord mesoscale variations on the record-averaged flux.

We define an effective eddy diffusivity to relate the record-averaged turbulence flux to the record-averaged vertical gradient

$$-[\overline{w'F'}] \equiv K_{\mathrm{eff}} \frac{\partial [F]}{\partial z}. \quad (12)$$

Due to nonstationarity of the mean flow, the effective diffusivity K_{eff} is different from the average of the subrecord diffusivities over the record, $[K_F]$. With stationarity flow, K_{eff} approaches $[K_F]$.

For temperature ($F = \theta$), we expect the correlation between $\tilde{K_\theta}$ and the vertical potential temperature gradient to be negative since increased stratification (increased gradient Richardson number) generally reduces turbulent mixing. As a result, nonstationarity within the record acts to reduce the eddy diffusivity for the time-averaged heat flux. In contrast, $\tilde{K_m}$ is expected to be positively correlated with mean vertical shear since enhanced shear (decreased gradient Richardson number) generally increases the turbulence. Therefore, the effective eddy diffusivity for momentum is enhanced by nonstationarity of the mean flow. With these arguments, the turbulent Prandtl number should increase with nonstationarity.

We evaluated Eq. 11 for both heat and momentum using 10-min intervals for the subrecords. These evaluations generally indicated that the second term on the right hand side of Eq. 11 due to subrecord variations (product of tilde terms), had the same sign as predicted by Businger (2005), resulting in enhanced momentum flux and reduced heat flux. However, averaged over all of the data, the extra term due to subrecord variations was only a few percent or less for both heat and momentum. Therefore, the effective diffusivities are close to the average of the subrecord diffusivities. If we restrict the analysis to values of the nonstationarity of the mean flow greater than unity, the momentum flux is augmented by 24% by the second term on the right hand side of Eq. 11 while this term decreases the heat flux by 21%.

Since the eddy diffusivity is a dimensional quantity, the generality of the above analysis cannot be easily assessed, although the eddy diffusivity is directly linked to ϕ through surface layer similarity theory (Eq. 5). Enhanced eddy diffusivity for momentum corresponds to reduced ϕ_m.

As an aside, the subrecord variation of the heat flux correlates better with variations of the shear than with variations of the stratification. Greater shear and shear-generation of turbulence leads to more downward heat flux. The sensitivity of the heat flux to shear is probably due to the much greater percentage-wise changes of the shear compared to the smaller percentage-wise changes of stratification. Due to this effect and the quadratic dependence of the gradient Richardson number on shear, the variation of the gradient Richardson number depends mainly on the shear.

We attempted to reduce the role of nonstationarity by evaluating the flux–gradient relationship over 10-min periods instead of 1-hr periods. The resulting flux–gradient relationship showed the expected increase of scatter compared to the 1-hr values. However, the average value of ϕ_m did not increase compared to the average of the 1-hr values, again suggesting that the influence of nonstationarity is not primarily an averaging problem.

6.2 Influence of profile distortion

The influence of nonstationarity on the Prandtl number appears to be partly related to distortion of the wind profile. In the present study, we use a simplified version of the term "curvature" defined to be the second derivative of $u(z)$ the wind profile, as has been done in previous geophysical studies. The dynamics of the turbulence depends on the second derivative and not the true curvature. The true curvature leads to dimensional problems in the denominator. We evaluate the simple curvature of the wind profile near the surface using second-order finite differencing based on the 1-m, 5-m and 10-m levels.

Typical wind profiles in the nocturnal boundary layer, where the shear decreases with height, are characterized by negative curvature of the wind profile. For the bin-averaged data from FLOSS (Sect. 2.6), the negative curvature of the wind profile decreases with increasing nonstationarity (Fig. 4). With strong nonstationarity of the mean flow, the magnitude of the negative curvature becomes small, corresponding to near-linear profiles, or becomes positive, with an inflection point at higher levels. Cases of positive curvature of the wind profile sometimes correspond to a nonstationary wind maximum within the lower part of the tower level. Other cases of positive curvature correspond to a near-stagnant cold layer in the lowest 5–10 m with more significant shear above the cold layer.

While inflection points are well studied in the roughness sublayer over rough canopies (Finnigan 2000), cases of strong stratification and weak winds can experience inflection points at higher levels, and associated instabilities may develop (e.g., Nappo 2002). Then, turbulence may be generated at larger values of the gradient Richardson number (Abarbanel et al. 1984; Grisogono 1994). With positive wind profile curvature, the wind shear increases with height from the ground surface, leading to maximum shear some distance above the surface, which in turn can promote shear instability (Newsom and Banta 2003). It is not possible from a single tower to unambiguously assess the possibility of inflection point instability.

For significant negative curvature of the wind profile, ϕ_m is bounded from below by unity (Fig. 5) with larger values occurring due to stability effects, as expected from sim-

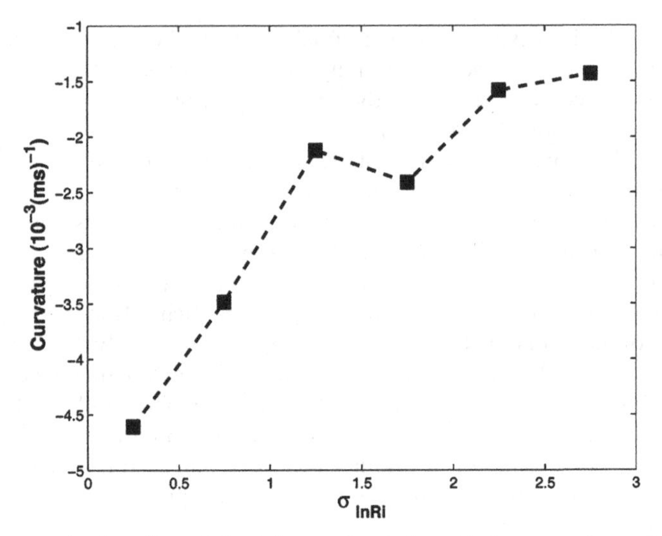

Fig. 4 Bin-averaged values (Sect. 2.6) of the simple wind profile "curvature" as a function of the nonstationarity of the mean flow

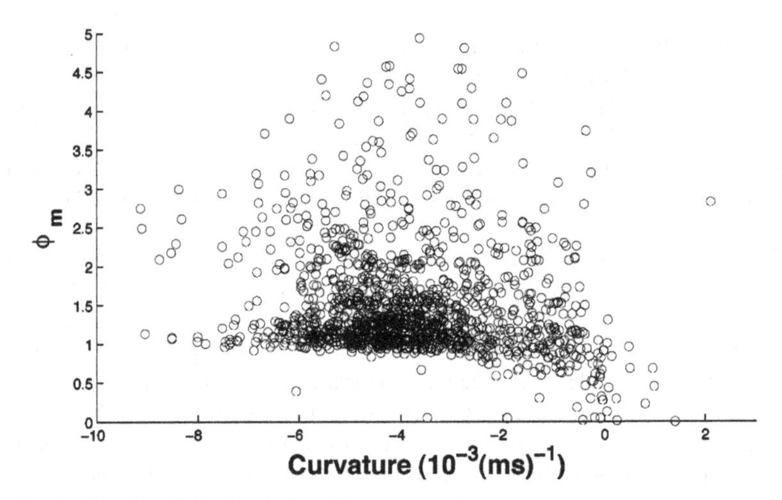

Fig. 5 ϕ_m as a function of simple wind-profile curvature

ilarity theory. However, as the negative curvature becomes weak or vanishes (Fig. 5), values of ϕ_m less than unity occur, even with significant stability, corresponding to the more efficient transfer of momentum than predicted by similarity theory. With positive curvature, the momentum transport is even more efficient and ϕ_m is less than unity for the majority of the records in spite of stable stratification.

With the usual cases of negative curvature of the wind profile, the shear near the surface is greater than that averaged over a deeper layer. With positive curvature of the wind profile, the shear near the surface is less than that averaged over a deeper layer and the calibration of Monin–Obukhov similarity is no longer appropriate. With this argument, the local shear near the surface underestimates the effective shear for positive curvature, leading to smaller ϕ_m. For the smaller CASES-99 dataset, ϕ_m decreases when the wind profile curvature becomes positive, although the impact of profile curvature is less defined than in FLOSSII.

Small negative or positive curvature occur primarily for strong nonstationarity of the mean flow, implying that such curvature is associated with disturbed wind profiles resulting from accelerations (decelerations) in the flow. Small negative or positive curvature of the wind profile have a much weaker influence on the heat transport such that the eddy Prandtl number increases with nonstationarity and profile curvature (Sect. 7).

Records with positive or weak negative curvature generally occur with weak winds, less than about 1.5 m s^{-1} at 2 m, but weak winds do not necessarily imply large deviations from the log-linear profiles. We can summarize that a subset of weak-wind cases correspond to strong nonstationary, large deviations from the log-linear profile, efficient momentum transport and large eddy Prandtl number.

7 Enhanced Prandtl number

The greater efficiency of the momentum transport relative to the heat transport, can be expressed in terms of the eddy Prandtl number

$$Pr \equiv \frac{K_m}{K_\theta} = \frac{\phi_h}{\phi_m}. \tag{13}$$

The correlation Prandtl number

$$R \equiv \frac{r_{wu}}{r_{w\theta}} \tag{14}$$

is of less practical application but less vulnerable to observational errors associated with the computation of vertical gradients. Here r_{wu} is the correlation between the vertical velocity and along-wind velocity fluctuations and $r_{w\theta}$ is the correlation between the vertical velocity and potential temperature fluctuations.

We have computed the eddy Prandtl number based on bin-averaged values of the fluxes and gradients for different intervals of the stability parameters or intervals of the nonstationarity (Sect. 2.6). The eddy-Prandtl number increases from near unity for nearly stationary flows to about 5 for strongly nonstationary flows (Fig. 6). The correlation Prandtl number (Fig. 6) also increases with increasing nonstationarity, although the increase is not as great as that for the eddy Prandtl number.

For the present datasets, the dependence of the Prandtl number on z/L is not well defined, as also found in Yagüe et al. (2001). Large self-correlation between the Prandtl number and the gradient Richardson number prevents evaluation of the physical significance of the increase of the Prandtl number with increasing gradient Richardson number (not shown). The eddy Prandtl number increases from a little greater than unity with normal negative curvature of the wind profile to about 2.5 for positive curvature of the wind profile (Fig. 7). The main increase is from negative to positive curvature. Inflection point instability and the relationship between the momentum flux and shear over a deeper layer may contribute to this dependence.

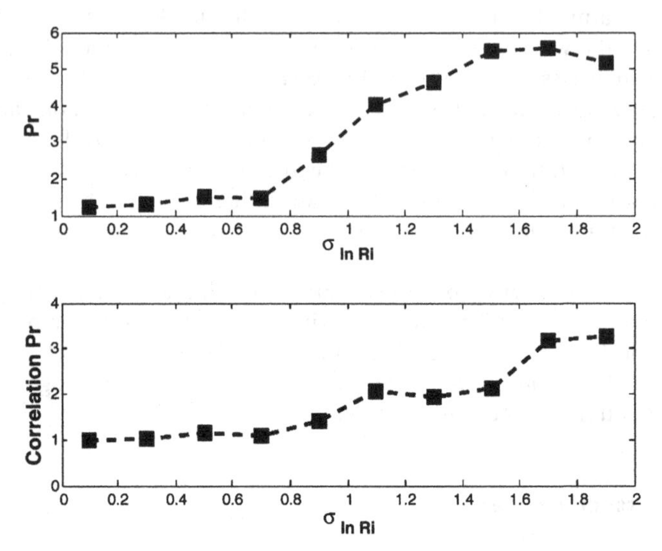

Fig. 6 The eddy-Prandtl number (upper panel) and correlation Prandtl number (lower panel) as a function of the nonstationarity of the mean flow. The correlations, fluxes and gradients are averaged over the records within fixed intervals of the nonstationarity of the mean flow and then the Prandtl numbers are computed from these averages

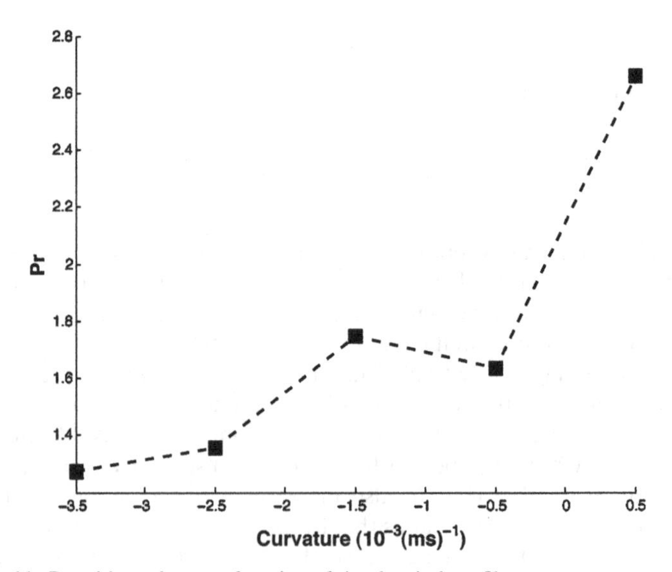

Fig. 7 The eddy-Prandtl number as a function of simple wind-profile curvature

8 Conclusions

The above study examined the influence of stability, nonstationary profile distortion and self-correlation on the flux-gradient relationship near the surface for weak and moderate stability ($z/L < 1$). The dependence of the nondimensional shear on z/L cannot be established as a physically viable relationship because large self-correlation dominates the total correlation and masks any physical relationship. In contrast, the physical significance of the dependence of the nondimensional temperature gradient on z/L is statistically supported by the data analysis, as is the physical significance of the dependence of the nondimensional shear on the local gradient Richardson number. In these two cases, the self-correlation is negative and does not contribute to the observed positive correlation between the nondimensional gradient and stability. The nondimensional gradients near the surface are also reasonably well correlated with the bulk Richardson number computed across the entire tower layer.

Nonstationarity of the mean flow, due to wave-like motions, meandering motions and common more complex signatures, increases the efficiency of the momentum transport in terms of reduced ϕ_m. The within-record variation of the gradient Richardson number was used to formally represent the nonstationarity of the mean flow. Such variations are related primarily to changes of wind shear, not changes in stratification. The nonstationarity of the mean flow appears to influence the turbulent flux–gradient relationship mainly through profile distortion. With significant nonstationarity of the mean wind, the usual negative curvature of the wind profile, where the mean shear decreases rapidly with height, often yields to more linear wind profiles. Nonstationarity of the mean wind may even occur with positive curvature where the wind shear increases with height at the surface with an inflection point at higher levels. The possibility of inflection point instability cannot be determined from the data. Small negative or positive curvature of the wind profile may correspond to smaller ϕ_m because the shear near the surface is smaller relative to the layer-averaged shear over a deeper layer.

The eddy diffusivity of heat increases with nonstationary distortion of the wind profile much more slowly (with more scatter) compared to the increase of the eddy diffusivity of momentum. As a result, the Prandtl number increases with increased nonstationarity and distortion of the mean wind profile. The increase of the Prandtl number with the Richardson number is dominated by self-correlation for these data.

Covariances between subrecord variations (Eq. 11) of the momentum eddy diffusivity and shear or thermal diffusivity and stratification (Businger 2005) contribute significantly to the record-averaged flux-gradient relationship only in the most nonstationary cases. The majority of the enhancement of the momentum flux cannot be explained by this mechanism. Reducing the averaging time from 1 h to a smaller value, to filter out part of the nonstationarity, only modestly reduces the influence of nonstationarity on the flux-gradient relationship. Therefore, the more efficient transfer of momentum with strong nonstationarity is not primarily due to averaging over nonstationarity of the mean flow, but rather due to an intrinsic change of turbulent transport in non-equilibrium turbulence, or, to an altered relation of the momentum flux to the distorted wind profiles.

The generality of the above results is not known. While the FLOSSII site was locally relatively flat, it is embedded in complex terrain. The systematic influence of external nonstationarity and profile distortion may be a common feature in stable flow over less ideal terrain. The present study was limited to weak and modest stability ($z/L < 1$), and the estimation of vertical gradients for stronger stability with significant nonstationarity becomes sensitive to the method of calculation of the vertical gradient, currently under investigation.

Acknowledgements I gratefully acknowledge the helpful detailed comments of the reviewers, the data processing and extensive comments of Dean Vickers, as well as the helpful comments of Danijel Belušić, Reina Nakamura and Joan Cuxart. I also acknowledge collection of the Alaska data by Mathew Sturm, collection of ASIT data by Jim Edson, and collection of FLOSS and CASES99 data by the NCAR ATD staff. This material is based upon work supported by Contract W911FN05C0067 from the Army Research Office and Grant 0107617-ATM from the Physical Meteorology Program of the National Sciences Program.

References

Abarbanel HDI, Holm DD, Marsden JE, Ratiu T (1984) Richardson number criterion for the nonlinear stability of three-dimensional stratified flow. Phys Rev Lett 52:2352–2355

Acevedo OC, Moraes OLL, Degrazia GA, Medeiros LE (2006) Intermittency and the exchange of scalars in the nocturnal surface layer. Boundary-layer meteorol 119:41–55

Andreas EL (2002) Parameterizing scalar transfer over snow and ice: a review. J Hydrometeorol 3:417–431

Anfossi D, Oettl D, Degrazia G, Boulart A (2005) An analysis of sonic anemometer observations in low wind speed conditions. Boundary-Layer Meteorol 114:179–203

Atlas D, Metcalf JI, Richter JH, Gossard EE (1970) The birth of "CAT" and microscale turbulence. J Atmos Sci 27:903–913

Banta RM, Pichugina YL, Brewer WA (2007) Turbulent velocity–variance profiles in the stable boundary layer generated by a nocturnal low-level jet. J Atmos Sci (In press)

Basu S, Porté-Agel F, Foufoula-Georgiou E, Vinuesa J-F, Pahlow M (2006) Revisiting the local scaling hypothesis in stably stratified atmospheric boundary-layer turbulence: an integration of field and laboratory measurements with large-eddy simulations. Boundary-Layer Meteorol 119:473–500

Beljaars ACM, Holtslag AAM (1991) Flux parameterization over land surfaces for atmospheric models. J Appl Meteorol 30:327–341

Businger J (2005) Reflections on boundary-layer problems of the past 50 years. Boundary-Layer Meteorol 116:149–159

Chimonas G (1984) Apparent counter-gradient heat gluxes generated by atmospheric waves. Boundary-Layer Meteorol 31:1–12

Chimonas G (2002) On internal gravity waves associated with the stable boundary layer. Boundary-Layer Meteorol 102:139–155

Chimonas G (2003) Pressure gradient amplification of shear instabilities in the boundary layer. Dyn Atmos Oceans 37:131–145

Cooper DI, Leclerc MY, Archuleta J, Coulter R, Eichinger WW, Kao CYJ, Nappo CJ (2006) Mass exchange in the stable boundary layer by coherent structures. Annu Rev Fluid Mech 136:113–131

Coulter RL, Doran JC (2002) Spatial and temporal occurrences of intermittent turbulence during CASES-99. Boundary-Layer Meteorol 105:329–349

Derbyshire H (1995) Stable boundary layers: Observations, models and variability part II: data analysis and averaging effects. Boundary-Layer Meteorol 75:1–24

Derbyshire S (1999) Boundary-layer decoupling over cold surfaces as a physical boundary-instability. Boundary-Layer Meteorol 90:297–325

Edson JB, Crofoot R, McGillis W, Zappa C (2004) Investigations of flux–profile relationships in the marine atmospheric boundary layer during CBLAST. 16th Symposium on Boundary Layers and Turbulence, 9–13 August 2004, Portland, Maine.

Etling D (1990) On plume meandering under stable stratification. Atmos Environ. 24A:1979–1985

Fernando HJS (2003) Turbulence patches in a stratified shear flow. Phys Fluids 15:656, 3164–3169

Finnigan JJ (1999) A note on wave-turbulence interaction and the possibility of scaling the very stable boundary layer. Boundary-Layer Meteorol 90:529–539

Finnigan JJ (2000) Turbulence in plant canopies. Ann Rev Fluid Mech 32:519–571

Finnigan JJ, Einaudi F (1981) The interaction between an internal gravity wave and the planetary boundary layer. Part II: effect of the wave on the turbulence structure. Quart J Roy Meteorol Soc 107:807–832

Finnigan JJ, Einaudi F, Fua D (1984) The interaction between an internal gravity wave and turbulence in the stably-stratified nocturnal boundary layer. J Atmos Sci 41:2409–2436

Grachev AA, Fairall CW, Persson POG, Andreas EL, Guest PS (2005) Stable boundary-layer scaling regimes, the Sheba data. Boundary-Layer Meteorol 116:201–235

Grisogono B (1994) A curvature effect on the critical Richardson number. Croatian Meteorol J 29:43–46
Herring JR, Métais O (1989) Numerical experiments in forced stably stratified turbulence. J Fluid Mech 202:97–115
Hicks BB (1976) Wind profile relationships from the 'Wangara' experiment. Quart J Roy Meteorol Soc 102:535–551
Hicks BB (1981) An examination of turbulence statistics in the surface boundary layer. Boundary-Layer Meteorol 21:389–402
Holtslag AAM, De Bruin HAR (1988) Applied modeling of the nighttime surface energy balance over land. J Appl Meteorol 27:689–704
Howell J, Sun J (1999) Surface layer fluxes in stable conditions. Boundary-Layer Meteorol 90:495–520
Kim J, Mahrt L (1992) Simple formulation of turbulent mixing in the stable free atmosphere and nocturnal boundary layer. Tellus 44A:381–394
Klipp C., Mahrt L (2004) Flux-gradient relationship, self-correlation and intermittency in the stable boundary layer. Quart J Roy Meteorol Soc 130:2087–2104
Kondo J, Kanechika O, Yasuda N (1978) Heat and momentum transfers under strong stability in the atmospheric surface layer. J Atmos Sci 35:1012–1021
Kristensen L, Jensen NO, Peterson EL (1981) Lateral dispersion of pollutants in a very stable atmosphere — the effect of the meandering. Atmos Environ 15:837–844
Lee S-M, Giori W, Princevac M, Fernando HJS (2006) Implementation of a stable PBL turbulence parameterization for the mesoscale model MM5: nocturnal flow in complex terrain. Boundary-Layer Meteorol 119:109–134
Lilly DK (1983) Stratified turbulence and the mesoscale variability of the atmosphere. J Atmos Sci 40:749–761
Mahrt L (1987) Grid-averaged surface fluxes. Mon Wea Rev 115:1550–1560
Mahrt L, Vickers D (2002) Contrasting vertical structures of nocturnal boundary layers. Boundary-Layer Meteorol 105:351–363
Mahrt L, Vickers D, Frederickson P, Davidson K, Smedman A-S (2003) Sea-surface aerodynamic roughness. J Geophys Res 108:1–9
Mahrt L, Vickers D (2005) Boundary-layer adjustment over small-scale changes of surface heat flux. Boundary-Layer Meteorol 116:313–330
Mahrt L, Vickers D (2006) Extremely weak mixing in stable conditions. Boundary-Layer Meteorol 119:19–39
McWilliams J (2004) Phenomenological hunts in two-dimensional and stably stratified turbulence. In: Federovich E, Rotunno R, Stevens B (eds) Atmospheric turbulence and mesoscale meteorology. Cambridge University Press, pp 35–49
Monti P, Fernando HJS, Chan W, Princevac M, Kowalewski T, Pardyjak E (2002) Observations of flow and turbulence in the nocturnal boundary layer over a slope. J Atmos Sci 59:2513–2434
Moraes OLL, Acevedo OC, Da Silva R, Magnago R, Siqueira AC (2004) Nocturnal surface-layer characteristics at the bottom of a valley. Boundary-Layer Meteorol 112:159–177
Nakamura R, Mahrt L (2005) A study of intermittent turbulence with CASES-99 tower measurements. Boundary-Layer Meteorol 114:367–387
Nappo CJ (2002) An introduction to atmospheric gravity waves. Academic Press, 276 pp
Newsom R, Banta R (2003) Shear-flow instability in the stable nocturnal boundary layer as observed by Doppler lidar during CASES99. J Atmos Sci 60:16–33
Ohya Y (2001) Wind-tunnel study of atmospheric stable boundary layers over a rough surface. Boundary-Layer Meteorol 98:57–82
Ohya Y, Uchida T (2003) Turbulence structure of stable boundary layers with a near-linear temperature profile. Boundary-Layer Meteorol 108:19–38
Panofsky HA, Dutton JA (1984) Atmospheric turbulence — models and methods for engineering applications. John Wiley and Sons, New York, 397 pp
Pardyjak E, Monti P, Fernando H (2002) Flux Richardson number measurements in stable atmospheric shear flows. J Fluid Mech 449:307–316
Poulos GS, Blumen W, Fritts D, Lundquist J, Sun J, Burns S, Nappo C, Banta R, Newsom R, Cuxart J, Terradellas E, Balsley B, Jensen M (2002) CASES-99: a comprehensive investigation of the stable nocturnal boundary layer. Bull Amer Meteorol Soc 81:757–779
Riley JJ, Lelong M-P (2000) Fluid motions in the presence of strong stable stratification. Annu Rev Fluid Mech 32:613–657
Salmond JA (2005) Wavelet analysis of intermittent turbulence in a very stable nocturnal boundary layer: Implications for the vertical mixing of ozone. Boundary-Layer Meteorol 114:463–488

Sorbjan Z (2006) Local structure of turbulence in stably stratified boundary layers. J Atmos Sci 63:526–537
Strang EJ, Fernando HJS (2001) Vertical mixing and transports through a stratified shear layer. J Phys Oceanog 31:2026–2048
Sukoriansky S, Galperin B, Perov V (2005) Application of a new spectral theory of stably stratified turbulence to the atmospheric boundary layer over sea ice. Boundary-Layer Meteorol 117:231–257
Sun J, Burns SP, Lenschow DH, Banta R, Newsom R, Coulter R, Frasier S, Ince T, Nappo C, Cuxart J, Blumen W, Lee X, Hu X-Z (2002) Intermittent turbulence associated with a density current passage in the stable boundary layer. Boundary-Layer Meteorol 105:199–219
Sun J, Lenschow DH, Burns SP, Banta RM, Newsom RK, Coulter S, Frasier S, Ince T, Nappo C, Balsley B, Jensen M, Mahrt L, Miller D, Skelly B (2004) Atmospheric disturbances that generate intermittent turbulence in nocturnal boundary layers. Boundary-Layer Meteorol 110:255–279
Ueda H, Mitsumoto S, Komori S (1981) Buoyancy effects on the turbulent transport processes in the lower atmosphere. Quart J Roy Meteorol Soc 107:561–578
van de Wiel BJH, Moene AF, Ronda RJ, De Bruin HAR, Holtslag AAM (2002) Intermittent turbulence in the stable boundary layer over land. Part II: a system dynamics approach. J Atmos Sci 59:2567–2581
van den Kroonenberg A, Bange J (2007) Turbulent flux calculation in the polar stable boundary layer: multiresolution flux decomposition and wavelet analysis. J Geophys Res (In press)
Van Ulden AP, Holtslag AAM (1985) Estimation of atmospheric boundary layer parameters for diffusion applications. J Clim Appl Meteorol 24:1197–1207
Vickers D, Mahrt L (1997) Quality control and flux sampling problems for tower and aircraft data. J Atmos Oceanic Tech 14:512–526
Vickers D, Mahrt L (2006) A solution for flux contamination by mesoscale motions with very weak turbulence. Boundary-Layer Meteorol 118:431–447
Yagüe C, Maqueda G, Rees JM (2001) Characteristics of turbulence in the lower atmosphere at Halley IV Station, Antarctica. Dyn Atmos Oceans 34:205–223
Yagüe C, Viana S, Maqueda G, Redondo JM (2006) Influence of stability on the flux-profile relationships for wind speed, ϕ_m, and temperature, ϕ_h, for the stable atmospheric boundary layer. Nonlin Proc Geophys 13:185–203
Zilitinkevich SS, Calanca P (2000) An extended similarity theory for the stably stratified atmospheric surface layer. Quart J Roy Meteorol Soc 126:1913–1923
Zilitinkevich S, Baklanov A, Rost J, Smedman AS, Lykosov V, Calanca P (2002) Diagnostic and prognostic equations for the depth of the stably stratified Ekman boundary layer. Quart J Roy Meteorol Soc 128:25–46

Chemical perturbations in the planetary boundary layer and their relevance for chemistry transport modelling

Adolf Ebel · Michael Memmesheimer · Hermann J. Jakobs

Abstract The role of perturbations of reactive trace gas concentration distributions in turbulent flows in the planetary boundary layer (PBL) is discussed. The paper focuses on disturbances with larger spatial scales. Sequential nesting of a chemical transport model is applied to assess the effect of neglecting subgrid chemical perturbations on the formation and loss of ozone, NO_x, peroxyacetyl nitrate (PAN) and HNO_3 calculated with a highly complex chemical mechanism. The results point to characteristic differences regarding the process of mixing of chemically reactive species in the PBL and lower troposphere.

Keywords Chemical perturbations · Chemistry transport model (CTM) · Damköhler number · Deposition · Emission · Mesoscale · Segregation · Turbulence

1 Introduction

Dynamical and chemical perturbations with scales smaller than the spatial and temporal resolution of a chemistry transport model (CTM) may lead to significant effects on the simulated temporal variation and spatial distribution of reactive species and cause errors of the calculated behaviour of atmospheric composition if they are not taken into account. This is particularly true in the planetary boundary layer (PBL). Yet parameterization of subgrid chemical processes in the complex dynamic and chemical system of the atmosphere in general and the PBL in particular is a tedious task and subject to fundamental ambiguities. It is therefore neglected in most CTM applications and usually only treated in special (simplified) model versions focusing on selected perturbation processes. Only recently a method of parameterizing effective chemical reaction coefficients has been suggested by Vinuesa and Vilà-Guerau

A. Ebel (✉) · M. Memmesheimer · H. J. Jakobs
Rhenish Institute for Environmental Research at the University of Cologne, Aachener Str. 209, 50931 Cologne, Germany
e-mail: eb@eurad.uni-koeln.de

Atmospheric Boundary Layers. A. Baklanov & B. Grisogono (eds.),
doi: 10.1007/978-0-387-74321-9_8, © Springer Science+Business Media B.V. 2007

de Arellan (2005, see Sect. 4). Existing studies (e. g. Komori et al. 1991; Kramm and Meixner, 2000; Vilà-Guerau de Arellano et al. 2004) mainly deal with smaller scale turbulent disturbances that are most difficult or impossible to resolve by observations as far as chemical fluctuations in space and time in the real atmosphere are concerned. In contrast, this paper emphasizes the effect of larger scale eddies in a range that, in principle, is observable through standard operational networks (O(5 km), O(1 h)).

The effect of chemical perturbations has been studied under various conditions for several decades. First studies were devoted to perturbed chemical transport in the laboratory (Damköhler 1940; Hawthorne et al. 1949). Comparing homogeneous and inhomogeneous reactive mixtures in turbulent flows in the laboratory, lower yields were found in the latter case (Toor 1969). Wind-tunnel experiments exploring the degree of segregation of atmospheric pollutants indicate preferably negatively correlated perturbations of the studied reactive gases (e. g. Builtjes 1981). Such a tendency to negative correlations was confirmed by a Lagrangian model study carried out by Komori et al. (1991). Chemical reactions of non-premixed reactants in a turbulent and diffusive medium were shown to be less efficient than those of perfectly mixed ones. Knowledge gained in the laboratory was applied to atmospheric conditions the first time by Donaldson and Hilst (1972), who pointed to the fact that the mixing of emitted gases in the vicinity of their sources is not fast enough and so the effects of inhomogeneous mixing are relevant. Such studies have preferably been devoted to reactive plumes by applying plume models (e.g. Georgopoulos and Seinfeld (1986); Vilà-Guerau de Arellano et al. 1990, 2004). Attempts have been made to reproduce the mixing of reactive species in the convective PBL with the help of large-eddy simulation (LES; Schumann 1989; Sykes et al. 1994) They show that the asymmetry between strongly confined updrafts and slow extended downdrafts lead to height dependent segregation affecting atmospheric chemistry with a different intensity at different levels.

Such studies have mainly been concerned with the impact of molecular diffusion and turbulent motions (eddies) of air on atmospheric chemistry, sometimes taking into account irregularities of emission sources. It is emphasized that other processes exist that cause perturbations of mixing ratios of atmospheric minor constituents. A general impression on how physical processes represented in chemical transport models by parameterizations, e.g. deposition, and specific choices of input data, e.g. land-use categories, may be obtained from the ensemble study of ozone modelling by Mallet and Sportisse (2006). Classifying the impact of mixing ratio perturbations on the budget of chemicals according to their generation, proceeding from purely dynamical to purely chemical and additional processes, and focusing on gas phase chemistry, one may differentiate between the following processes: (a) pure turbulent mixing of species with very large life and residential times (e.g. Stull 1988); (b) turbulent mixing with chemistry leading to the modulation of eddy fluxes (e.g. Kramm and Meixner 2000); (c) only chemical perturbations of binary and ternary reactions (e.g. Damköhler 1940; Stockwell 1995); (d) reaction rate perturbations in the case of temperature dependence (Finlayson-Pitts and Pitts 1986); (e) perturbations of the radiation field leading to variable actinic fluxes, e.g. changing photolysis rates in the presence of clouds (Vilà-Guerau de Arellano et al. 2005) or through fluctuations of total ozone (Wetzel and Slusser 2005).

Furthermore, differences of daytime and nighttime chemistry should result in a noticeable time dependence of chemical perturbation effects. Additional specific perturbation processes may be identified in heterogeneous chemical systems and for cloud chemistry that are out of the scope of this study. In this context it should be noted that heterogeneous reactions and phase transitions in a complex gas-particle system may lead to feedbacks on turbulent motions as shown in a fundamental study by Elperin et al. (1998).

In addition to turbulence (eddies) there is the irregular distribution of emission sources (Krol et al. 2000) with fluctuating emissions and transport in and by clouds as causes of mixing ratio perturbations of minor atmospheric constituents through transport. The role of earth surface properties has also to be mentioned. Orography, land type, vegetation and surface temperature play a decisive role for the generation of chemical perturbations. Several of these issues are addressed in the following sections. We start with general considerations about the interaction of dynamical and chemical perturbations in Sect. 2. The role of deposition, i.e. perturbations at the lower boundary, is briefly treated in Sect. 3. Sect. 4 contains results of nested mesoscale simulations regarding the change of calculated chemical production with increasing model resolution. Conclusions and an outlook for future work needed are given in Sect. 5.

2 General aspects

The relative importance of turbulent mixing and chemical transformation can be characterized by the Damköhler number D, i.e. the ratio of a characteristic mixing time scale τ_t and the chemical lifetime τ_{ch} of a minor constituent (Damköhler 1940):

$$D = \tau_{\mathrm{t}}/\tau_{\mathrm{ch}}. \tag{1}$$

If the lifetime of a reactive species is long in comparison to the time scale of the mixing process under consideration ($D \ll 1$) efficient mixing is possible and irregular changes of the concentrations of reactive species (briefly "chemical eddies") are less effective in the transformation process. The contrary is true when $D \gg 1$. This is schematically shown in Fig. 1 for various mixing conditions represented by typical time scales. Usually only vertical mixing is considered as the relevant turbulent process (time scales between about 10 and 5,000 s for unstable to stable stratification; the first three lines from left to right in Fig. 1). Yet it is emphasized here that mesoscale horizontal mixing is an important factor for the generation of chemical perturbations. This may easily be demonstrated with routine measurements from operational networks (e.g. EMEP, Netherlands, North-Rhine Westphalia). Therefore, an example of mesoscale eddies with larger time scales (12–24 h, stagnant anticyclonic conditions for chemistry and transport anticipated) is also exhibited in Fig. 1. Lifetimes of three inorganic and two organic compounds are marked in order to indicate that chemical yields of short-lived reactants such as NO_2 (the darkened area over bar 2) and NO are more sensitive to turbulence than relatively long-lived ones such as ozone. This is confirmed by observations and models. Yet if one takes into account horizontal fluctuations of the flow also ozone and other long-lived species will show noticeable sensitivity to eddy activity as confirmed, for instance, by operational measurements of ozone and NO_x (Memmesheimer et al. 2005).

It is common to decompose chemical parameters as is done for physical ones, that is into mean and fluctuating terms. Usually it is assumed that ensemble averaging is possible in a fluid (and that ergodic conditions hold, i.e. that ensemble averaging is approximately equivalent to time and/or spatial averaging). For simplicity we will

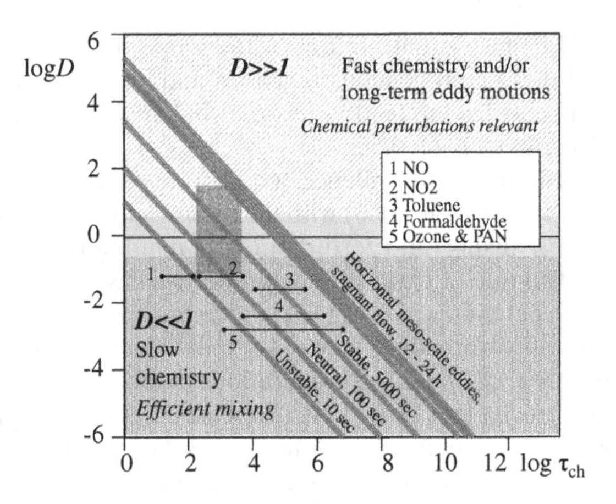

Fig. 1 Damköhler number D as function of chemical lifetime τ_{ch} for different turbulent time scales representing different PBL stability conditions. Horizontal bars indicate approximate lifetimes of reactive species in the PBL. Time in seconds

formally use this approach though such conditions appear to be rare in the real atmosphere as far as chemical mixtures are concerned. The question of averaging methods is comprehensively discussed by Kramm and Meixner (2000) showing that Hesselberg averaging generally gives more correct results than Reynolds averaging in the PBL, but that both methods converge if the effect of density fluctuations becomes small. The reader is referred to this paper. For the sake of brevity and with the aim of discussing the principle effects of eddy mixing we leave open the question of averaging method and apply notations that are common for the Reynolds method of averaging. A tentative justification for this is the fact that we deal with relatively large eddies for which density fluctuations are assumed to be less important. For our purposes it is important to hint to the formal difference if one applies temporal and a simple form of spatial averaging separately. This has to do with the inhomogeneity and anisotropy of eddy perturbations in mixed realistic PBL tracer flows. It is noticed that the problem of averaging is closely related to the question of representativity of atmospheric measurements and the difference of local (point) and volume values of concentrations of a chemical species. Indicating averages and perturbations by $x = \bar{x} + x'$ (ensemble), $x = \langle x \rangle + x^*$ (time) and $x = [x] + x''$(space) one finds for binary reactions proportional to the product $c_i c_j$ of the concentrations of the species i, j:

$$\overline{c_i c_j} = \bar{c}_i \bar{c}_j + \overline{c'_i c'_j}, \tag{2a}$$

$$\langle c_i c_j \rangle = \langle c_i \rangle \langle c_j \rangle + \left\langle c_i^* c_j^* \right\rangle, \tag{2b}$$

$$[c_i c_j] = [c_i] [c_j] + \left[c''_i c''_j \right]. \tag{2c}$$

Averaging the time average spatially one has

$$[\langle c_i c_j \rangle] = [\langle c_i \rangle] [\langle c_j \rangle] + \left[\left\langle c_i^* c_j^* \right\rangle \right] + [\langle c_i \rangle^* \langle c_j \rangle^*]. \tag{3}$$

It is obvious that the last term in Eq. 3 will to a great deal be determined by processes and structures causing spatial irregularities, e.g. by the irregular distribution of sources or inhomogeneous deposition. Consider a simple version of the continuity equation of a constituent $c_i = C$,

$$\frac{\partial C}{\partial t} + \frac{\partial (u_l C)}{\partial x_l} = P - L + E, \tag{4}$$

with t time, u_l the wind component in direction x_l, P production, L loss and E emission in the interior of the model domain (internal emission) of C. Sedimentation of matter (aerosols) would require an additional vertical flux gradient term in this equation (Elperin et al. 1998; Kramm and Meixner 2000). We restrict ourselves to gaseous compounds. Assuming homogeneous production of C through a second-order reaction of species A and B ($= k_P AB$) and chemical loss through a reaction of C with D ($= k_L CD$) and photolysis (rC) (where k_P, k_L and r are rate constants of the chemical reactions and the photolytic rate constant for C, respectively) and averaging Eq. 4 (ensemble average applied for formal simplicity) one obtains

$$\frac{\partial \bar{C}}{\partial t} + \frac{\partial \bar{u}_l \bar{C}}{\partial x_l} = -\frac{\partial \overline{u_l' C'}}{\partial x_l} + k_P(\bar{A}\bar{B} + \overline{A'B'}) - k_L(\bar{C}\bar{D} + \overline{C'D'}) - r\bar{C} + \bar{E}. \tag{5}$$

Many studies hint at the importance of the correlation of the subgrid fluctuations A', B' and C' for the averaged continuity Eq. 5 at least for three decades (e.g. Donaldson and Hilst 1972; Builtjes and Talmon 1987). For the averaging procedure it is common to assume that fluctuations of the chemical (k) and photolytic rate constants (r) can be neglected but which is certainly not true in the general case (e.g. Elperin et al. 1998). Here, k may depend on air temperature, and r, for instance, on cloudiness that obviously exhibit fluctuations in the real atmosphere. The perturbation method can also be expanded to third- and higher-order reactions, e.g. $L = -k_3 c_i c_j c_k = -(\bar{k}_3 + k_3') \prod_{s=i,j,k} (\bar{c}_s + c_s')$for perturbed third-order loss reactions (e. g. Hellmuth 2005). This would lead to higher moments in the set of continuity equations for a reactive gas mixture and a large number of additional unknowns in the partial differential equation system of chemical mechanisms. Subgrid effects on simulated purely chemical production and loss as discussed in Sect. 4 have mainly to be attributed to the neglect of the k', r' and X' terms originating from the decomposition in Eq. 4.

The efficiency of chemical perturbation effects depends on the intensity of segregation S, i.e. the degree of mixing of two or more components, which is defined for the perturbations of the concentrations of two species by

$$S_{ij} = \frac{\overline{c_i' c_j'}}{\bar{c}_i \bar{c}_j}. \tag{6}$$

Reformulating the chemical reaction terms in Eq. 5 using S (e.g. $k(\bar{A}\bar{B} + \overline{A'B'}) = k\bar{A}\bar{B}(1 + S_{AB})$ it is evident that the efficiency of chemical perturbations is controlled by segregation intensity. In PBL, wind-tunnel and LES experiments values between 0 (fluctuations not correlated, well mixed) and -1 (highly segregated) have been found (Vilà-Guerau de Arellano et al. 1993b; Builtjes and Talmon 1987; Schumann 1989; Kramm and Meixner 2000). It seems that S usually decreases with height in the upper PBL (Schumann 1989; Vilà-Guerau de Arellano et al. 1993b).

Integration of the set of partial differential equations given by Eq. 5 for a specific reactive mixture requires knowledge of the perturbation terms. The most widely used method of dealing with the problem in atmospheric chemistry transport modelling is the first-order closure (flux-gradient relationship, K theory) neglecting the correlation of perturbations of reactive species. Their treatment is calling for second- or higher-order closure (Verver et al. 1997; Kramm and Meixner 2000; Hellmuth 2005). A complete perturbation treatment in a CTM also requires the definition of fluctuation terms for boundary conditions. This fact seems to be neglected in most perturbation studies. That it does play a role is clearly evident when flux conditions are applied at the lower boundary, e.g. for the vertical flux F_{iz} of species i with surface concentration c_{oi}, deposition velocity w_{Di} (Eq.9) and surface emission E_{Si}

$$F_{iz} = w_{Di}c_{oi} + E_{si}, \tag{7a}$$

$$\bar{F}_{iz} = \bar{w}_{Di}\bar{c}_{oi} + \overline{w'_{Di}c'_i} + \bar{E}_{si}. \tag{7b}$$

3 Chemical perturbations and deposition

Chemical perturbations affect the eddy transport properties of a flow in the ABL particularly in the vertical direction in the PBL and may thus cause deviations of vertical chemical tracer profiles from those of inert species. Thereby deposition fluxes are modified (Kramm and Meixner 2000). Vilá-Guerau de Arellano et al. (1993a) using a one-dimensional model with a simplified chemical system, only taking into account ozone production and loss through nitrogen oxides, photolysis of NO_2, and turbulent fluctuations of all species showed that the deposition/emission of a reactive tracer (NO) near the ground is also crucial for the formation of the vertical profile of the concentration under the influence of turbulent mixing. Ozone with a relatively long lifetime is a species less affected by chemical fluctuations in such a system.

The deposition velocity w_{Di} in Eq. 7a is strongly coupled to the wind field thus experiencing perturbations through turbulence. This is best seen when the usual resistance formulation is used for a species i:

$$w_{Di} = \frac{1}{R_a + R_b + R_{ci}}, \tag{8}$$

where R_a is the aerodynamic resistance, R_b is the quasi-laminar sublayer resistance and R_c is the overall surface (including canopies, grass, sand, water, etc.) resistance. Note that R_a changes with PBL conditions as described by the friction velocity u_*,the roughness length z_0 and stability parameters such as the Obukhov length L_*(also a function of u_*).Since all mentioned parameters are controlled by the velocity of the flow, R_a fluctuates with the fluctuating wind speed. The same is true for $R_b = f(u_*)$. The overall surface resistance R_c can be composed from the resistances of various processes (e. g. Zhang et al. 2003): stomatal resistance R_{st}, specific resistance $R_{sp,i}$ of species i, in-canopy aerodynamic resistance R_{ac}, resistance of the underlying ground R_g and resistance to cuticle uptake R_{cu}. Evidently R_c does not exhibit an essential dependence on flow characteristics, but may be modulated by changing temperature and radiation. It shows strong spatial variability and can thus induce significant spatial perturbations in the chemical system not only on small scales (for which the described parameterizations are usually derived), but also on larger ones that are of special interest for this study.

4 Subgrid effects in mesoscale simulations

In this section we focus on the analysis of subgrid effects that occur in standard simulations with complex chemistry transport models not treating chemical perturbations explicitly as is common for most CTM applications. We apply the method of sequential nesting (Jakobs et al. 1995) with gradual increase of the horizontal resolution. The expectation is that with increasing resolution more and more smaller scale structures and processes are resolved. Comparison of results obtained with coarse and fine resolution can thus be used to assess the impact of unresolved eddy structures in coarser grid simulations. As already mentioned, this approach allows for the application of atmospheric chemistry systems in full complexity. To our knowledge all specified models that attempt to explicitly introduce chemical perturbation terms through second- and higher-order closure have been confined to a very limited number of reactions in the case of practical applications since information needed for defining sensible closure assumptions for such systems is quite limited at present. The complex model allows the separation of the chemical production/loss term and of the deposition term of the chemical mechanism for specific species to be chosen. It also offers the possibility of studying mechanisms of eddy generation in concentration distributions other than pure turbulence that underlies the concept of the Damköhler number.

The EURAD model system has been applied taking into account only gas phase reactions. The description of the model may be found in Memmesheimer et al. (2004). K parameterization is employed for vertical transport calculations following Holtslag and Nieuwstadt (1986) as described by Hass (1991). The model was applied to a domain covering Berlin, Germany, and a photo-smog episode between 21 and 27 July 1994 was simulated (Memmesheimer et al. 1999). Calculations of ozone, NO_x, PAN and HNO_3 budgets were carried out and discussed by Weber (1999) using the method of space–time integration introduced by Memmesheimer et al. (1997). The budgets can advantageously be employed to asses the role of chemical and dynamical subgrid perturbations for chemical yields and deposition fluxes, in particular.

4.1 Results of simulations

Changes of chemical production/loss and dry deposition velocities of ozone, NO_x, PAN and HNO_3 with increasing resolution are depicted in Fig. 2. Relative variations are shown for easier comparison. It is evident from panel (a) that in the height range between 80 m and 2000 m the calculated yields (of O_3,PAN, HNO_3) and losses (NO_x) decrease by about 20–25 % for the four analysed species when the horizontal resolution is decreased from 54 km (one grid box of the coarse resolution simulation) to 2 km (729 grid boxes). This confirms the expected overestimation of chemical activity when (horizontal) perturbations of concentrations are only coarsely resolved. The relative change decreases with finer resolution indicating that a large part of the resolvable and/or more efficient spatial fluctuations is already covered by the first nesting step. It is briefly noted that the reduction of calculated chemical production/loss is not fully reflected by the concentrations of the species since vertical and horizontal transport is also modified by changing resolution. This is especially true for ozone with significant non-local production and long lifetime. Concentrations of species that are more controlled by local sources and conditions show a stronger, but also reduced response, to increasing resolution of chemical perturbations.

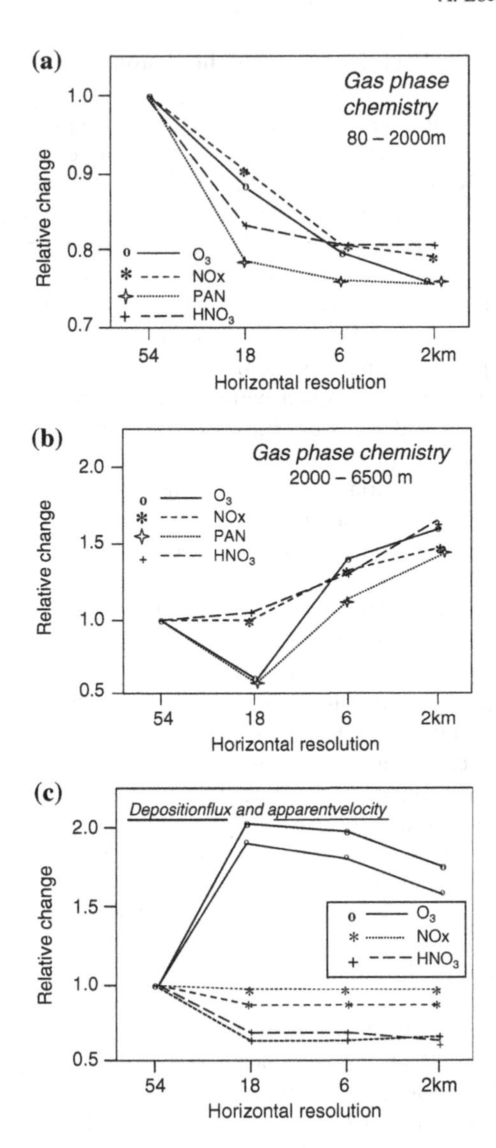

Fig. 2 (**a**) Relative change of chemical production of ozone, PAN and HNO_3 and loss of NO_x with increasing resolution. Calculated with the EURAD model system (Memmesheimer et al. 2004) for the layer from about 80 m to 2,000 m altitude. (**b**) The same for the height range from about 2,000 m to 6,500 m. (**c**) Relative change of deposition flux (thick lines) and deposition velocity (thin lines) of ozone, NO_x and HNO_3. Berlin region, episode 21–27 July 1994

When analysing changes of chemical yields in the lowest model layer (0–80 m) a more complex behaviour is found for all substances, and is interpreted as an effect of the heterogeneous emission source distribution for NO in particular. A peculiar result not reported to date in the literature is the apparent relative increase of the simulated chemical reaction intensity with finer resolution (starting at 18 km in this case) above about 2,000 m, at levels mainly in the lower free troposphere (about 2,000–6,500 m, Fig. 2b). There the efficiency of chemistry ranges between only 2.5% (O_3, NO_x) and 6% (PAN, HNO_3) of the respective values found for the lower height range during the studied episode. Nevertheless the systematic increase found for all analyzed species for a grid size of less than 18 km is significant. Such behaviour may be expected for premixed flows with positive correlation of chemical perturbations, i.e. positive segregation intensity. We speculate that this kind of premixing is possible in air masses subject to long-range horizontal transport.

4.2 Discussion of chemical production/loss

In the PBL a similar relative change of simulated net chemical production or loss (in case of NO_x) is found for ozone and NO_x on the one hand and PAN and HNO_3 on the other hand. Regarding the Damköhler numbers (Fig. 1) one would expect closer similarity between O_3 and PAN (as is the case in the free troposphere) with nearly identical range of chemical lifetimes, whereas HNO_3 should show a behaviour similar to an inert tracer, i.e. no change of chemical production, due to its extremely long chemical lifetime (20 days to 11 years neglecting the removal by aerosols and precipitation as is done in the present study; Finlayson-Pitts and Pitts 1986). These findings point to two processes that appear to be essential for the formation of chemical perturbations in addition to pure turbulence. The first one is cross-correlation of the concentrations of reactive species in the atmosphere with complex chemistry. In the case of O_3 and NO_x it can be quite strong as may be seen from the O_x concept ($O_x = O_3 + NO_2$) of ozone chemistry. Under the anticyclonic conditions of the simulated case there is a tendency towards the conservation of O_x due to the titration by NO ($O_3 + NO \rightarrow NO_2 + O_2$). This implies enhanced negative correlation of O_3 and NO_2 and strong coupling of the change of chemical productivity (reduction of ozone formation induces a reduction of NO_2 and thus NO_x formation).

The other process is the generation of perturbations by irregular lower boundary conditions determined by land-use characteristics (deposition, low-level temperature fluctuations) and surface emissions (see Eq. 3); Orography, another relevant factor, is of minor importance for the Berlin case. The impact of boundary conditions appears to be reflected in the similarity of net PAN and HNO_3 production changes with resolution. Both species show a strong sensitivity when the drastic change of urban (54 km) to mixed (18 km and less) land-use conditions occurs where only reduced sensitivity to resolution refinement is found. The role of land-use changes is further discussed below in the context of deposition.

The situation is different in the free troposphere, where NO emitted from the ground is less active and chemical coupling between HNO_3 and NO_2 becomes important. This probably causes the similarity between the relative change of net chemical production and loss of nitric acid and NO_x, respectively. The reactions are slow so that air masses with nearly persistent concentration ratios of admixtures may be expected as they are characteristic for premixed states. O_3 and PAN show a behaviour expected for negative segregation intensity when changing the grid size from 54 km to 18 km, and then an increase with increasing resolution typical for premixing. The increase is stronger than for HNO_3 and NO_x. We are not sure how to explain the transition from one to the other type of mixing. It might be an artefact or an accidental result for the specific simulated case, yet it could also be an indication of scale dependence of mixing characteristics exhibiting predominantly positively correlated components at shorter spatial scales and predominant negative correlation at larger scales.

4.3 Discussion of deposition

Vilá Guerau de Arellano et al. (1993a) and Kramm et al. (1995) dealing with simplified chemical systems, namely $NO–NO_2–O_3$ and $NO–NO_2–O_3/HNO_3–NH_3–NH_4NO_3$, respectively, showed that turbulent dry deposition fluxes of reactive species are modified by chemistry in the PBL. Regarding the vertical resolution (order of some 10 m) of the lowest levels of commonly employed CTMs this process cannot explicitly be resolved and would need to be parameterised if there were a reliable method to do so. Yet regarding the impact of horizontal irregularities on eddy deposition fluxes the nesting method can well be used to analyse the dependence of the flux (and to a certain degree of deposition velocity) on the scales resolved by a model. This is of specific interest in the framework of this study since flux irregularities can be a source of perturbations of the concentration fields of reactive species and thus contribute to the reduction of chemical production in the PBL by macro-turbulence.

The relative change of dry eddy deposition flux is exhibited in Fig. 2c for ozone, nitric monoxide plus dioxide and nitric acid. The variation found for PAN resembling that of O_3 is not shown since the estimated values are less reliable than for the other gases. The major features of the difference between the different analyzed species may be attributed to differences regarding the change of the resistance to dry deposition of the individual compounds. Using the estimates of surface resistance for different land types by Walcek et al. (1986) and Chang et al. (1987) for dry summer conditions one finds high values for urban conditions (Berlin, single box, resolution of 54 km). Increasing the resolution (first nest, 18 km) the land type will be a mixture of urban, suburban, forest, and agriculture causing a decrease of resistance by about 50%. With a further increase in resolution (6 and 2 km) more and more land-use categories such as water and coniferous forest are resolved leading to a gradual increase of the average surface resistance and thus a decrease of deposition in comparison with the 18-km resolution. In contrast to ozone nitric acid does not experience significant surface resistance so that the change with resolution has to be solely attributed to a change of aerodynamic and (to a lesser extent) sublayer resistance (R_a and R_b, respectively, in Eq. 8). In this specific case only minor changes with resolution are found when changing from purely urban to mixed land-use conditions. The same is true for NO_x, but at a reduced level. This compound is characterized by main concentrations at night during the episode, so that the flux is preferably controlled by nighttime resistance conditions. These are relatively homogeneous for non-urban areas around Berlin and allow somewhat higher fluxes than for purely urban land-use. Together with the counteracting effect of aerodynamic resistance that is evident from the behaviour of the HNO_3 flux, only a slight deviation from the flux found for purely urban conditions is obtained exhibiting minor variations with increasing resolution. The contribution of the aerodynamic resistance to total ozone deposition resistance can be assumed to be small (de Miguel and Bilbao 1999) since ozone is a daytime product and therefore mainly experiences convective mixing. Nyogi et al. (2003) observed differences of daytime and nighttime deposition velocities of ozone up to an order of magnitude, mainly attributing this behaviour to variations of surface (canopy) resistance.

Relative changes of apparent average deposition w^*_{Di} are also exhibited in Fig. 2c. They are calculated employing the relation $\bar{F}_i = w^*_{Di}\bar{c}_i$ (symbols as Eq. 7, concentration averages $\bar{c}_i$ of the lowest layer), and closely follow the curves drawn for the flux, since the average concentrations of the studied gases only slightly vary with resolution when mixed land-use conditions prevail (decrease of ozone around 7%, decrease of NO_x and HNO_3 varying from 4% after the first nesting step to 12% with 2-km grid size). It is noted that in contrast to "apparent" deposition velocities "true" average deposition velocities would require the estimation of surface concentrations and extraction of the fluctuation term introduced in Eq. 7b. Our model does not provide such data in its standard version.

Summarising the discussion we may state with reference to the simulated episode and its specific regional characteristics that three different types of responses to scale refinement are represented in Fig. 2c: (a) a daytime type (ozone) showing a pronounced sensitivity to land-use changes with varying surface resistance and reduced efficiency of aerodynamic resistance, (b) a nighttime type (NO_x) with reduced variability of surface resistance and efficient aerodynamic resistance, (c) a low surface resistance type (HNO_3) predominantly controlled by aerodynamic resistance. It remains to be seen how such a classification of deposition behaviour will work under varying weather and land-use conditions.

A rather detailed model study of the change of the calculated vertical deposition velocity of ozone by feeding more detailed information about the land use into a CTM was recently carried out by Miao et al. (2006). Their results indicate a similar change of the parameter with resolution, but of lesser magnitude, since the basic run already exhibits higher horizontal resolution in contrast to our analysis starting with a single coarse grid box. In addition, those authors demonstrate the dependence of the results on meteorological conditions.

4.4 Segregation

Finally it is noted that the average deviations of the O_3, PAN, HNO_3 and NO_x concentrations from those of the coarse grid simulation gradually decrease with increasing height. The deviations become small and irregular around about 2,000 m, i.e. roughly the top of the convective PBL. This coincides with the finding that transition from a state with negative correlations of chemical tracer fluctuations to a premixed state with positive correlation occurs above the mixing layer (Fig. 2a, b). As a consequence the negative segregation intensities in the PBL tend to approach zero around the top of the layer. The trend to a reduction of segregation in the PBL with increasing height is in accordance with the already mentioned results of Schumann (1989) and Vilà-Guerau de Arellano et al. (1993b). Yet, comparing the values characterizing the state of segregation in the boundary layer for different combinations of compounds and resolution at individual levels a rather unsystematic behaviour is found for the simulated episode. Using, for instance, O_3 and NO the segregation estimate in the middle PBL (around -0.5) does not change much with resolution, whereas O_3/NO_x and PAN/NO_x reveal a change to larger negative values (about -0.4 to -0.8). This is taken as an additional hint for an interdependence of various compounds, in particular NO_x and O_3, which in this case seems to increase with grid refinement.

Based on results of LES simulations with simplified chemistry Vinuesa and Vilà-Guerau de Arellano (2005) suggest defining effective reaction rates $k_{\mathrm{eff}} = k(1 + S_{ij})$ (see Eq. 6) for the averaged chemical terms in Eq. 5. Considering the fact that large chemical mechanisms such as the one used in the EURAD model lead to complex chemical interrelationships and that our simulations do not show a clear segregation relationship for different combinations of reactive species, it seems that a straightforward use of this attractive and, in principle, sensible parameterization of turbulent effects is not possible in the case of macro-turbulent perturbations. Regarding the interdependence of chemical processes a combination of different segregation intensities may help to overcome this problem. More comprehensive tests, then carried out in the framework of this study, would be needed to explore such a possibility.

5 Concluding remarks

We have shown in the preceding section that CTM simulations can well be used to demonstrate the role of mesoscale turbulent perturbations of atmospheric flows, and reactive trace gases contained within such flows, for the average behaviour of the contaminated PBL. As for smaller scale turbulence, overestimation of chemical production/loss is obtained in the PBL due to inappropriate representation of mixing effects in the simulated case. Other studies dealing with this effect seem to be rare or absent. Applying the Damköhler concept (Eq. 1) beyond the range of turbulent scales, for which it was originally developed and has exclusively been used so far, it is found that it also holds for mesoscale turbulence. Yet a modification of the definition of lifetimes for reactive species is required taking into account other processes influencing the residence time in the PBL (e.g. deposition) besides chemistry. Two major categories of perturbation terms have been addressed, namely purely chemical (terms with $A'B', k'A'B', A'B'C'$, etc. after decomposition of Eq. 4) and the more commonly discussed mixed dynamic-chemical ones (terms $u'C'$, etc). Emission and land-type heterogeneities have been invoked as possible causes of the perturbations in addition to dynamical eddies. Also free tropospheric behaviour appears to be different, pointing to specific segregation characteristics, possibly premixing on larger scales, i. e. during long-range transport, above the PBL. As mentioned in Sect. 1 other causes such as temporal and spatial fluctuations of the radiation field and impacts of clouds on boundary-layer eddy effects should be addressed more comprehensively in future studies. Methods, which could be applied, are correlation and spectral analysis as well as sensitivity tests with parameter variation similar to ensemble modelling (Mallet and Sportisse 2006).

The relevance of purely chemical eddy effects has been demonstrated in an indirect way by increasing the horizontal resolution of a complex CTM in a specific domain (Berlin). The change of deposition flux with grid box size has also been studied. By analyzing ozone, NO_x and nitric acid three different types of scale dependence regarding the modification of deposition by changing surface properties with increasing resolution could be identified, namely modifications mainly by surface resistance (ozone), aerodynamic resistance (nitric acid) and a mixture of both (NO_x). Such differences should be reflected in respective concentration perturbations in the lower PBL.

It is stressed that the EURAD model was applied to an intensive photo-smog episode and a rather polluted domain. Nevertheless the results are believed to be of general importance showing that air quality estimates derived for pollution episodes may exhibit a significant degree of uncertainty if mesoscale chemical perturbations are neglected. Of course, one may wonder why they are usually not addressed in model studies going beyond basic research of this phenomenon. There are two main reasons for this. The first one is the complexity of turbulent impacts on chemistry regarding the considerable diversity of possible turbulent interactions in large chemical mechanisms. They are difficult to handle and would require a larger number of speculative closure assumptions. The other one is the fact that it is a tedious and in many cases impossible task to validate simulated eddy and segregation effects in the real atmosphere and their control by land-use and emission irregularities. Nevertheless more intense research and model development is needed to cope with this deficiency in chemistry transport modelling. A necessary next step that could not yet be done in the framework of this study is a more rigorous use of the complex chemistry transport model through extension of the perturbation analysis to other trace substances, in particular volatile organic compounds, and an extension of work to other cases with different meteorological and geographical conditions.

Acknowledgements The paper is dedicated to Professor Sergej S. Zilitinkevich in honour of his 70th birthday and his outstanding, still continuing contributions, to innovative PBL research. The authors are grateful for fruitful and helpful exchange of ideas with G. Kramm, University of Alaska Fairbanks, and O. Hellmuth, Institute for Tropospheric Research, Leipzig. They thank A. A. M. Holtslag and J. Vilà-Guerau de Arellano, University of Wageningen, Netherlands, for encouraging insights into their work on chemistry and turbulence. Helpful comments and suggestions by the reviewers regarding the paper in general and interpretation of results, in particular, are gratefully acknowledged. Emission data used for the simulations were provided by the IER of the University of Stuttgart and EMEP. The European Centre for Medium-Range Weather Forecast (ECMWF) and the German Weather Service (DWD) gave access to global meteorological analyses. The numerical calculations were supported by the computer centre of the University of Cologne (ZAIK/RRZK) and the Research Centre Juelich (NIK/ZAM). A part of the work was funded by the German Federal Ministry of Education and Research (BMBF) under grant 07tfs10/LT1-C1.

References

Builtjes PJH (1981) A comparison between chemically reacting plume models and wind tunnel experiments. Paper presented at the12th ITM on Air Pollution and Its Application, Palo Alto, USA

Builtjes PJH, Talmon AM (1987) Macro- and micro-scale mixing in chemically reactive plumes. Boundary-Layer Meteorol 41:417–426

Chang JS, Brost RA, Isaksen ISA, Madronich S, Middleton P, Stockwell WR, Walcek CJ (1987) A three-dimensional Eulerian acid deposition model: physical concepts and formulation. J Geophys Res 92:14681–14700

Damköhler G (1940) Influence of turbulence on the velocity of flames in gas mixtures. Z Elektrochem 46:601–626

de Miguel A, Bilbao J (1999) Ozone dry deposition and resistances onto green grasland in summer in central Spain. J Atmos Chem 34:321–338

Donaldson C du P, Hilst GR (1972) Effect of inhomogeneous mixing on atmospheric photochemical reactions. Environ Sci Technol 6:812–816

Elperin T, Kleeorin N, Rogachevskii I (1998) Effect of chemical reactions and phase transitions on turbulent transport of particles and gases. Phys Rev Lett 80:69–72

Finlayson-Pitts BJ, Pitts JN (1986) Atmospheric chemistry. John Wiley and Sons, New York, 1098 pp

Georgopoulos PG, Seinfeld JH (1986) Mathematical modeling of turbulent reacting plumes. Atmos Environ 20:1791–1807

Hass H (1991) Description of the EURAD chemistry transport model version 2 (CTM2). Mitteilungen des Instituts für Geophysik und Meteorologie der Universität zu Koeln, no. 83, 100 pp

Hawthorne WR, Weddell DS, Hottel HC (1949) Mixing and combustion in turbulent gas jets. Paper presented at 3rd symp. on combustion, flame and explosion phenomena, Maryland, USA

Hellmuth O (2005) Conceptual study on nucleation burst evolution in the convective boundary layer – Part I: modelling approach. Atmos Chem Phys Disc 5:11413–11487

Holtslag AAM, Nieuwstadt FTM (1986) Scaling the atmospheric boundary layer. Boundary-Layer Meteorol 36:201–209

Jakobs HJ, Feldmann H, Hass H, Memmesheimer M (1995) The use of nested models for air pollution studies: an application of the EURAD model to a SANA episode. J Appl Meteorol 34:1301–1319

Komori S, Hunt JCH, Kanzaki T, Murakami Y (1991) The effects of turbulent mixing on the correlation between two species and on concentration fluctuations in non-premixed reacting flows. J Fluid Mech 228:629–659

Kramm G, Meixner FX (2000) On the dispersion of trace species in the atmospheric boundary layer: a re-formulation of the governing equations for the turbulent flow of the compressible atmosphere. Tellus 52A:500–522

Kramm G, Dlugi R, Dollard GJ, Foken T, Mölders N, Müller H, Seiler W, Sievering H (1995) On the dry deposition of ozone and reactive nitrogen species. Atmos Environ 29:3209–3231

Krol MC, Molemaker MJ, Vilà-Guerau de Arellano J (2000) Effects of turbulence and heterogeneous emissions on photochemically active species in the convective boundary layer. J Geophys Res 105:6871–6884

Mallet V, Sportisse B (2006) Uncertainty in a chemistry-transport model due to physical parameterizations and numerical approximations: an ensemble approach applied to ozone modelling. J Geophys Res 111, D01302, doi:10.1029/2005JD006149

Memmesheimer M, Roemer M, Ebel A (1997) Budget calculations for ozone and its precursers: seasonal and episodic features based on model simulations. J Atmos Chem 28:283–317

Memmesheimer M, Jakobs HJ, Tippke J, Ebel A, Piekorz G, Weber M, Geiss H, Jansen S, Wickert B, Friedrich R, Schwarz U, Smiatek G (1999) Simulation of a summer-smog episode in July 1994 on the European and urban scale with special emphasis on the photo-oxidant plume of Berlin. In: Borrell PM, Borrell P (eds) Proceedings of the EUROTRAC Symposium '98, WIT Press, Southampton, pp 591–595

Memmesheimer M, Friese E, Ebel A, Jakobs HJ, Feldmann H, Kessler C, Piekorz G (2004) Long-term simulations of particulate matter in Europe on different scales using sequential nesting of a regional model. Int J Environ Poll 22:108–132

Memmesheimer M, Friese E, Jakobs HJ, Kessler C, Feldmann H, Piekorz G, Ebel A (2005) OZURMI: Lokal geprägte Ozonspitzenwerte in Nordrhein-Westfalen – Ursachen und Minderungspotential für ein ausgewähltes Gebiet im Kölner Süden. Report, Landesumweltamt Nordrhein-Westfalen, Essen, November 2005, 274 pp

Miao J-F, Chen D, Wyser D (2006) Modelling subgrid scale dry deposition velocity of O_3 over the Swedish west coast with MM5-PX model. Atmos Environ 40:415–429

Niyogi DDS, Alapaty K, Raman S (2003) A photosynthesis-based dry deposition modelling approach. Water, Air Soil Pollut 144:171–194

Schumann U (1989) Large-eddy simulation of turbulent diffusion with chemical reactions in the convective boundary layer. Atmos Environ 23:1713–1727

Stockwell WR (1995) Effects of turbulence on gas-phase atmospheric chemistry: calculation of the relationship between time scales for diffusion and chemical reaction. Meteorol Atmos Phys 57:159–171

Stull RB (1988) An introduction to boundary layer meteorology. Kluwer Academic Publishers, Dordrecht, 666 pp

Sykes R, Parker S, Henn D, Lewellen W (1994) Turbulent mixing with chemical reactions in the planetary boundary layer. J Appl Meteorol 33:825–834

Toor HL (1969) Turbulent mixing of two species with and without chemical reactions. Ind Eng Chem Fundam 8:655–659

Verver GHI, van Dop H, Holtslag AAM (1997) Turbulent mixing of reactive gases in the convective boundary layer. Boundary-Layer Meteorol 85:197–222

Vilà-Guerau de Arellano J, Talmon AM, Builtjes PJH (1990) A chemically reactive plume model for the NO–NO_2–O_3 system. Atmos Environ 24A: 2237–2246

Vilà-Guerau de Arellano J, Duynkerke PG, Builtjes PJH (1993a) The divergence of the turbulent diffusion flux due to chemical reactions in the surface layer: NO–O_3–NO_2 system. Tellus 45B:23–33

Vilà-Guerau de Arellano J, Duynkerke PG, Jonker PJ, Builtjes PJH (1993b) An observational study on the effects of time and space averaging in photochemical models. Atmos Environ 27:353–362

Vilà-Guerau de Arellano J, Dosio A, Vinuesa J-F, Holtslag AAM, Galmarini S (2004) The dispersion of chemically reactive species in the atmospheric boundary layer. Meteorol Atmos Phys 87:23–28. DOI 10.1007/s0073-003-0059-2

Vilà-Guerau de Arellano J, Kim S-W, Barth MC, Patton EG (2005) Transport and chemical transformations influenced by shallow cumulus over land. Atmos Chem Phys 5:3219–3231

Vinuesa J-F, Vilà-Guerau de Arellano J (2005) Introducing effective reaction rates to account for inefficient mixing in the convective boundary layer. Atmos Environ 39:445–461

Walcek CJ, Taylor GR (1986) A theoretical method for computing vertical distributions of acidity and sulphate production within growing cumulus clouds. J Atmos Sci 43:339–355

Weber M (1999) Die Auswirkung der Nestung auf die Konzentrationen und Budgets ausgewählter Spurenstoffe während einer Sommersmogepisode. Diploma thesis, University of Cologne, Institute for Geophysics and Meteorology 146 pp

Wetzel MA, Slusser JR (2005) Mesoscale distributions of ultraviolet spectral irradiance, actinic flux, and photolysis rates derived from multispectral satellite data and radiative transfer models. Opt Eng 44(4):041006

Zhang L, Brook JR, Vet R (2003) A revised parameterization for gaseous dry deposition in air-quality models. Atmos Chem Phys 3:2067–2082

Theoretical considerations of meandering winds in simplified conditions

Antônio G. O. Goulart · Gervásio A. Degrazia · Otávio C. Acevedo · Domenico Anfossi

Abstract The influence of turbulence on the meandering phenomenon is investigated. The study, based on the three-dimensional Navier–Stokes equations, shows that when the turbulent fluxes can be neglected an asymptotic solution results. This solution reproduces a horizontal wind oscillation with an infinite relaxation time. When there is turbulent forcing, on the other hand, a transition occurs to a new order, characterized by a spatial reorganization, leading to a wind field with a well-defined direction.

Keywords Atmospheric turbulence · Meandering · Navier–Stokes equations · Stable boundary layer

1 Introduction

Meandering of the horizontal mean wind vector is an important and complex phenomenon associated with turbulence in the Planetary Boundary Layer (PBL). Meandering behaviour in low wind speed conditions is characterized by low frequency oscillations of the horizontal wind, which are responsible for the presence of negative lobes in the observed autocorrelation functions (Hanna 1983; Anfossi et al. 2005). Anfossi et al. (2005) investigated in detail the shape of the autocorrelation function under meandering conditions, showing that

A. G. O. Goulart
Centro de Tecnologia de Alegrete, Unipampa/UFSM,
Alegrete, RS, Brazil

G. A. Degrazia · O. C. Acevedo (✉)
Departamento de Física, Universidade Federal de Santa Maria,
Santa Maria, RS, Brazil
e-mail: acevedo@pesquisador.cnpq.br

D. Anfossi
C. N. R., Istituto di Scienze dell'Atmosfera e del Clima,
Torino, Italy

Atmospheric Boundary Layers. A. Baklanov & B. Grisogono (eds.),
doi: 10.1007/978-0-387-74321-9_9,

the classical formulation proposed by Frenkiel (1953) appropriately represents the observations. Recently, Oettl et al. (2005), employing the two-dimensional averaged Navier–Stokes equations, proposed a physically based theory of meandering flow for low wind speeds. According to this study, the meandering phenomenon appears when there is, in the Navier–Stokes equations, an equilibrium between the Coriolis force and the horizontal pressure gradient.

In this study, we present a physical derivation employing the three-dimensional Reynolds-averaged Navier–Stokes equations for the PBL. Differently from Oettl et al. (2005) and Anfossi et al. (2005), we investigate how the presence of turbulent forcing in the Navier–Stokes equations modifies the structure of the meandering flow. Therefore, the aim of this study is to employ the physical model described by the three-dimensional Navier–Stokes equations to investigate how the existence of turbulent momentum fluxes affects the meandering phenomenon.

2 Theoretical development

The theoretical derivation starts from the three-dimensional Reynolds-averaged Navier–Stokes equations for the PBL. Neglecting viscosity terms, this equation describing the three components of the mean wind can be written as (Holton 1992):

$$\begin{cases} \dfrac{\partial \bar{u}}{\partial t}+\bar{u}\dfrac{\partial \bar{u}}{\partial x}+\bar{v}\dfrac{\partial \bar{u}}{\partial y}+\bar{w}\dfrac{\partial \bar{u}}{\partial z}=f_c\bar{v}-\dfrac{1}{\bar{\rho}}\dfrac{\partial \bar{p}}{\partial x}-\left[\dfrac{\partial \overline{(u'u')}}{\partial x}+\dfrac{\partial \overline{(u'v')}}{\partial y}+\dfrac{\partial \overline{(u'w')}}{\partial z}\right], & \text{(1a)}\\[2ex] \dfrac{\partial \bar{v}}{\partial t}+\bar{u}\dfrac{\partial \bar{v}}{\partial x}+\bar{v}\dfrac{\partial \bar{v}}{\partial y}+\bar{w}\dfrac{\partial \bar{v}}{\partial z}=-f_c\bar{u}-\dfrac{1}{\bar{\rho}}\dfrac{\partial \bar{p}}{\partial y}-\left[\dfrac{\partial \overline{(u'v')}}{\partial x}+\dfrac{\partial \overline{(v'v')}}{\partial y}+\dfrac{\partial \overline{(v'w')}}{\partial z}\right], & \text{(1b)}\\[2ex] \dfrac{\partial \bar{w}}{\partial t}+\bar{u}\dfrac{\partial \bar{w}}{\partial x}+\bar{v}\dfrac{\partial \bar{w}}{\partial y}+\bar{w}\dfrac{\partial \bar{w}}{\partial z}=-\dfrac{1}{\bar{\rho}}\dfrac{\partial \bar{p}}{\partial z}+g-\left[\dfrac{\partial \overline{(u'w')}}{\partial x}+\dfrac{\partial \overline{(v'w')}}{\partial y}+\dfrac{\partial \overline{(w'w')}}{\partial z}\right], & \text{(1c)} \end{cases}$$

where $\bar{u}$, $\bar{v}$ and $\bar{w}$ denote the mean wind components in the x, y and z directions averaged over some time interval, $\bar{\rho}$ is the mean density, u', v' and w' are the velocity fluctuations, $\bar{p}$ is the mean pressure, f_c is the Coriolis parameter, and g is the acceleration due to gravity.

From a mathematical point of view Eq. 1 cannot be analytically solved. However, we assume here that all the horizontal gradients of the wind velocity components and of pressure can be taken as constant. This simplification leads to

$$\frac{\partial \bar{u}}{\partial t}=-a_1\bar{u}+b_1\bar{v}+d_1\bar{w}+c_1, \tag{2a}$$

$$\frac{\partial \bar{v}}{\partial t}=-a_2\bar{v}+b_2\bar{u}+d_2\bar{w}+c_2, \tag{2b}$$

$$\frac{\partial \bar{w}}{\partial t}=-a_3\bar{w}+b_3\bar{u}+d_3\bar{v}+c_3, \tag{2c}$$

where

$$a_1 = \frac{\partial \bar{u}}{\partial x}, \quad a_2 = \frac{\partial \bar{v}}{\partial y}, \quad a_3 = \frac{\partial \bar{w}}{\partial z},$$

$$b_1 = f_c - \frac{\partial \bar{u}}{\partial y}, \quad b_2 = -\left(f_c + \frac{\partial \bar{v}}{\partial x}\right), \quad b_3 = -\frac{\partial \bar{w}}{\partial x},$$

$$d_1 = -\left(\frac{\partial \bar{u}}{\partial z}\right), \quad d_2 = -\frac{\partial \bar{v}}{\partial z}, \quad d_3 = -\frac{\partial \bar{w}}{\partial y},$$

$$c_1 = -\frac{1}{\bar{\rho}}\frac{\partial \bar{p}}{\partial x} - \left[\frac{\partial \overline{(u'u')}}{\partial x} + \frac{\partial \overline{(u'v')}}{\partial y} + \frac{\partial \overline{(u'w')}}{\partial z}\right],$$

$$c_2 = -\frac{1}{\bar{\rho}}\frac{\partial \bar{p}}{\partial y} - \left[\frac{\partial \overline{(u'v')}}{\partial x} + \frac{\partial \overline{(v'v')}}{\partial y} + \frac{\partial \overline{(v'w')}}{\partial z}\right],$$

$$c_3 = -\frac{1}{\bar{\rho}}\frac{\partial \bar{p}}{\partial z} + g - \left[\frac{\partial \overline{(u'w')}}{\partial x} + \frac{\partial \overline{(v'w')}}{\partial y} + \frac{\partial \overline{(w'w')}}{\partial z}\right].$$

The magnitude of $\bar{w}$ characterizing subsidence can be as high as $0.1\,\mathrm{m\,s^{-1}}$ (Stull 1988), so in fair weather conditions, the terms $d_1\bar{w}$ and $d_2\bar{w}$ are much smaller than the terms $a_1\bar{u}$, $b_1\bar{v}$, c_1, $a_2\bar{v}$, $b_2\bar{u}$ and c_2 in (2a) and (2b). Therefore, the terms containing vertical gradients of the wind speed (d_1 and d_2) can be neglected. Furthermore, assuming hydrostatic balance, the terms $a_3\overline{w}$, $b_3\bar{u}$ and $d_3\bar{v}$ and the vertical turbulent flux divergence terms can be neglected, leading to:

$$\frac{\partial \bar{u}}{\partial t} = -a_1\bar{u} + b_1\bar{v} + c_1, \tag{3a}$$

$$\frac{\partial \bar{v}}{\partial t} = -a_2\bar{v} + b_2\bar{u} + c_2, \tag{3b}$$

$$\frac{\partial \bar{w}}{\partial t} = 0. \tag{3c}$$

From Eq. 3a we obtain

$$\bar{v} = \frac{1}{b_1}\left(\frac{\partial \bar{u}}{\partial t} + a_1\bar{u} - c_1\right) \tag{4}$$

and differentiating with respect to time gives,

$$\frac{\partial^2 \bar{u}}{\partial t^2} = -a_1\frac{\partial \bar{u}}{\partial t} + b_1\frac{\partial \bar{v}}{\partial t}. \tag{5}$$

From Eqs. 3b and 4 we obtain

$$\frac{\partial \bar{v}}{\partial t} = -\frac{a_2}{b_1}\left(\frac{\partial \bar{u}}{\partial t} + a_1\bar{u} - c_1\right) + b_2\bar{u} + c_2,$$

and introducing the above expression in (5), the following results

$$\frac{\partial^2 \bar{u}}{\partial t^2} + (a_1 + a_2)\frac{\partial \bar{u}}{\partial} + (a_1a_2 - b_1b_2)\,\bar{u} = a_2c_1 + b_1c_2. \tag{6}$$

Defining

$$B = a_1 + a_2, \tag{7a}$$

$$C = (a_1 a_2 - b_1 b_2), \tag{7b}$$

$$D = a_2 c_1 + b_1 c_2 \tag{7c}$$

we obtain

$$\frac{d^2\overline{u}}{dt^2} + B\frac{d\overline{u}}{dt} + C\overline{u} = D \tag{8}$$

Equation 8 has a known analytical solution, with three cases according to the values of the roots r_1 and r_2 from the auxiliary equation $r^2 + Br + C = 0$.

We only consider the case in which $B^2 - 4C < 0$ that results in oscillatory behaviour, i.e. $(a_1 + a_2)^2 < -4b_1 b_2$. Setting $r_1 = -p + qi$ and $r_2 = -p - qi$, where p is associated with the horizontal flow divergence:

$$p = \frac{B}{2} = \frac{1}{2}\left(\frac{\partial \overline{u}}{\partial x} + \frac{\partial \overline{v}}{\partial y}\right), \tag{9a}$$

$$q = \frac{\sqrt{-B^2 + 4C}}{2}. \tag{9b}$$

The solutions are:

$$\bar{u}(t) = e^{-pt}\left(\alpha_1 \cos(qt) + \alpha_2 \cos(qt)\right) + \frac{D}{C} \tag{10}$$

and

$$\bar{v}(t) = e^{-pt}\left[\frac{-p\alpha_1 + q\alpha_2 + a_1\alpha_1}{b_1}\cos(qt) + \frac{-p\alpha_2 - q\alpha_1 + a_1\alpha_2}{b_1}\sin(qt)\right.$$
$$\left. + \left(\frac{D}{C}\right)\frac{a_1}{b_1} + \frac{c_1}{b_1}\right] \tag{11}$$

where

$$\alpha_1 = u_0 - \frac{D}{C} \tag{12}$$

and

$$\alpha_2 = \frac{1}{q}\left[v_0 b_1 - (a_1 - p)\, u_0 - \frac{D}{C} p + c_1\right]. \tag{13}$$

The analytical solutions (10) and (11) for $\bar{u}(t)$ and $\bar{v}(t)$, respectively, exhibit oscillatory characteristics similar to those associated with wind meandering. In fact, the Navier–Stokes equations containing all information about the distinct phenomenological aspects associated with fluid flows supply this particular solution that is capable of describing the meandering behaviour.

If the horizontal and vertical gradients of the turbulent momentum fluxes can be disregarded, a scale analysis allows the derivation of the following simplifications for Eqs. 10 and 11 (Oettl et al. 2005):

$$\bar{u}_m(t) = \alpha_1 e^{-pt} \cos(qt), \tag{14a}$$

$$\bar{v}_m(t) = -\alpha_1 e^{-pt} \sin(qt). \tag{14b}$$

These equations can be written in an analytical functional form as

$$U(t) = \alpha_1 e^{-pt-iqt} \quad (15)$$

where $\overline{U}(t)$ is the mean meandering wind velocity.

In particular, if we compute the normalized autocorrelation function of Eq. 15 we find

$$R(\tau) = e^{-p\tau}(\cos(q\tau) + i \sin(q\tau)), \quad (16)$$

and by taking the real part of Eq. 16, the following autocorrelation function is obtained.

$$R(\tau) = e^{-pt} \cos(q\tau). \quad (17)$$

Equation 17 presents the same functional form as the autocorrelation function proposed by Frenkiel (1953). Anfossi et al. (2005) employed the Frenkiel (1953) classical mathematical expression to reproduce autocorrelation functions observed during meandering periods, which presented strong negative lobes. Figure 1 shows experimental data from a dataset obtained over flat terrain in Sweden (Anfossi et al. 2005) exhibiting meandering autocorrelation functions that are represented by the functional form given by Eq. 17.

The simplified solutions (14a) and (14b) can be assumed for an asymptotic situation, in which the magnitude of the horizontal mean wind (15) is very low. As a consequence of the zero eddy viscosity hypothesis (Oettl et al. 2005), the Navier–Stokes equations, in this situation, can be considered as a Reynolds-stress-free equation. The meandering behaviour described by the simplified solutions (14a) and (14b), which allow the derivation of the Frenkiel autocorrelation function, will be called the *asymptotic meandering solution*. It is interesting here to identify how the presence of turbulence in the solutions (10) and (11) modifies the state described by the asymptotic meandering solution.

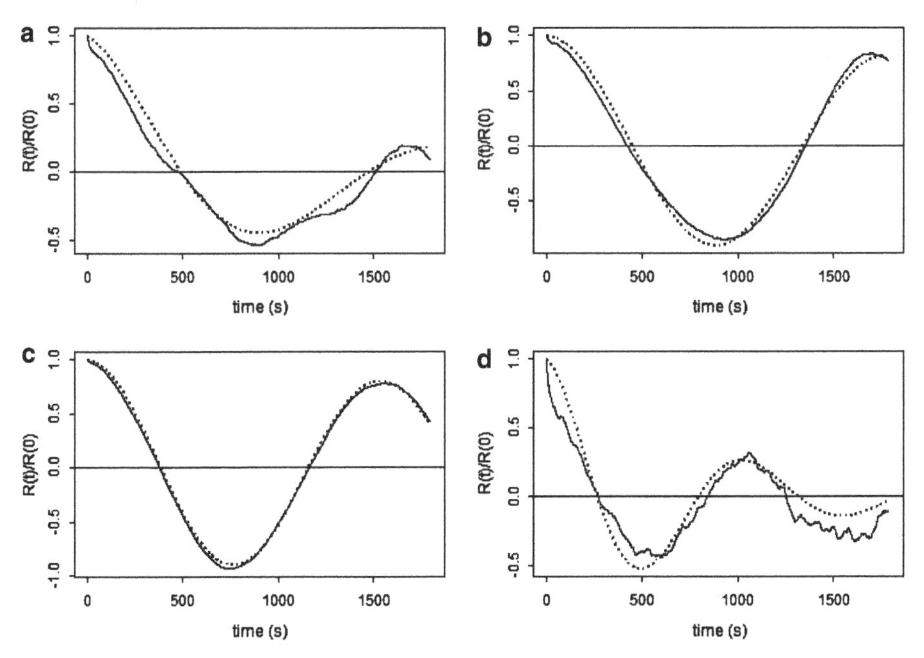

Fig. 1 Four examples of autocorrelation functions under meandering conditions. In all panels, the solid line represents observed autocorrelation functions and the dotted line is the fitting from Frenkiel expression (Eq. 17)

3 Turbulence and meandering occurrence

To investigate the effect of the Reynolds-stress terms on the meandering phenomenon, we consider their role in the general solutions (10) and (11). Therefore, a flux-gradient relationship is employed with the horizontal turbulent momentum flux parameterized as:

$$\overline{u'u'} = -K_m \frac{\partial \bar{u}}{\partial x}, \tag{18a}$$

$$\overline{v'v'} = -K_m \frac{\partial \bar{v}}{\partial y} \tag{18b}$$

where K_m is the eddy viscosity.

For low wind speed conditions, we assume here that all horizontal turbulent momentum fluxes are identical ($\overline{u'u'} = \overline{u'v'} = \overline{v'v'}$) and that $\partial\overline{u}/\partial x = \partial\overline{v}/\partial y$. With these assumptions and considering Eq. 9, the parameterization for the Reynolds-stress terms can be written as:

$$\overline{u'u'} = \overline{u'v'} = \overline{v'v'} = -K_m p. \tag{19}$$

The substitution of (19) in the solutions (10) and (11) allows the investigation of the interaction between meandering and turbulence. The parameter controlling this interaction is the product $K_m p$, showing that the solution behaviour depends both on the surface stability (which determines K_m) and on the horizontal wind divergence (which controls p).

Analysis of the behaviour of these solutions (Fig. 2) shows that, for very small values of the product $K_m p(K_m p = 10^{-7}\text{m}^2\text{s}^{-2}$ in Fig. 2) associated, for example, with very stable conditions, the general solutions (10) and (11) are identical to the asymptotic meandering solutions (14a) and (14b) that were derived from the Reynolds-stress-free Navier–Stokes equation.

On the other hand, for larger values of the product $K_m p$, the simplified solutions (14a) and (14b) are rather distinct from the general solutions given by Eqs. 10 and 11 ($K_m p = 10^{-4}\,\text{m}^2\text{s}^{-2}$ in Figs. 3 and 4). Thus, for the case with higher turbulent intensity, the larger

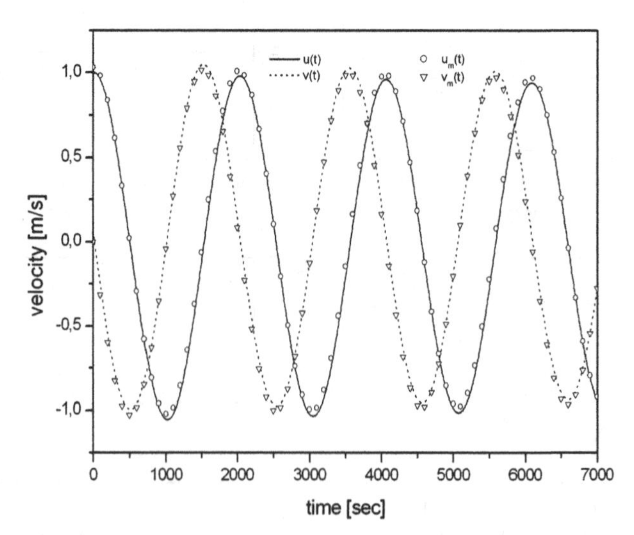

Fig. 2 Time series of velocity components $\overline{u}$ (solid line) and $\overline{v}$ (dotted line), calculated from Eqs. 10 and 11 with $K_m p = 10^{-7}\text{m}^2\text{s}^{-2}$, and the corresponding asymptotic meandering solutions for $\overline{u}$ (circles) and $\overline{v}$ (triangles), from Eqs. 14a and b

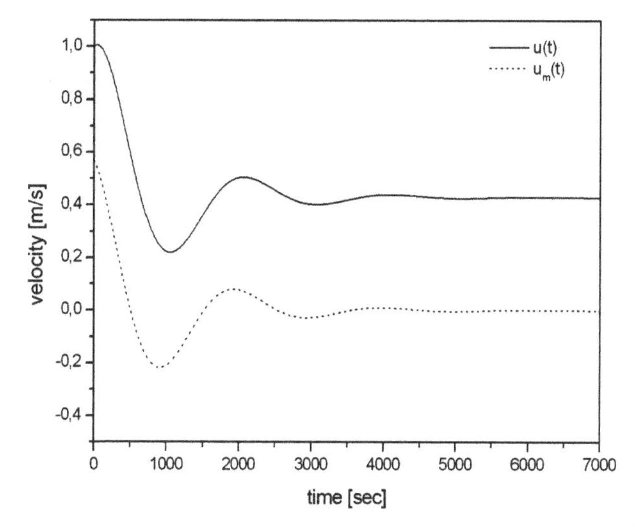

Fig. 3 Time series of $\overline{u}$ component (solid line) determined from Eq. 10, with $K_m p = 10^{-4}\,\mathrm{m^2 s^{-2}}$ and the corresponding asymptotic meandering solution from Eq. 14a

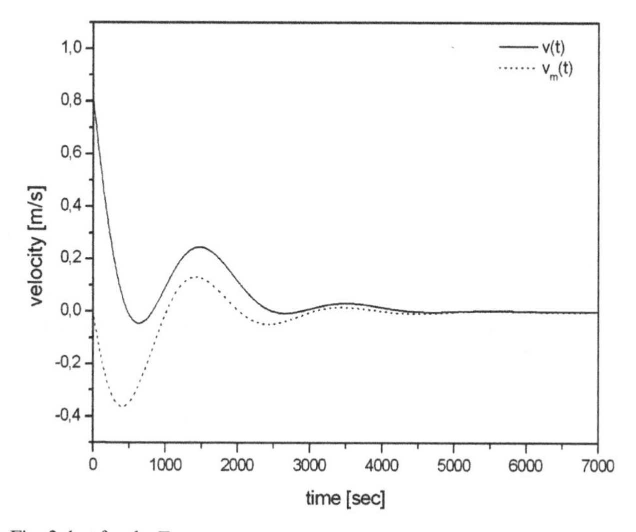

Fig. 4 Same as Fig. 2, but for the $\overline{v}$ component

Reynolds stress mitigates against the coincidence between the complete solutions (10) and (11) and the asymptotic meandering solutions given by (14a) and (14b). In fact, only for very particular cases (those when the Reynolds stress is neglectably small), can we use the asymptotic meandering solution as a surrogate for the general solutions (10) and (11).

From the above considerations, we may conclude that under very small values of the horizontal wind divergence as well as under strong stable stability, it is impossible to modify the state of meandering characterized by a permanent oscillation with constant amplitude and infinite relaxation time. Only under these very restrictive conditions does the meandering phenomenon not decay, and can be described by the simplified solutions (14a), (14b) and by Frenkiel's relation (17).

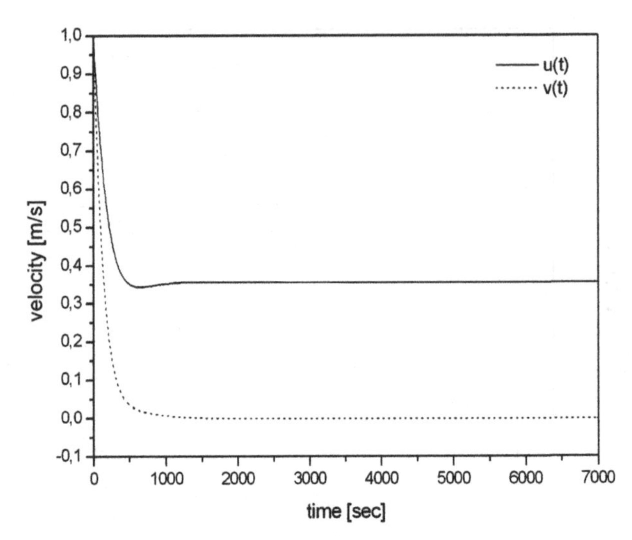

Fig. 5 Time series of $\overline{u}$ (solid line) and $\overline{v}$ (velocity components), determined from Eqs. 10 and 11 respectively, with $K_m p = 7.5 \times 10^{-4} \mathrm{m}^2/\mathrm{s}^2$

Figures 3 and 4 show that from the general solution, for initial times, the components $\overline{u}$ and $\overline{v}$ of the horizontal mean wind exhibit an oscillatory behaviour, characterizing the meandering phenomenon. However, for larger times, the magnitude of the horizontal velocity divergence cannot be neglected and acts as a turbulent forcing, generating momentum fluxes that disconnect $\overline{u}\,(t)$ and $\overline{v}\,(t)$, breaking the oscillatory characteristic of the meandering motion and establishing a precise direction for the mean wind, which becomes one-dimensional. Increasing the turbulent forcing (given by the product $K_m p$) even more ($K_m p = 7.5 \times 10^{-4}\,\mathrm{m}^2\mathrm{s}^{-2}$ in Fig. 5) attenuates rapidly the oscillatory character. Therefore, Fig. 5 shows that when the turbulent forcing is larger, the flow converges more rapidly to a one-dimensional structure.

For a well-developed meandering phenomenon, the presence of an external turbulent forcing causes the oscillatory movement to decay. This means that turbulent action provokes the vanishing of the meandering phenomenon and the transition from non-precise to precise mean wind direction.

4 Conclusion

The present study employs the Navier–Stokes equations to investigate the influence of the Reynolds-stress terms on the meandering phenomenon. The analysis shows that when the turbulent forcing can be neglected, the Navier–Stokes equations provide an asymptotic meandering solution that describes a non-decaying horizontal wind oscillation. Only for this very particular state, free of turbulent fluxes, and presenting an infinite relaxation time, can the general solutions (10) and (11) converge to simplified solutions (14a) and (14b) and the Frenkiel relation (Eq. 17) can be derived.

On the other hand, for increasing values of the turbulent forcing, the presence of the horizontal Reynolds-stress terms makes the horizontal wind components u and v become disconnected from each other, which breaks the undulating behaviour associated with the meandering phenomenon. In fact, the general Navier–Stokes solutions given by (10) and

(11) demonstrate that the action of turbulence transforms the geometry of the flow field. In this case, the turbulent forcing imposes a finite relaxation time and changes a two-dimensional flow to one-dimensional. This new order of the flow, generated by the turbulence, leads to a different spatial symmetry, establishing a precise mean wind direction.

The results described here refer to a mathematical solution, which show that the presence of turbulent stresses plays a role in destroying the oscillatory behaviour typical of meandering flow. A number of simplifications are assumed, which may not allow this description to completely apply to realistic cases. Anyhow, the general behaviour expected by the system has been reproduced by this solution. Further developments are necessary, applying to more general situations. The results of the present study will, therefore, serve as a first approximation, and it is expected that similar interactions between turbulence and meandering exist even when more realistic systems are considered.

Acknowledgements This study was supported by Brazilian Research Agencies Conselho Nacional de Desenvolvimento Científico e Tecnológico (CNPq) and Fundação de Amparo à Pesquisa do Estado do Rio Grande do Sul (FAPERGS).

References

Anfossi D, Oettl D, Degrazia GA, Goulart A (2005) An analysis of sonic anemometer observations in low wind speed conditions. Boundary-Layer Meteorol 114: 179–203

Frenkiel FN (1953) Turbulent diffusion: mean concentration distribution in a flow field of homogeneous turbulence. Adv Appl Mech 3: 61–107

Hanna SR (1983) Lateral turbulence intensity, and plume meandering during stable conditions. J Appl Meteorol 22: 1424–1430

Holton JR (1992) An introduction to dynamic meteorology. Academic Press, 510 pp

Oettl D, Goulart A, Degrazia GA, Anfossi D (2005) A new hypothesis on meandering atmospheric flows in low wind speed conditions. Atmos Environ 39: 1739–1748

Stull RB (1988) An introduction to boundary-layer meteorology. Kluwer Academic Publishers, Boston 666 pp

Aerodynamic roughness of the sea surface at high winds

Vladimir N. Kudryavtsev · Vladimir K. Makin

Abstract The role of the surface roughness in the formation of the aerodynamic friction of the water surface at high wind speeds is investigated. The study is based on a wind-over-waves coupling theory. In this theory waves provide the surface friction velocity through the form drag, while the energy input from the wind to waves depends on the friction velocity and the wind speed. The wind-over-waves coupling model is extended to high wind speeds taking into account the effect of sheltering of the short wind waves by the air-flow separation from breaking crests of longer waves. It is suggested that the momentum and energy flux from the wind to short waves locally vanishes if they are trapped into the separation bubble of breaking longer waves. At short fetches, typical for laboratory conditions, and strong winds the steep dominant wind waves break frequently and provide the major part of the total form drag through the air-flow separation from breaking crests, and the effect of short waves on the sea drag is suppressed. In this case the dependence of the drag coefficient on the wind speed is much weaker than would be expected from the standard parameterization of the roughness parameter through the Charnock relation. At long fetches, typical for the field, waves in the spectral peak break rarely and their contribution to the air-flow separation is weak. In this case the surface form drag is determined predominantly by the air-flow separation from breaking of the equilibrium range waves. As found at high wind speeds up to $60\,\mathrm{m\,s^{-1}}$ the modelled aerodynamic roughness is consistent with the Charnock relation, i.e. there is no saturation of the sea drag. Unlike the aerodynamic roughness, the geometrical surface roughness (height of short waves) could be saturated or even suppressed when the wind speed exceeds $30\,\mathrm{m\,s^{-1}}$.

Keywords Breaking wind waves · High wind conditions · Sea drag · Separation of the air flow

---

V. N. Kudryavtsev
Nansen International Environmental and Remote Sensing Center (NIERSC), Saint-Petersburg, Russia
Nansen Environmental and Remote Sensing Center (NERSC), Bergen, Norway

V. K. Makin (✉)
Royal Netherlands Meteorological Institute (KNMI), 3730AE, De Bilt, The Netherlands
e-mail: makin@knmi.nl

Atmospheric Boundary Layers. A. Baklanov & B. Grisogono (eds.),
doi: 10.1007/978-0-387-74321-9_10,

1 Introduction

A strong local enhancement of the surface stress above breaking waves was reported in a number of laboratory experiments (e.g., Banner and Melville 1976; Kawamura and Toba 1988; Banner 1990; Melville 1996; Giovanangeli et al. 1999; Reul et al. 1999). It has been argued that the air-flow separation (AFS) from the crest of breaking waves is responsible for this enhancement, which may in turn significantly contribute to the total form drag of the wavy surface. Kudryavtsev and Makin (2001) (hereinafter KM01), and Makin and Kudryavtsev (2002) (hereinafter MK02) proposed a model, which takes into account the impact of the AFS on the sea surface drag. They showed that the contribution of the AFS to the form drag rapidly increases with the wind speed, and at wind speeds 20–25 $\mathrm{m\,s^{-1}}$ the AFS supports about half of the total surface wind stress.

Wave breaking manifests itself in the form of white caps—an observable phenomenon on the sea surface. At high wind speeds and young seas white caps are formed very intensively that suggests that the AFS may play a dominant role in supporting the form drag of the sea surface. Donelan et al. (2004) investigated the aerodynamic roughness of the water surface at extreme wind speeds in laboratory conditions. They observed a saturation of the surface drag coefficient at the wind speed exceeding 33 $\mathrm{m\,s^{-1}}$. As a plausible mechanism the separation of the air flow from continually breaking wave crests was suggested to explain this fact. A similar mechanism as the limiting regime of the form drag was also suggested by KM01. The experimental finding by Donelan et al. (2004) is similar to that found by Powell et al. (2003) in the open sea under hurricane wind speeds. Though both datasets indicate the saturation of the drag coefficient at the wind speed above 30–35 $\mathrm{m\,s^{-1}}$, the physics lying behind this phenomenon could be quite different. In laboratory conditions this phenomenon, as was suggested by Donelan et al. (2004), could be explained by the saturation of the aerodynamic roughness due to the air-flow separation. While in the open field, where waves are not so short and steep as in laboratory conditions, a plausible mechanism is the impact of the sea droplets and the foam on the air-flow dynamics. Recent theoretical studies by Makin (2005), Bye and Jenkins (2006), and Kudryavtsev (2006) offer the physical grounds for the efficiency of this mechanism.

An adequate description of the exchange processes at the sea surface at high wind speeds is very important for the storm surge and the hurricane prediction. For example, the sensitivity study of the tropical cyclone model performed by Emanuel (1995) showed that cyclones cannot attain their observed intensity with the traditional parameterizations of the surface exchange coefficients, and to obtain that it is necessary to reduce the ratio of the drag coefficient to the enthalpy transfer coefficient. Makin (2005) and Kudryavtsev (2006) explored the impact of the sea droplets on the surface drag through the effect of the buoyancy force on the turbulent mixing. They showed that the efficiency of such a mechanism is sufficient to suppress the drag coefficient. However, both of these model approaches were based on the Charnock's parameterization of the aerodynamic roughness of the sea surface. Deviation of the aerodynamic roughness at high wind speeds from the Charnock relation may significantly affect the result.

The main goal of the present paper is focused on the aerodynamic roughness of the sea surface at high wind conditions. In this context, the present study complements the study by Makin (2005) and Kudryavtsev (2006), where the validity of the Charnock relation for the description of the aerodynamic roughness of the sea surface at high wind conditions was postulated. On the other hand, this study is essentially based on the KM01 and MK02 model, which is extended here to the case of high wind conditions when the intensive breaking of waves becomes a dominant surface feature. One may anticipate that at such conditions the

AFS from breaking wave crests on one hand will dominate the form drag, and on the other hand will reduce the form drag due to sheltering of some fraction of the sea surface. The latter results from the fact that the shorter waves could be sheltered by longer waves: being trapped in the separation bubble induced by the longer wave they do not extract momentum from the air flow and thus locally do not contribute to the form drag. In the present paper we present a model description of this effect and analyze its significance for high wind conditions when the wind seas are essentially undeveloped.

2 Form drag at intensive wave breaking

In KM01 the form drag τ_f of the sea surface was presented as a sum of the wave-induced stress τ_w (correlation of the surface pressure with the slope of the regular streamlined wavy surface) and the AFS stress τ_s, describing the action of the pressure drop on the surface slope discontinuity that models the wave breaking front: $\tau_f = \tau_w + \tau_s$. The spectral density of the form drag supported by the surface waves with the wavenumber from $\mathbf{k}$ to $\mathbf{k} + d\mathbf{k}$ reads:

$$d\tau_f(\mathbf{k}) = d\tau_w^0(\mathbf{k}) + d\tau_s^0(\mathbf{k}), \tag{1}$$

where the components of the form drag are

$$d\tau_w^0(\mathbf{k}) = c_\beta u_*^2 \cos^3\theta k^{-2} B(\mathbf{k}) d\mathbf{k}, \tag{2}$$

$$d\tau_s^0(\mathbf{k}) = c_s u_*^2 \cos^3\theta k^{-1} \Lambda(\mathbf{k}) d\mathbf{k}. \tag{3}$$

In these equations c, k and θ are the phase velocity, the wavenumber and its direction; u_* is the friction velocity; c_β is the growth rate coefficient defined as $c_\beta = 1.5\kappa^{-1}\ln(\pi/(kz_c))$; $B(\mathbf{k})$ is the saturation spectrum; $\Lambda(\mathbf{k})d\mathbf{k}$ is the length of wave breaking fronts per unit area; $c_s = \varepsilon_b\gamma/\kappa^2 \ln^2(\varepsilon_b/(kz_c))$ is the separation stress coefficient; $\kappa = 0.4$ is the von Karman constant; $\varepsilon_b = 0.5$ is the characteristic steepness of the breaking wave; $\gamma \sim 1$ taken here as 0.75 is an empirical constant relating the pressure drop in the separation bubble to the airflow velocity; $z_c = z_0 \exp(\kappa c/(u_* \cos\theta_b))$ is traditionally referred to as the height of the critical layer, and z_0 is the surface roughness parameter. Integration of (1) over all $\mathbf{k}$ at specified $B(\mathbf{k})$ and $\Lambda(\mathbf{k})$ gives the total surface form drag τ_f. The solution of the momentum conservation equation

$$u_*^2 = \tau_f + \tau_v, \tag{4}$$

where τ_v is the viscous surface stress

$$\tau_v = \frac{1}{\kappa d}\ln\left(\frac{d\nu}{z_0 u_*}\right) u_*^2, \tag{5}$$

(ν is the molecular viscosity and $d = 10$ is the molecular sub-layer constant) provides the drag coefficient of the sea surface. It was shown that the model results are consistent with the observations at low and moderate wind speeds.

KM01 and MK02 restricted their analysis to low and moderate wind conditions, when the fraction of the sea surface covered (or sheltered) by the separation bubbles is relatively small. However, at high wind speeds this assumption may lose its validity. Let us assume that the surface waves are quasi-monochromatic with the wavenumber k_p and the probability of the wave crest breaking P_p. Then the total length of breaking crests per unit surface is

$$L_p \equiv \Lambda(\mathbf{k})d\mathbf{k} = \frac{k_p}{2\pi} P_p. \tag{6}$$

The air flow separates from the breaking crest and reattaches to the surface on the up-wind slope of the downwind wave, closer to its crest. Thus the individual breaking crest with the length l_i shelters the area q^i proportional to $q^i \sim 2\pi/k_p l_i$, and the total fraction of the sea surface sheltered by all breaking crests $q_p = \sum q^i$ is

$$q_p = 2\pi/k_p L_p = P_p. \tag{7}$$

The action of the pressure drop inside the separation bubble on the breaking front was already included in the definition of the separation stress. Therefore the sheltered surface area should be excluded from the wave-induced momentum flux, and the expression for the form drag now reads:

$$\tau_f^p = \tau_s^p + (1 - q_p)\tau_w^p.$$

KM01 and MK02 assumed that $q_p \ll 1$ and therefore the form drag is simply the sum of τ_w and τ_s, Eq. 1. The assumption is reasonable for low and moderate wind conditions when wave breaking occurs relatively rare, but it certainly breaks down at high and extreme wind speeds when the intensive wave breaking becomes a dominant surface feature.

Let us consider the extreme case when slow waves with $c_p \ll u_{10}$, where u_{10} is the wind speed at the reference level of 10 m height, are so steep that each of their crest breaks: $P_p = 1$ and $q_p = 1$. In this case the surface stress is fully supported by the AFS: $u_*^2 = \tau_s^p$. Then, taking into account the expression (3) for the AFS stress, and L_p defined by (6) at $P_p = 1$, we have the following equation for the sea-surface roughness parameter z_0:

$$z_0/h_p = \frac{1}{2}\exp\left(-\kappa\sqrt{\frac{2\pi}{\varepsilon_b \gamma}}\right), \tag{8}$$

where $h_p = 2\varepsilon_b/k_p$ is the height of the breaking wave. At $\varepsilon_b = 0.5$ in the range of γ from 0.25 to 1 the roughness parameter varies from $z_0/h_p = 0.03$ to $z_0/h_p = 0.1$, i.e. approximately from $1/30$ to $1/10$ of the height of the roughness element. This estimate is consistent with the classical empirical knowledge (Monin and Yaglom 1971).

For the real wind seas the surface waves are not narrow, and we introduce the cumulative fraction of the sheltered surface

$$q(k) = 2\pi \int_{k_1<k} \cos\theta k_1^{-1} \Lambda(\mathbf{k}_1) d\mathbf{k}_1 \tag{9}$$

describing the cumulative contribution of breaking wind waves to the sheltered zones. We suggest that there is a cascade sheltering, i.e. the AFS from the breaking crest of longer waves shelters the shorter waves and thus prevents the wave-induced momentum flux to these shorter waves that are trapped in the sheltered zone. In terms of the cumulative fraction of the sheltered surface, the wave-induced component of the form drag can be written as:

$$d\tau_w(\mathbf{k}) = (1 - q(k)) d\tau_w^0(\mathbf{k}), \tag{10}$$

where $d\tau_w^0(\mathbf{k})$ is the wave-induced momentum flux described by (2).

One may anticipate that at high wind speeds, when the wave breaking of waves of different scales is strongly intensified, the AFS from the long breaking waves may shelter the shorter breaking waves. In other words, there is an overlapping of the sheltered area generated by the AFS from the breaking crest of different wave scales. This may lead to the fact that the fraction of the sheltered zones will be more than unity $q(k) > 1$ at some wavenumber exceeding a threshold value $k > k_o$, where k_o is the overlapping wavenumber defined as the solution of the equation $q(k_o) = 1$. Precise description of the statistics of the overlapping sheltered areas is out of the scope of the present study. Instead, in order to take into account this effect on a qualitative level, we simply assumed that the AFS from the breaking crest of waves with the wavenumber $k > k_o$ do not contribute to the total AFS stress τ_s since their separation bubbles are absorbed by the separation bubbles induced by longer waves. Thus the AFS stress can be written as

$$d\tau_s(\mathbf{k}) = h(k_o - k)d\tau_s^0(\mathbf{k}), \tag{11}$$

where $d\tau_s^0(\mathbf{k})$ is the AFS stress described by (3), and $h(x)$ is the Heaviside step-function: $h(x) = 0$ at $x < 0$ and $h(x) = 1$ at $x > 0$.

3 Surface drag

The experiment by Reul et al. (1999) revealed the intensive vortex inside the separation bubble that produces near the surface a counter flow with the velocity of about 20% of the wind velocity in the free stream. This presumes that contrary to the "regular surface", the viscous surface stress inside the sheltered area could be negative. However, taking into account that the stress is proportional to the square of the wind speed, we shall ignore this negative contribution, which is small (of order $0.04q$) relative to the non-separated fraction of the surface. Therefore, the total surface stress in the case of intensive wave breaking at high wind speeds reads

$$\int (1 - q(k))d\tau_w^0(\mathbf{k}) + \int_{k<k_b} h(k_o - k)d\tau_s^0(\mathbf{k}) + (1 - q(k_b))\tau_v = u_*^2, \tag{12}$$

where $k_b \simeq 2\pi/0.15$ rad m^{-1} is the wavenumber of the shortest breaking wave, which provides the AFS. As discussed by KM01, the generation of parasitic capillaries by shorter breaking waves prevents the air-flow separation. Notice, that q in (12) is limited by the value 1, i.e. $q(k) = \min(q(k), 1)$. If $q(k_b) \ll 1$ the model described by KM01 is retrieved.

To complete the problem one needs to define $\Lambda(\mathbf{k})$ describing the length of the wave breaking front. In the equilibrium range of the spectrum, KM01 defined this quantity following the approach proposed by Phillips (1985). The quantity $\Lambda(\mathbf{k})$ defines the spectral energy loss $D(\mathbf{k})$ due to wave breaking

$$D(\mathbf{k}) = bg^{-1}c^5\Lambda(\mathbf{k}), \tag{13}$$

where b is an empirical constant of order $b \sim 10^{-1} - 10^{-2}$ (see, for example, references in KM01, and the discussion by Babanin and Young (2005) for more details; we mention here that the exact value of b is not important for the present study as b appears in the expression for the separation stress, Eq. 20 below in the text, in combination with other constants, and only their combined value is relevant for the model results). Since in the equilibrium range D is proportional to the wind energy input $I(\mathbf{k})$

$$D(\mathbf{k}) \sim I(\mathbf{k}) = \beta\omega g k^{-4}B(\mathbf{k}) \tag{14}$$

then, combining (13) and (14), the following equation for $\Lambda(\mathbf{k})$ is obtained

$$\Lambda(\mathbf{k}) \sim b^{-1}\beta k^{-1}B(\mathbf{k}). \tag{15}$$

In developed seas, the main contribution to the total length of wave breaking fronts comes from the shortest breaking waves (Phillips 1985), and the role of dominant waves (waves in the spectral peak) in supporting the form drag is negligible (see KM01 for more details).

The present study is focused on high wind conditions, when wind seas are most likely undeveloped. MK02 gave an estimate of the impact of the AFS from the dominant waves on the sea drag for young seas. Adopting the threshold level approach proposed by Longuet-Higgins (1957), they found quite a strong effect of the AFS from dominant breaking waves on the sea drag, the impact being stronger the younger (and thus the steeper) are the seas. However, as found by Makin et al. (2004), the threshold approach is not universal, at least if the threshold value for the breaking wave steepness is assumed to be a universal constant.

In order to avoid an uncertainty with a choice of the threshold level, we shall define here $\Lambda(\mathbf{k})$ for the dominant waves in the manner similar to the equilibrium range. In stationary conditions, the energy balance equation for developing wind waves reads (e.g., Komen et al. 1994):

$$c_g \frac{\partial}{\partial x} E(\mathbf{k}) = N(\mathbf{k}) + I(\mathbf{k}) - D(\mathbf{k}), \tag{16}$$

where $E(\mathbf{k})$ is the wave energy spectral density, $N(\mathbf{k})$, $I(\mathbf{k})$, $D(\mathbf{k})$ are the energy sinks describing the non-linear four-wave interactions, wind energy input and dissipation due to wave breaking. As a well established fact (Komen et al. 1994), we note that the non-linear wave interactions (term N on the right-hand-side of (16)) provides the development of wind seas, i.e. the shifting of the spectral peak towards the low frequency with the increasing fetch. Therefore, in the vicinity of the spectral peak the energy balance Eq. 16 could be approximately reduced to

$$c_g \frac{\partial}{\partial x} E(\mathbf{k}) \simeq N(\mathbf{k}). \tag{17}$$

Then, in the vicinity of the peak, the wind energy input and dissipation due to wave breaking should be also balanced, at least in the order of magnitude

$$I(\mathbf{k}) \simeq D(\mathbf{k}) \tag{18}$$

(see Komen et al. (1994), their Figure 3.9). Referring to (14) and (18), we may expect that the expression for $\Lambda(\mathbf{k})$ in the spectral peak domain should have the same form (15) as for the equilibrium range. Thus we suggest that the spectral density of wave breaking fronts is defined by (15) in the full spectral range. Note, that the length of breaking fronts plays a crucial role in the present study defining both the sheltered area (9) and the separation stress (3). The validity of the adopted parameterization is restricted most likely by moderate wind speeds (< 20 m s^{-1}), and is questionable for strong or hurricane wind speeds. According to (15), in the equilibrium range of the wave spectrum $\Lambda(\mathbf{k})$ is a growing function of the wind speed, which at very high wind speeds should inevitably be saturated. From dimensional reasoning the length of breaking fronts should be saturated at

$$\Lambda(\mathbf{k}) \sim k^{-1}. \tag{19}$$

On the other hand, at very strong wind speeds generated foam and spume droplets result in the fact that the near surface layer becomes a two-phase "liquid", whose properties (e.g., density), differ significantly from the air. The interaction of this two-phase "liquid" with waves as well as the formation of the wave-induced and separation stress are not investigated so far. We leave these problems beyond the scope of the present study, relying on that extrapolation of (15) to high wind speeds will give the right trend in model results.

With the use of (15) Eq. 3 can be rewritten as

$$d\tau_s^0(\mathbf{k}) = c_s b^{-1} u_*^2 \cos^3\theta k^{-2} \beta(\mathbf{k}) B(\mathbf{k}) d\mathbf{k}. \tag{20}$$

Thus, the equation for the sea surface drag (12) is the governing equation of the model, where the spectral density of the wave-induced $d\tau_w^0(\mathbf{k})$ and separation $d\tau_w^0(\mathbf{k})$ stress is defined by (2) and (20) correspondingly. The effect of the surface sheltering on the wave-induced and separation stress is taken into account through the cumulative sheltered area $q(k)$ defined by (9) and the Heaviside function $h(k_o - k)$ centered around the wavenumber k_o and determined as a solution of the equation $q(k_0) = 1$.

In this paper, as well as in KM01 and MK02, the saturation spectrum is defined as a sum $B(\mathbf{k}) = B_p(\mathbf{k}) + B_{eq}(\mathbf{k})$ of the dominant wave spectrum $B_p(\mathbf{k})$ with the shape proposed by Donelan et al. (1985), and the equilibrium spectrum $B_{eq}(\mathbf{k})$ with the shape proposed by Kudryavtsev et al. (1999)

$$B_{eq}(\mathbf{k}) = \alpha \left[\frac{\beta_\nu(\mathbf{k}) + (\beta_\nu^2(\mathbf{k}) + 4I_{pc}/\alpha)^{1/2}}{2} \right]^{1/n}, \tag{21}$$

where I_{pc} is the rate of the parasitic capillaries generation (which is vanished at $k < 2(g/T)^{1/2}$; T is the surface tension), $\beta_\nu = \beta - 4\nu k^2/\omega$ is the effective wave growth rate, which is the difference between the wind wave growth rate and the rate of viscous dissipation, α and n are the spectral parameters defined here as reported in Kudryavtsev et al. (2003) after the wave spectrum validation against the radar data. Similar to the wave-induced momentum flux we suggest that there is no wind energy input to short waves inside the sheltered area, i.e. the wind wave growth rate defining the shape of the spectrum (21) is

$$\beta(\mathbf{k}) = c_\beta[1 - q(k)](u_*/c)^2 \exp(-\varphi^2). \tag{22}$$

If for some scale of short waves $q(k) = 1$, i.e. they are totally covered by separation bubbles induced by longer waves, the wind energy flux to these waves vanishes, and thus their energy also vanishes. This corresponds to the suggestion by Donelan et al. (2004) that at high wind speeds the outer flow separating from continually breaking waves does not "see" the troughs of the long waves and is unable to generate small-scale roughness there.

4 Model results

4.1 Comparison with the laboratory experiment

Donelan et al. (2004) investigated in laboratory conditions the sea-surface drag at very high wind speeds. They found that the drag coefficient at wind speeds exceeding 30 m s^{-1} reaches saturation. The separation of the air flow from continually breaking dominant waves was suggested as the most plausible mechanism explaining this effect. The Donelan et al. (2004) experiment is simulated here by using the model described above.

The model wind speed dependence of the drag coefficient $C_{d10} = u_*^2/u_{10}^2$ is shown in Fig. 1a. The calculation of C_{d10} for infinite fetch is also shown. The wind speed dependence of C_{d10} below and above 20 m s^{-1} is quite different for limited and infinite fetch. For limited fetch the drag coefficient has a tendency for saturation, while for infinite fetch it continues to increase linearly. The model calculation of C_{d10} for limited fetch without accounting for the effect of sheltering is given in Fig. 1a by the dashed line. As expected the exclusion of this effect results in the overestimation of C_{d10}. Sheltering leads to the suppression of the surface drag. Model calculations are close to laboratory measurements by Donelan et al. (2004), their Fig. 2, that are shown in Fig. 1a by open circles (momentum budget method) and diamonds (Reynolds stress method). Though a systematic shift between the measurements

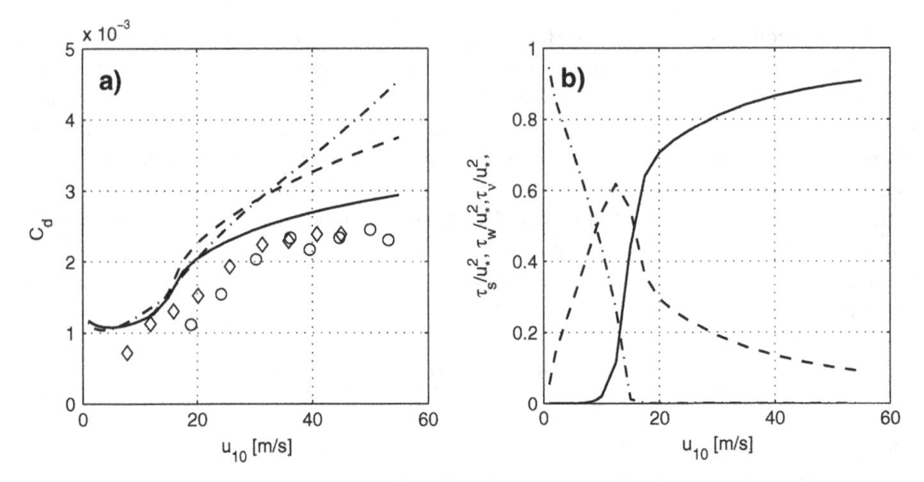

Fig. 1 (**a**) Drag coefficient C_{d10} versus wind speed u_{10}. Solid line, full model, limited fetch 10 m; dashed line, limited fetch 10 m, sheltering is switched off; dashed-dotted line, full model, infinite fetch. Laboratory measurements by Donelan et al. (2004) compiled from their Fig. 2 are shown by open circles (momentum budget method) and diamonds (Reynolds stress method) (**b**) Stress partitioning versus wind speed u_{10}. Solid line, stress due to separation; dashed line, wave-induced stress; dashed-dotted, viscous stress

and the model results is clearly observed, on the whole the model reproduces correctly a saturation trend in the distribution of C_{d10} at high wind speeds. We may consider this model result as an analogue of the C_{d10} saturation at high wind speeds revealed by Donelan et al. (2004) and interpret it as a result of the air-flow separation from continually breaking waves. The contribution of different components of the surface stress (viscous, wave-induced, and AFS) to the total stress is shown in Fig. 1b. At low wind speed $u_{10} < 5\,\mathrm{m\,s^{-1}}$ the viscous stress τ_ν dominates the stress, while at moderate wind speed of $5 < u_{10} < 15\,\mathrm{m\,s^{-1}}$ that is the wave-induced stress τ_w. At higher wind speeds the impact of the AFS strongly increases, and at $u_{10} > 25\,\mathrm{m\,s^{-1}}$ separation plays the crucial role in supporting stress providing the major part of the total stress (stresses in Fig. 1b are normalized on u_*^2, so that the total stress equals 1). This model result once again supports the suggestion by Donelan et al. (2004) that the separation from breaking waves leads to the saturation of the drag coefficient.

The role of sheltering by the AFS is further explained by Fig. 2, which shows the wind speed dependence of the total sheltered area $q(k_b)$ and the cumulative sheltered area $q(k)$. The total sheltered area is strongly wind speed dependent. At $u_{10} > 30\,\mathrm{m\,s^{-1}}$ the AFS from breaking waves of different scales covers more than 75% of the surface area. As follows from Fig. 2b the most part of the sheltered area is produced by the AFS from waves of the spectral peak (the wavenumber of the spectral peak k_p for the corresponding wind speed is shown by vertical dashed lines, the lowest wavenumber corresponds to the highest wind speed). At highest wind speed the AFS from the spectral peak quenches the form drag from shorter scales for both the wave-induced and the AFS component, and thus is responsible for the total surface stress.

Donelan et al. (2004) investigated also the C-band (5.3 GHz) radar scattering for HH and VV polarizations at high wind speeds. Their measurements for HH polarization are shown in Fig. 3a by open circles (the signal at VV polarization is very similar to HH and thus not shown here). Note that the measured values of the radar cross-section are multiplied by the factor 2.2×10^{-2} to fit the plot for the saturation spectrum. As it follows from this plot,

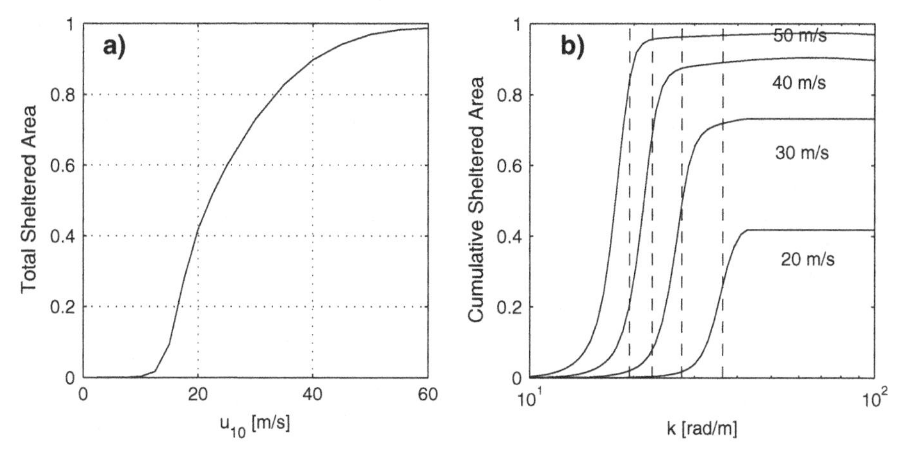

Fig. 2 Wind speed dependence of (**a**) the total sheltered area $q(k_b)$; and (**b**) the cumulative sheltered area $q(k)$. Dashed lines, the wavenumber of the spectral peak k_p for the corresponding wind speed, the lowest wavenumber corresponds to the highest wind speed

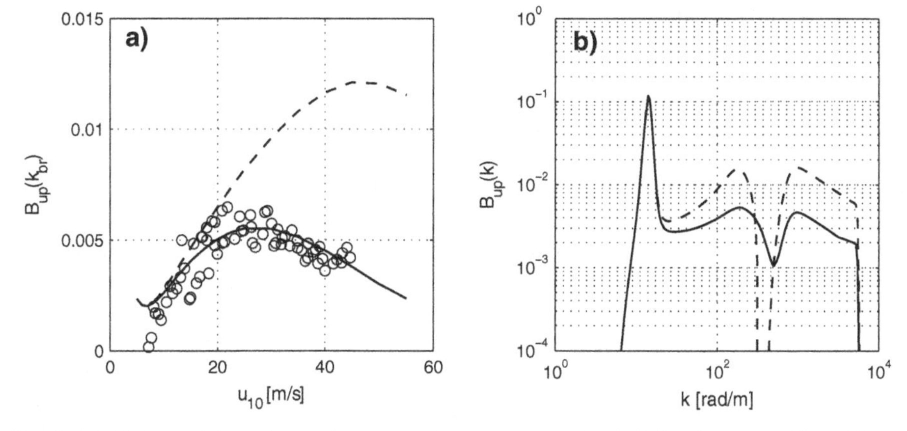

Fig. 3 (**a**) Wind speed dependence of the saturation spectrum in the wind direction at the C-band Bragg wavenumber. Solid line, full model; dashed line, sheltering is switched off. Open circles are measurements of the C-band radar cross section for HH polarization compiled from Donelan et al. (2004), their Fig. 5. Notice, that the measured values of the radar cross section are multiplied by factor 2.2×10^{-2}. (**b**) Spectral shape of the wave spectrum at $u_{10} = 30$ m s^{-1}. Solid line, full model; dashed line, sheltering is switched off

the radar signal associated with the geometric roughness of the centimetric waves reaches the maximum at the wind speed where the drag coefficient reaches the saturation level, and then decreases with the increase in the wind speed. As suggested, at very high wind speeds the air flow separating from continually breaking dominant wave crests no longer "sees" the troughs of these waves, and thus does not generate the small-scale roughness there, reducing the overall microwave reflectivity.

Qualitatively this mechanism is included in the wave spectrum model (21) through the reduction of the wind wave growth rate according to (22) due to sheltering by the AFS from dominant waves. Figure 3b illustrates the significance of the impact of the sheltering effect on the shape of the wave spectrum for laboratory conditions at $u_{10} = 30$ m s^{-1}. First, we note that the growth rate c_β depends on the aerodynamic roughness z_0 of the sea surface (see notation for c_β in the text below Eq. 2). This is because the wind wave growth rate at the wavenumber k is proportional to the wind velocity squared at $z = \pi/k$ relative to the phase velocity. Therefore, the larger is z_0 the smaller is the growth rate. The dashed line in Fig. 3b shows the spectral shape when the effect of sheltering is not taken into account. The spectral gap in the vicinity of $k \sim (g/T)^{1/2}$ is caused by the weak wind energy input due to the high aerodynamic surface roughness (see Fig. 1a, dashed line). On the contrary, in the capillary range the spectral density does not vanish since these waves are parasitic capillaries, i.e. they are not dependent on the direct wind energy input to this spectral range. The solid line in Fig. 3b shows the spectral shape resulting from the full model. There is a dual effect of the AFS sheltering on the wave spectrum. On one hand it reduces the aerodynamic roughness (compare the solid and dashed lines in Fig. 1a) and thus the growth rate coefficient c_β is increased. As a result the spectral gap at $k \sim (g/T)^{1/2}$ is less pronounced. On the other hand sheltering of short waves by the AFS from breaking crests of longer waves decreases the effective growth rate that results in the reduction of the spectral level of short waves.

Figure 3a shows the wind speed dependence of the up-wind spectral level at the Bragg wavenumber (5.3 GHz) corresponding to the radar measurements by Donelan et al. (2004). Similar to observations, the spectrum of Bragg waves reaches maximum at 25 m s^{-1} and then decreases with increasing wind speed. This is because sheltering of the short Bragg wave by the AFS from breaking crests of longer waves decreases the effective growth rate of the Bragg wave and that results in the reduction of its spectral level. When sheltering is switched off the spectral level is considerably overestimated.

4.2 Aerodynamic and geometrical roughness at high winds

The experimental data and the model simulations indicate that the AFS from continually breaking dominant waves can be considered as a plausible mechanism explaining the saturation of the surface drag coefficient at high wind speeds. A question arises however as to how this mechanism works in real field conditions characterized by much longer fetches than in the laboratory? And can it explain a similar saturation and further reduction of the drag coefficient with increasing of the wind speed revealed by Powell et al. (2003) in tropical cyclones?

Figure 4 shows the model calculation of C_{d10} (in terms of the Charnock parameter $\alpha = z_0 g/u_*^2$) in a wide range of fetch from approximately 1m to 10^6 m and the wind speed from 10 m s^{-1} to 50 m s^{-1}. In the range of fetch of practical interest ($X > 10^3$ m) and at the wind speed $u_{10} > 20$ m s^{-1} the model Charnock parameter appears to be almost independent of the fetch and the wind speed and approximately equal to $\alpha = 0.014$. At short fetches typical for the laboratory conditions α is considerably reduced. This regime of the air flow and the water surface interaction is described in the previous section.

Independence of α from the fetch and the wind speed in the range of long fetches is a surprising fact. To obtain a deeper insight, the contribution of different stress components to the total surface stress and the cumulative contribution of breaking wind waves to the sheltered area for various wind speeds at fetch 10^5 m are shown in Fig. 5. Unlike laboratory conditions, the separation stress τ_s and the wave-induced momentum flux τ_w contribute equally to the total stress at high wind speeds. Moreover, as follows from Fig. 5b, unlike very short fetches the dominant waves in this case do not contribute significantly to the sheltered area and to the AFS stress. At $X > 10^3$ m breaking of waves from the equilibrium range produces most of the AFS, and since they extract also most of the wave-induced momentum flux, we may conclude that the self-consistent interaction of the air flow with the equilibrium range of wind waves totally defines the form drag of the sea surface, in close relation to the Charnock prediction. No saturation or levelling off of C_{d10} (or suppression of the aerodynamic

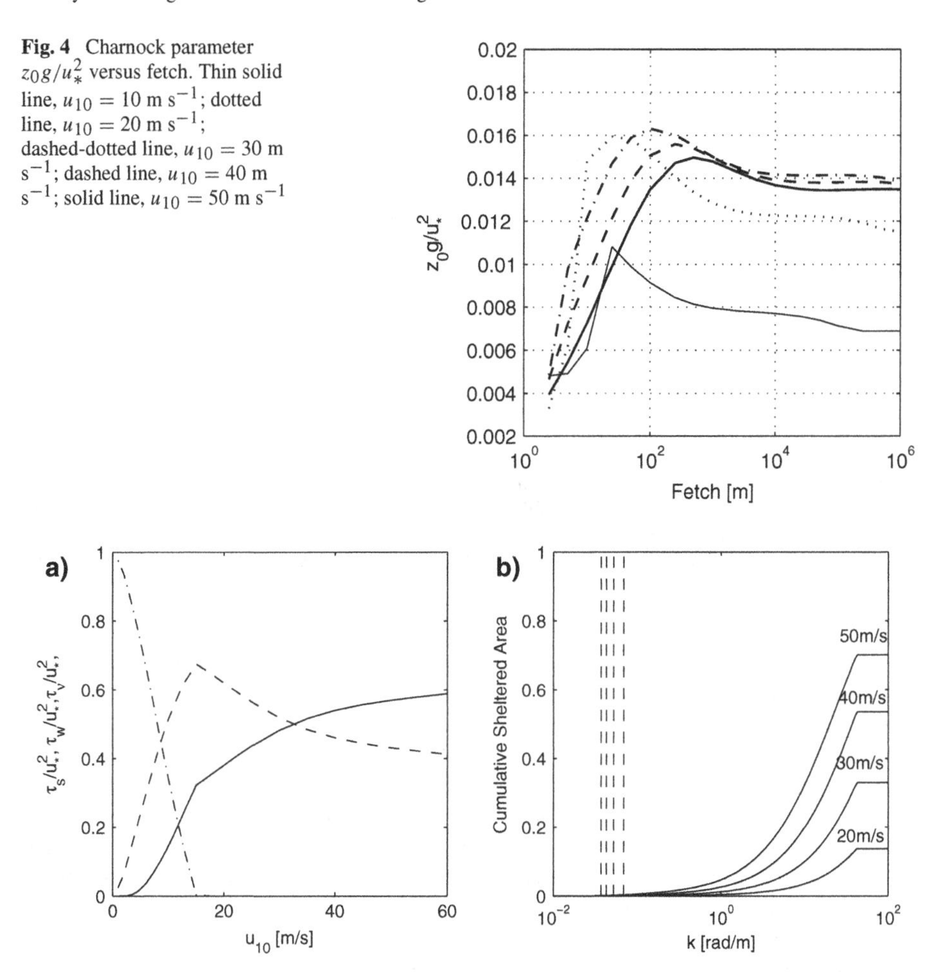

Fig. 4 Charnock parameter $z_0 g/u_*^2$ versus fetch. Thin solid line, $u_{10} = 10$ m s^{-1}; dotted line, $u_{10} = 20$ m s^{-1}; dashed-dotted line, $u_{10} = 30$ m s^{-1}; dashed line, $u_{10} = 40$ m s^{-1}; solid line, $u_{10} = 50$ m s^{-1}

Fig. 5 (**a**) Stress partitioning versus wind speed. Solid line, stress due to separation; dashed line, wave-induced stress; dashed-dotted line, viscous stress. (**b**) Wind speed dependence of cumulative sheltered area $q(k)$. Dashed lines, the wavenumber of the spectral peak k_p for the corresponding wind speed, the lowest wavenumber corresponds to the highest wind speed. Fetch 10^5 m

roughness) is to be anticipated at high wind speeds if the wind fetch is long enough. Most probable that other mechanisms such as spray effects are responsible for the reduction of the drag coefficient in the field as observed by Powell et al. (2003) and showed by Makin (2005) and Kudryavtsev (2006).

Figure 6a shows a behaviour of the geometrical surface roughness, which is related to the saturation spectrum in the wind direction at the Bragg wavenumber in C- and L-band radar signals at high wind speeds. Unlike the aerodynamic roughness, the geometrical surface roughness demonstrates the apparent saturation around $u_{10} \simeq 30$ m s^{-1} with the following suppression (in C-band) at higher wind speeds. This behaviour is very similar to that revealed in laboratory conditions by Donelan et al. (2004) (compare with their Fig. 5). A physical mechanism leading to this behaviour is sheltering of short waves by the AFS from breaking longer waves. The sheltered area for C- and L-band roughness is shown in Fig. 6b. At highest wind speeds the sheltered area reaches 70% of the sea surface. According

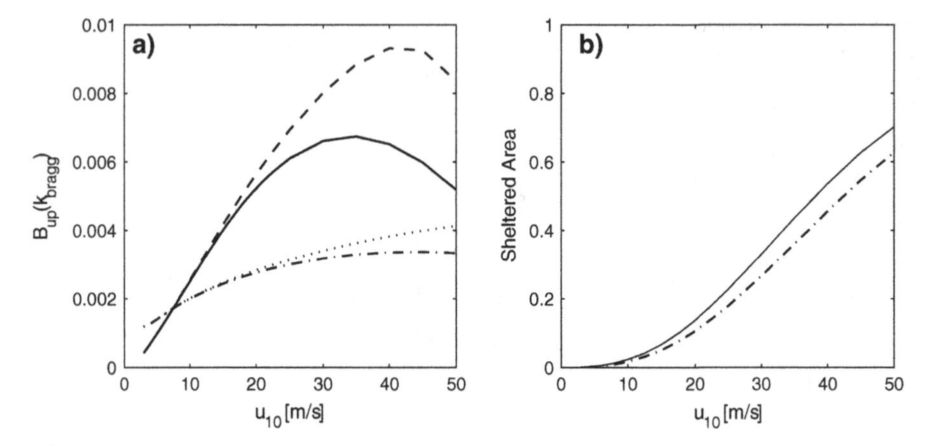

Fig. 6 (**a**) Wind speed dependence of the saturation spectrum in the wind direction at the Bragg wavenumber. Solid line, radar C-band (wavelength 0.06m), full model; dashed line, the same but sheltering is switched off; Dashed-dotted line, radar L-band (wavelength 0.20m); dotted line, the same but sheltering is switched off. (**b**) Wind speed dependence of the total sheltered area $q(k_b)$. Solid line, radar C-band; dashed-dotted line, L-band. Fetch 10^5 m

to the model, inside these areas the wind energy input to the short waves is suppressed, and thus their spectrum level is reduced. The model calculations of the wave spectral level without accounting for the sheltering effect are shown in Fig. 6a and demonstrate the significance of this mechanism.

Notice, that levelling off of the C-band radar signal at high wind speeds in the field conditions was revealed by Donnelly et al. (1999). This effect is included in the improved empirical ocean C-band backscatter model CMOD-5 (Hersbach 2003). Though the mechanism of radar scattering at high wind speeds is more complicated than the Bragg scattering model prediction (see, e.g., the discussion by Kudryavtsev et al. (2003) on the effect of wave breaking), our results clearly indicate that the effect of levelling off of the radar backscatter at high wind speeds results from the suppression of the geometrical surface roughness, and there is no need to invoke the suppression of the aerodynamic surface roughness to explain this phenomenon.

5 Conclusion

The current understanding of the physical processes at the sea surface at extreme wind conditions is based on few experiments performed in the field (Powell et al. 2003) and in the laboratory (Donelan et al. 2004). In both the saturation of the sea surface drag coefficient C_{d10} at very high wind speeds was revealed. One theoretical attempt to investigate this problem is based on the description of the sea droplets impact on the turbulent momentum flux and thus on the sea drag and was performed by Makin (2005), Bye and Jenkins (2006) and Kudryavtsev (2006). Though these studies showed strong potential ability of this mechanism in levelling off and further reduction of the drag coefficient at high wind speeds, the question remains what is the role of the surface roughness in the aerodynamic friction of the ocean surface at high wind speeds.

In this context, the laboratory study by Donelan et al. (2004) provides a good opportunity to answer this question. As concluded by Donelan et al. (2004) the saturation of the drag coefficient in laboratory conditions could be explained by the saturation of the aerodynamic roughness due to the air flow separation from continually breaking waves. The role of water droplets in that study was not investigated (only the presence of droplets at highest winds was mentioned). However, one may anticipate that the impact of droplets on the air-flow dynamics at short fetches should be much weaker than in real field conditions, as (i) the range of breaking waves generating spume droplets is quite narrow (in laboratory conditions at highest wind speed the dominant wavelength was about 0.8 m) and thus the concentration of water droplets should be significantly lower that in field conditions, and (ii) generation of the vertical spread of the droplets and their possible influence on turbulence is confined to the internal boundary layer, which depth is about 3% of the fetch. So, the air flow with suspended droplets in laboratory conditions strongly differs from the air flow above ocean waves, where droplets can be transported by turbulence far away from the surface and thus strongly affecting the dynamics of the boundary layer.

In this paper based on the model developed by KM01 and MK02 we have investigated the effect of the surface roughness on the surface drag at high wind speeds. In the model, wind waves and the atmospheric boundary layer represent a coupled system. Waves provide the surface friction velocity through the form drag, while the energy input from the wind to waves depends on the friction velocity and the wind speed. Here, we extended the model to high wind speeds taking into account the effect of sheltering of short wind waves by the AFS from breaking crests of longer waves. It is suggested that the momentum and energy flux from the wind to short waves, which are trapped into the separation bubble of breaking longer waves, is locally vanished. On one hand, this leads to the reduction of the form drag due to the exclusion of contributions from these areas to the form drag supported by these short waves through the wave-induced momentum flux. On the other hand, in these sheltered areas the short waves do not receive energy from the wind and that reduces the wind wave growth rate, which defines the shape of the wave spectrum.

At short fetches, typical for laboratory conditions, and strong wind speeds steep dominant wind waves break very frequently and provide the major part of the total form drag through the AFS from breaking crests. At $u_{10} > 30$ m s^{-1} this contribution attains 90%, so that the effect of the short waves on the drag coefficient C_{d10} is considerably suppressed. This is a limiting regime when the aerodynamic roughness at short fetches is defined by the height of the dominant waves. In this case, the dependence of the drag coefficient on the wind speed is much weaker than would be anticipated from the standard parameterization of the roughness scale through the Charnock relation. This result is similar to the saturation of the drag coefficient at $u_{10} > 30\,\mathrm{m\,s^{-1}}$ revealed experimentally by Donelan et al. (2004). According to the model, sheltering of the surface by the AFS from the dominant breaking waves prevents the energy flux to short waves, which in turn restrains their wind growth rate and leads to the reduction of the spectral level of the wave spectrum at the wind speed $u_{10} > 30\,\mathrm{m\,s^{-1}}$. This phenomenon was also found by Donelan et al. (2004) in the signature of radar measurements.

At long fetches representing the field conditions the spectral contribution of the wave components to the form drag is significantly changed as compared to the short fetches. Unlike the laboratory condition, waves of the spectral peak in the field are not so steep, thus they break rarely and their contribution to the AFS is weak. In this case, the surface form drag is determined predominantly by the AFS from breaking of the equilibrium range waves. Since the wave-induced momentum flux is supported to a large extent also by these waves, the aerodynamic roughness at high wind conditions becomes independent from the fetch and the wind speed. As shown, at high wind speeds up to $60\,\mathrm{m\,s^{-1}}$ the model aerodynamic roughness is consistent with the Charnock relation, i.e. no saturation of C_{d10} at high wind speeds can be explained by this mechanism if the fetch is long enough. Unlike the aerodynamic roughness, the geometrical surface roughness (height of short wind waves) could be saturated or even suppressed when the wind speed exceeds $30\,\mathrm{m\,s^{-1}}$. The effect is similar to that occurring at short fetches, and its origin is sheltering by the AFS from longer breaking waves that restraints the short wave growth rate. This mechanism can explain the effect of saturation of the C-band radar backscatter at high wind speeds found by Donnelly et al. (1999) in the field, which is adopted in the empirical geophysical backscatter model CMOD-5 (Hersbach 2003).

Acknowledgements The Expert Visit grant ESP.NR.NREV. 981938 and the Collaborative Linkage Grant ESP.NR.NRCLG 982529 by Public Diplomacy Division, Collaborative Programmes Section, NATO are gratefully acknowledged.

References

Babanin AV, Young IR (2005) Two-phase behaviour of the spectral dissipation of wind waves. In: Proceedings ocean waves measurement and analysis, fifth int sym WAVES 2005, 3–7 July, 2005, Madrid, Spain, Edge B, Santas JC, (eds) paper no. 51, 11

Banner ML (1990) The influence of wave breaking on the surface pressure distribution in wind wave interaction. J Fluid Mech 211:463–495

Banner ML, Melville WK (1976) On the separation of air flow above water waves. J Fluid Mech 77:825–842

Bye JAT, Jenkins AD (2006) Drag coefficient reduction at very high wind speeds. J Geophys Res 111, C033024, doi:10.1029/2005JC003114

Donelan MA, Hamilton J, Hui WH (1985) Directional spectra of wind generated waves. Phil Trans R Soc Lond, Ser A 315:509–562

Donelan MA, Haus BK, Reul N, Plant WJ, Stiassnie M, Graber HC, Brown OB, Saltzman ES (2004) On the limiting aerodynamic roughness of the ocean in very strong winds. Geophys Res Lett 31, L18306, doi:10.1029/2004GL019460

Donnelly WJ, Carswell JR, McIntosh RE, Chang PS, Wilkerson J, Marks F, Black PG (1999) Revised ocean backscatter models at C and Ku band under high-wind conditions. J Geophys Res 104:11485–11497

Emanuel KA (1995) Sensitivity of tropical cyclones to surface exchange coefficients and a revised steady-state model incorporating eye dynamics. J Atmos Sci 52:3969–3976

Giovanangeli JP, Reul N, Garat MH, Branger H (1999) Some aspects of wind-wave coupling at high winds:an experimental study. In: Wind-over-wave couplings. pp 81–90, Clarendon Press, Oxford 356 pp

Hersbach H (2003) CMOD5 an improved geophysical model function for ERS C-band scatterometry,/ Techn. Mem./, ECMWF, Reading, UK

Kawamura H, Toba Y (1988) Ordered motions in the turbulent boundary layer over wind waves. J Fluid Mech 197:105–138

Komen GJ, Cavalery L, Donelan M, Hasselmann K, Hasselmann S, Janssen PAEM (1994) Dynamics and modelling of ocean waves. Cambridge University Press, 540 pp

Kudryavtsev VN (2006) On effect of sea drops on atmospheric boundary layer. J Geophys Res 111: C07020, doi:10.1029/2005JC002970

Kudryavtsev VN, Makin VK (2001) The impact of air-flow separation on the drag of the sea surface. Boundary-Layer Meteorol 98:155–171

Kudryavtsev VN, Makin VK, Chapron B (1999) Coupled sea surface-atmosphere model 2. Spectrum of short wind waves. J Geophys Res 104:7625–7639

Kudryavtsev V, Hauser D, Caudal G, Chapron B (2003) A semi-empirical model of the normalized radar cross-section of the sea surface. Part 1: The background model. J Geophys Res 108(C3):8054, doi:10.1029/2001JC001003

Lonquet-Higgins MS (1957) The statistical analysis of a random moving surface. Phil Trans R Soc Lond, Ser A 249:321–387

Makin VK (2005) A note on drag of the sea surface at hurricane winds. Boundary-Layer Meteorol 115:169–176

Makin VK, Kudryavtsev VN (1999) Coupled sea surface- atmosphere model 1. Wind over waves coupling. J Geophys Res 104:7613–7623

Makin VK, Kudryavtsev VN (2002) Impact of dominant waves on sea drag. Boundary-Layer Meteorol 103:83–99

Makin VK, Caulliez G, Kudryavtsev VN (2004) Drag of the water surface at limited fetch: laboratory measurements and modelling. In: Geophysical research abstracts 1016, EGU04-A-00113, EGU General Assembly, April 25–31, 2004, Nice, France

Melville WK (1996) The role of surface-wave breaking in air-sea interaction. Ann Rev Fluid Mech 28:279–321

Monin AS, Yaglom AM (1971) Statistical fluid mechanics, vol 1. Cambridge: MIT Press 769 pp

Reul N, Branger H, Giovanangeli JP (1999) Air flow separation over unsteady breaking waves. Physics of Fluids 11:1959–1961

Phillips OM (1985) Spectral and statistical properties of the equilibrium range in wind generated gravity waves. J Fluid Mech 156:505–531

Powell MD, Vickery PJ, Reinhold TA (2003) Reduced drag coefficient for high wind speeds in tropical cyclones. Nature 422:279–283

Modelling dust distributions in the atmospheric boundary layer on Mars

Peter A. Taylor · P-Y. Li · Diane V. Michelangeli · Jagruti Pathak · Wensong Weng

Abstract A time and height dependent eddy diffusion model is used to investigate possible scenarios for the size distribution of dust in the lower atmosphere of Mars. The dust is assumed to either have been advected from a distant source or to have originated locally. In the former case, the atmosphere is assumed to initially contain dust particles with sizes following a modified gamma distribution. Larger particles are deposited relatively rapidly while small particles are well mixed up to the maximum height of the afternoon boundary layer and are deposited more slowly. In other cases, a parameterization of the dust source at the surface is proposed. Model results show that smaller particles are rapidly mixed within the Martian boundary layer, while larger particles ($r > 10\,\mu m$) are concentrated near the ground with a stronger diurnal cycle. In all simulations we assume that the initial concentration or surface source depend on a modified gamma function distribution. For small particles (cross-sectional area weighted mean radius, $r_{eff} = 1.6\,\mu m$) distributions retain essentially the same form, though with variations in the mean and variance of the area-weighted radius, and the gamma function can be used to represent the particle size distribution reasonably well at most heights within the boundary layer. In the case of a surface source of larger particles (mean radius $50\,\mu m$) the modified gamma function does not fit the resulting particle size distribution. All results are normalised by a scaling factor that can be adjusted to correspond to an optical depth for assumed particle optical scattering properties.

Keywords Aeolian process · Diurnal cycle · Dust · Mars boundary layer · Phoenix mission

P. A. Taylor (✉) · P.-Y. Li · D. V. Michelangeli · J. Pathak · W. Weng
CRESS, York University, 4700 Keele St, Toronto, Ontario, M3J 1P3, Canada
e-mail: pat@yorku.ca

Atmospheric Boundary Layers. A. Baklanov & B. Grisogono (eds.),
doi: 10.1007/978-0-387-74321-9_11,

1 Introduction

Phoenix, a project led by Professor Peter Smith of the University of Arizona and the first NASA Scout mission, will be launched in August 2007 and should land on Mars at approximately 70° N in May 2008, during local summer. An important component of the instrumentation will be a vertically pointing, dual wavelength (532 nm and 1,064 nm) lidar, provided by the Canadian Space Agency and built by MDA Space Missions and Optech Inc.. Members of the Canadian Science team at York and Dalhousie Universities will be responsible for planning lidar activities and the interpretation of the data. Other members of the Canadian Phoenix Science team have been involved in dust, aerosol and cloud modelling.

The present paper describes a relatively simple, one-dimensional, unsteady eddy diffusion model for dust and presents results for sample dust scenarios. These can be used to study possible size and height distributions of dust in the Martian boundary layer and are being used, in conjunction with the Dalhousie lidar model, in refining lidar operational specifications and in developing data analysis strategies for the Phoenix mission. After landing in May 2008, Phoenix measurements of lidar backscatter from dust should allow us to determine boundary-layer dust layering and boundary-layer depth, as it does on Earth (Stull 1988), and in conditions without heavy dust loading it should be able to provide information on water ice clouds.

The atmosphere of Mars is somewhat different from that on Earth with the biggest difference being much lower near-surface air density ($0.01-0.02\,\mathrm{kg\,m^{-3}}$) and pressure (6–12 hPa). Surface pressures depend strongly on surface elevation and season, since there are large elevation differences relative to the zero geopotential level (of order ± 10 km) and because part of the CO_2 atmosphere freezes out at the poles in winter. Larsen et al. (2002) has argued, based on Viking and Mars Pathfinder data, that despite these differences the atmospheric boundary layers of Earth and Mars obey the same scaling laws, including Monin–Obukhov similarity. Despite a lower solar constant ($\approx 590\,\mathrm{Wm^{-2}}$), the much reduced heat capacity of the air leads to a relatively large diurnal cycle of near-surface temperature (maybe from $-80°$ C to $-10°$ C at the potential Phoenix landing sites). Such large surface temperature variations suggest the possibility of a relatively deep afternoon convective boundary layer, but it should be remembered that water vapour plays a smaller role in Martian meteorology than on Earth and deep convection associated with latent heat release and cumulous clouds is absent. Dust and clouds can play a significant role in the radiative heat budget (see, for example, Korablev et al. 2005) and improving our knowledge of their vertical distribution is a key component of the Phoenix meteorological mission.

The turbulence mechanisms and aeolian processes by which small particles of dust, sand and snow are lifted from the surface and suspended in the air have been studied on Earth for many years—see, for example, Bagnold (1941) and Shao (2000) and there has been considerable research on dust storms and dust devils on Mars (e.g., Toigo et al. 2003; Pankine and Ingersoll 2004). Dust-related processes are not fully understood, but models exist for both saltation (bouncing near the surface) and suspension modes. Recent field studies on Earth include the 2001 Aerosol Characterization Experiment, ACE-Asia (Hong et al. 2004) where lidar profile measurements of dust emanating from Asia were made from Jeju Island, Korea, and the Bodélé Field Experiment (BoDEx 2005, see Washington et al. 2006) in the Sahara.

When modelling the distribution of suspended particles an eddy diffusion approach is often adopted (e.g., Xiao and Taylor 2002) but Lagrangian simulation

(Taylor et al. 2002) and large-eddy simulation approaches are also used (see Shao 2000). In the present study we will use a simple eddy diffusivity approach, but we are fully aware of its limitations. Time and height dependent eddy diffusivities are obtained from a separate, uncoupled boundary-layer dynamics model. This is appropriate for light dust loading but coupling of the boundary-layer and dust models to include radiative effects of the predicted dust distribution is planned in the future.

At and close to the surface the situation is complex since particles interact with the surface in several different ways. The saltation process allows for an airborne particle to strike the surface and bounce back into the air, and/or to transfer its energy to other particles initially lying on the surface so that they are ejected into the air (saltation bombardment or sandblasting). This is especially important for small particles where there are large cohesive forces to be overcome in order to lift them from the surface. Well above the surface the time scales of turbulence are generally large compared to the inertial time scale of the particle ($|w_p/g|$, where w_p is the particle settling velocity and g is gravitational acceleration) and particles can be considered to be following the turbulent motions in the air but with a superimposed settling velocity. Near the surface (say in the lowest few metres) the particle inertial time scale and the time scale of the turbulent eddies can be similar and particle inertia may need to be taken into account (see Taylor et al. 2002). We have not included these near-surface effects, and will focus on heights, $z > 1$ m above the local surface—assuming that the lidar will be mounted at about that level, and that analysis of lidar returns will only start at ranges of about 50 m.

In addition to particles being raised from the surface due to surface wind shear stress in sustained high wind situations there is also a strong possibility that, in highly convective conditions, dust devils (Rennó et al. 2000) and convective plumes may provide an important mechanism for lifting particles and distributing them through the boundary layer. In this paper our focus is on the time and height variations of the particle size distribution of dust in the boundary layer and we choose to use a simple one-dimensional unsteady model in which surface wind shear stress is the only dust lifting mechanism represented. Lifting due to dust devils could be the source of an initial distribution and may be included explicitly in future studies (c.f. Newman et al. 2002a). They are however intermittent and very localised sources, and are not readily represented even in a three-dimensional model.

A significant uncertainty is the source of the dust at a particular location. Has it been lifted from the surface nearby or has it been advected over long distances from a remote source region? The latter is often assumed to be the case for major dust storms on Mars (Newman et al. 2002 a, b). Small (of order 1 μm radius) particles have low settling velocities (100 m sol^{-1}) and if mixed to large heights (10 km) can travel long distances, even without sustained turbulence maintaining them in suspension. Some authors, in particular Murphy et al. (1990), have simulated the decay of dust storms, assuming an initial distribution (by particle size and height) of dust, and dry deposition to the surface.

Our model is one dimensional, and so strictly speaking applies to situations that are completely horizontally homogeneous. We can however note that, in the absence of radiative coupling, the dust dispersion equation (the 3-D version of Eq. (5) below) is linear in N and with appropriate horizontal boundary conditions can be used to represent the height and time variation of horizontal average dust concentrations. Our dust computations will generally run for about 15 or 30 Martian days (sols) to produce a relatively regular diurnal cycle except for some residual amplitude variations due to net emission or removal of dust at the surface. The proposed landing sites for Phoenix are relatively homogeneous but high wind storm events are unlikely to be sustained in the same location for 15 or 30 sols. We can, however, envisage the results as approximately representing a Lagrangian column moving with a strong wind region, or a high dust concentration area. The primary aim is to explore a range of dust situations that may be encountered during the Phoenix mission. We are also collaborating with the Mars general circulation model group at York University with a view to incorporating similar, though lower resolution, dust treatment in the models developed by Moudden and McConnell (2005).

At the present time we have not linked our dynamics and dust models, and so cannot include the feedback effects associated with the radiative impacts of dust. The present cases can be viewed as an initial study, most appropriate for low dust concentrations and low optical depths.

2 Settling velocity

The settling velocity for a dust particle, w_p, is a function of particle and air densities (ρ_d, ρ_a), size and shape, gravitational acceleration (g) and properties of the fluid within which it is falling. The Stokes settling velocity, appropriate to small spherical particles of radius r, with low Reynolds number, is

$$w_s = 2\,gr^2(\rho_d/\rho_a)/(9\nu) = 2\,gr^2\rho_d/(9\eta), \tag{1}$$

where η and $\nu(=\eta/\rho_a)$ are the molecular dynamic and kinematic viscosities for CO_2 respectively. Since η is generally independent of air density the second form of Eq. (1) suggests the use of a constant w_s for particles of a given size. For low Reynolds number, this gives Stokes settling velocities ranging, on Mars, from approximately $0.0002 - 0.2\,m\ s^{-1}$ for spherical particles with radius from 1 to 30 μm, if we take $\eta = 1.0 \times 10^{-5}\,kg\ m^{-1}\ s^{-1}$ as a value appropriate to about 200 K, $g = 3.72\,m\ s^{-2}$ and dust particle density $\rho_d = 2.5 \times 10^3\ kg\,m^{-3}$. The value of η given above is based on the use of Sutherland's formula (see Chapman and Cowling 1964) with $\eta = 1.38 \times 10^{-5}$ kg $m^{-1}\ s^{-1}$ at T = 273 K.

For small particles and low air densities however the Cunningham slip-flow correction leads to significant increases in settling velocity above that predicted by Eq. (1) that need to be considered for dust on Mars. The Cunningham correction arises when the particle size is comparable to the mean free path of the air molecules and the continuum hypothesis begins to break down. Pruppacher and Klett (1997) discuss this in the context of water droplets in the Earth's atmosphere while Murphy et al. (1990) stress the importance of the factor for dust on Mars. The correction is given in the form,

$$w_c = w_s(1 + \alpha Kn), \tag{2}$$

where Kn (= λ/r) is the Knudsen number with λ representing the mean free path of the air (CO_2) molecules. There are various empirical fits to α based on settling velocity data. We follow Murphy et al. (1990) and use

$$\alpha = 1.246 + 0.42\exp(-0.87/Kn). \tag{3}$$

Results with other formulae are very similar for the range of particle sizes that we will be concerned with.

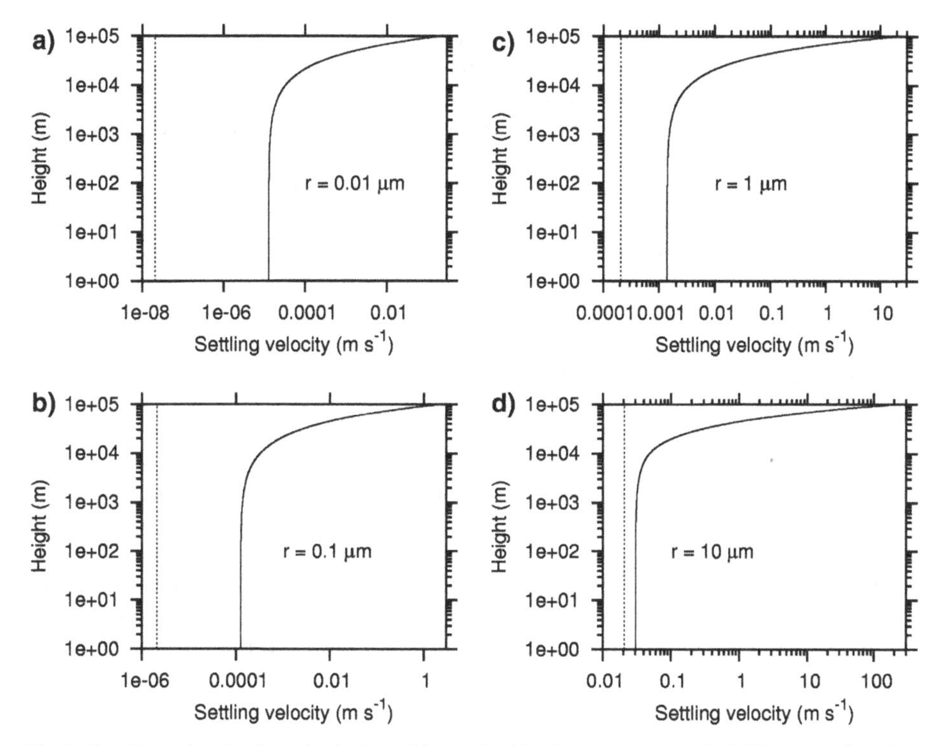

Fig. 1 Settling velocities for spherical particles in the Martian atmosphere. Solid line, with Cunningham slip correction; Dotted line, basic Stokes law result

The mean free path is given via the kinetic theory of gases as,

$$\lambda = 2\eta/[\mathrm{p}(8\,\mathrm{M}/\pi \mathrm{RT})^{1/2}], \tag{4}$$

where M is the molecular weight ($0.044\,\mathrm{kg\ mol^{-1}}$ for the Martian atmosphere) and R is the universal gas constant ($8.313\,\mathrm{J\ mol^{-1}\ K^{-1}}$)—see Seinfeld and Pandis (1998). For the Martian atmosphere at temperature, T = 200 K, and with $\eta = 1.0 \times 10^{-5}$ kg $\mathrm{m^{-1}\ s^{-1}}$, we would then have $\lambda = 3.64\,\mu$m at pressure p = 675 Pa.

Plots of the resulting settling velocity (w_c) as a function of height for a range of particle sizes are given in Fig. 1, together with the Stokes velocity, w_s. In these plots we have assumed an isothermal atmosphere (T = 200 K) with $p = p_0 \exp(-z/h)$, h = RT/Mg = 10.312 km and $p_0 = 675$ Pa (with corresponding $\rho_a = 0.0176\,\mathrm{kg\ m^{-3}}$). For particles of radius 1 μm there is a factor 6.7 increase relative to w_s, even at the surface, but the settling velocity ($0.0014\,\mathrm{m\ s^{-1}}$) is still small. By a height of just over 20 km the settling velocity has however risen to $0.01\,\mathrm{m\ s^{-1}}$ ($864\,\mathrm{m\ day^{-1}}$), and will be effective in removing particles from these heights over time if they are not replenished in some way. Overall, however, these particles would take a long time to settle out and removal processes may involve cloud microphysics as well as gravitational settling (see Michelangeli et al. 1993). The correction factor is less significant for large particles but remains important for radii up to about 10 μm.

The results discussed above have assumed that the particles are spherical with $\rho_d = 2.5 \times 10^3\,\mathrm{kg\ m^{-3}}$. As, for example, Murphy et al. (1990) have pointed out, the assumption of spherical particles may well be incorrect. Also the dust density may be lower than that assumed, which is based on terrestrial values, and we would not be too surprised if our settling velocity calculations are an overestimate. Density changes are easy to accommodate, and Murphy et al. provide an additional correction applicable to oblate spheroidal, disk-like particles should additional calculations be required. Although η decreases (and hence w_s increases) as temperature decreases, λ and α will also decrease. Numerical simulations with a lapse rate of $2\,K\,km^{-1}$ have shown that the temperature variations have negligible effect on w_c. In the model we will take $w_p = w_c$.

3 Turbulent mixing and simple analytic solutions

For high wind situations, surface sources and particles with moderate settling velocities, confined to the lower part of the Martian boundary layer, simple models of eddy diffusion in the neutrally stratified surface boundary layer can provide a starting point for the analysis. We should however note that, because of much lower air density, the boundary layer is even less likely to be in neutral stratification on Mars than on Earth and so these initial idealised models are for reference and illustration only. For uniform size particles with fixed settling velocity, w_p, in horizontally homogeneous conditions, a tendency term, downward settling and upward turbulent diffusion are represented by,

$$\partial N/\partial t - \partial(w_p N)/\partial z = \partial(K_s\, \partial N/\partial z)/\partial z, \tag{5}$$

where $N(z,t)$ is the particle number density (# m^{-3}, where # indicates number of particles). If we consider steady-state situations and assume a constant w_p, then on integration with eddy diffusivity, $K_s = \kappa z u_*$, where $\kappa = 0.4$ is the von Karman constant, and with constant friction velocity, u_*, we obtain the standard power-law profile (as in Xiao and Taylor 2002),

$$N = N_r (z/z_r)^{(-w_p/\kappa u_*)}, \tag{6}$$

where N_r is a reference value of the number density at height z_r. This is an appropriate formulation for the lowest 10–50 m of the atmospheric boundary layer (see, for example, Garratt 1994) on Earth or Mars but if $w_p/\kappa u_* < 1$ integrating the concentration profile (Eq. (6)) vertically gives an infinite number of particles, as noted by Xiao and Taylor (2002). If one is interested in a deeper layer, say up to several kilometres in the atmosphere one could modify the eddy diffusivity formulation to use,

$$K_s = l(z) v_s(z) \tag{7}$$

with $1/l(z) = 1/\kappa z + 1/\Lambda$. The mixing-length formulation for $l(z)$ was originally proposed by Blackadar (1962) and has been used by many others since (see Garratt 1994). For the turbulent velocity scale, v_s, we initially set this equal to the surface friction velocity, u_*.

With this formulation, steady state conditions and a constant w_p would lead, on integration of Eq. (5), to a number concentration profile in the form,

$$\ln(N(z)/N_r) = -(w_p/\kappa u_*)[\ln(z/z_r) + (\kappa/\Lambda)(z - z_r)]. \tag{8}$$

This gives the standard power law Eq. (6) for small z with a singularity at $z = 0$, and an exponential decay for large z. The singularity in relation to integrals from z_r to

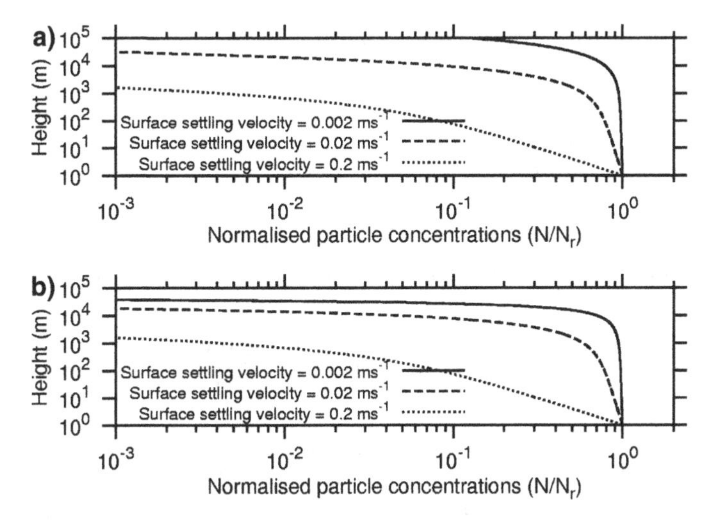

Fig. 2 Normalised particle concentrations in a simple, steady-state, model. Typical profiles for particles with (surface) settling velocities of 0.002, 0.02 and 0.2 m s^{-1}. N_r is at $z_r = 1$ m; (**a**) with $w_p = w_c(r, 0)$, (**b**) with $w_p = w_c(r, z)$

infinity with $w_p/\kappa u_* < 1$ is no longer a problem with finite values of Λ. Figure 2 illustrates particle concentration profiles with $\Lambda = 100$ m and $u_* = 1$m s^{-1}. We consider particles with surface settling velocities of 0.002, 0.02 and 0.2 m s^{-1}, which correspond very approximately to particles with radii 1, 10 and 30 μm, taking the Cunningham slip correction into account. We take $z_r = 1$ m, the approximate height of the Phoenix lidar. While Eq. (8) is a highly simplified formulation, the solutions shown in Fig. 2a with $w_p = w_c(0)$ emphasise the fact that, with a residual value for eddy diffusivity at large heights and a balance between upward diffusion and downward settling, we can maintain concentrations of small particles ($r < 1\,\mu$m) to significant heights. If we include height variations of w_c and set $w_p = w_c(z)$ we must numerically solve the steady state version of Eq. (5) with a balance between settling and diffusion, i.e.,

$$-w_c(z)N = K_s(z)\partial N/\partial z. \tag{9}$$

Solutions for this case are given in Fig. 2b and show a more rapid decrease of N with height, especially for small particles above 1 km. Reducing K_s will reduce the size of particle that can be maintained in suspension at relatively high concentrations but mesoscale or GCM models that maintain eddy diffusivities of order 100 m^2 s^{-1} or more above the boundary layer will resist removal of fine particles ($r < 1\,\mu$m) via settling. Using Eq. (8) as a basis for scaling we could argue that reductions in N with height scale with $zw_p/(u_*\Lambda)$ or zw_p/K_s and that substantial particle concentrations can be maintained to heights of order K_s/w_p. With $K_s = 100$ m^2 s^{-1} and $w_p = 0.002$ m s^{-1} ($r \approx 1\,\mu$m) this indicates a height of 50 km. This is confirmed by Fig. 2a, but note that including increases in w_c with height leads to a more rapid decrease in concentration with height.

In a diurnal cycle, eddy diffusivity will be time and height dependent and calculation requires numerical integration of Eq. (5). Here we will assume an eddy diffusivity of the form,

$$K_s(z,t) = l(z,t)v_s(z,t), \tag{10}$$

with

$$1/l(z,t) = \phi(\zeta)/[\kappa(z+z_0)] + 1/\Lambda(t).$$

In the near-surface layer we assume that Monin–Obukhov similarity can be applied and use the Businger-Dyer form of the dimensionless concentration gradient

$$\begin{aligned} \phi(\zeta) &= 1 + 5\zeta && \text{if } \zeta \geq 0, \text{or} \\ &= (1 - 16\zeta)^{-1/2} && \text{if } \zeta < 0, \end{aligned}$$

where $\zeta = z/L(t)$ and L is the Obukhov length. For large heights we set the maximum l as $\Lambda(t) = 0.1\ H(t)$, where H is the boundary-layer depth. We take $z_0 = 0.01\,m$, for dust, heat and momentum. This is a subjective estimate, based on images from Rover and Pathfinder sites, plus MRO high rise pictures of potential Phoenix lander sites. The following empirical form for the velocity scale, $v_s(z, t)$, is used in this study

$$v_s(z,t) = 0.5u_*(t)\{1 - \tanh[5(z - H(t))/H(t)]\} \tag{11}$$

based on a subjective fit to results obtained with our version of Savijaarvi's (1999) Mars boundary-layer model using the surface energy budget as the lower boundary condition. Note that $v_s \approx u_*$ at $z=0$, $v_s = u_*/2$ at $z=H$ and decreases towards zero at large z. The variables $1/L(t)$, $H(t)$ and u_* (t) are represented by slightly smoothed approximations to predictions of the diurnal variations over a sol (i.e., a Mars day = 24.66 Earth hours), computed with Savijaarvi's 1-D boundary-layer model for 70° N on Mars at areocentric longitude $L_s = 90$ degrees (summer). Computations were made with a range of values for U_g. The initial temperature profile had $T = 210\,K$ at the surface and a lapse rate of $2\,K\ km^{-1}$. Results for the 5th sol, by which time the boundary-layer model results are essentially periodic, are shown in Figs. 3 and 4 for geostrophic winds of $20\,m\ s^{-1}$ and $35\,m\ s^{-1}$. Despite its Earth connotation we still use the term "geostrophic" for Mars to indicate the wind for which there is a balance between Coriolis force and pressure gradient. Mars atmospheric model simulations, global and mesoscale, generally predict lower wind values, of order $10\,m\ s^{-1}$ but the Phoenix site will be in a strong baroclinic zone associated with the North polar cap and high winds are possible. In these and other plots we represent time in Mars hours where 1 Mars hour = (24.66/24) Earth hours. Seconds remain as Earth seconds. At this time of year and at this latitude the sun remains over the horizon all the time, but we will still refer to 0000 h as "night". In Fig. 3a we have also indicated the minimum threshold surface friction velocity of $1.3\,m\ s^{-1}$ used here for particles to be lifted from the surface under normal, maintained wind conditions. This is not exceeded in circumstances with $U_g = 20\,m\,s^{-1}$ while with $U_g = 35\,m\,s^{-1}$ the surface friction velocity exceeds this critical value in the middle of the sol when surface heating is strong and near-surface winds are strongest and most turbulent. The modelled boundary-layer heights reach about 3 km in late afternoon and shrink to less than 1 km at "night". Figure 4 shows the corresponding eddy diffusivities for dust, as computed from Eqs. (10) and (11), as they vary with height and time.

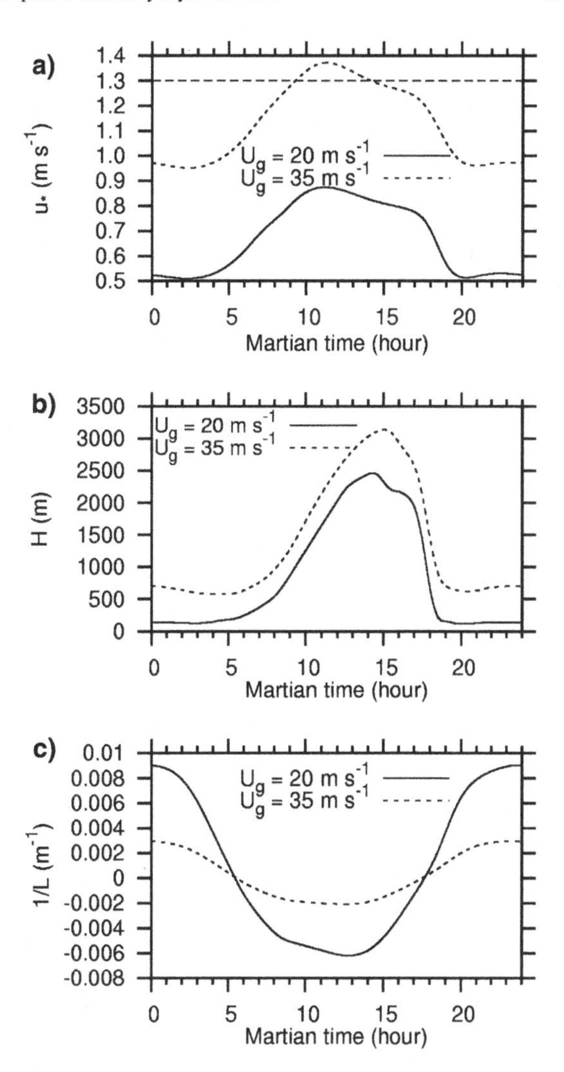

Fig. 3 Modelled time variations of boundary-layer parameters u_*, H and 1/L over a sol. Solid lines $U_g = 20\,m\,s^{-1}$, dashed line $U_g = 35\,m\,s^{-1}$. The minimum threshold friction velocity is also shown in the u_* plot

4 Surface boundary conditions

Considering a single size of particle with number density N (# m^{-3}), the net flux of N (# $m^{-2}\,s^{-1}$, positive upwards) at a given height is,

$$F = -w_p N - K_s \partial N/\partial z. \tag{12}$$

As the surface boundary condition we follow the suggestion of Shao (2000), and assume the following form:

$$F_0 = -\,w_d N \qquad \text{if } u_* \le u_{*t} \tag{13a}$$

$$-\,w_d N + \alpha_g (u_*/u_{*t})^3 (u_*/u_{*t} - 1) \qquad \text{if } u_* > u_{*t} \tag{13b}$$

where w_d is a deposition velocity. The lower boundary could be the surface itself, with an appropriate treatment of K_s, but assuming that the net flux is independent of

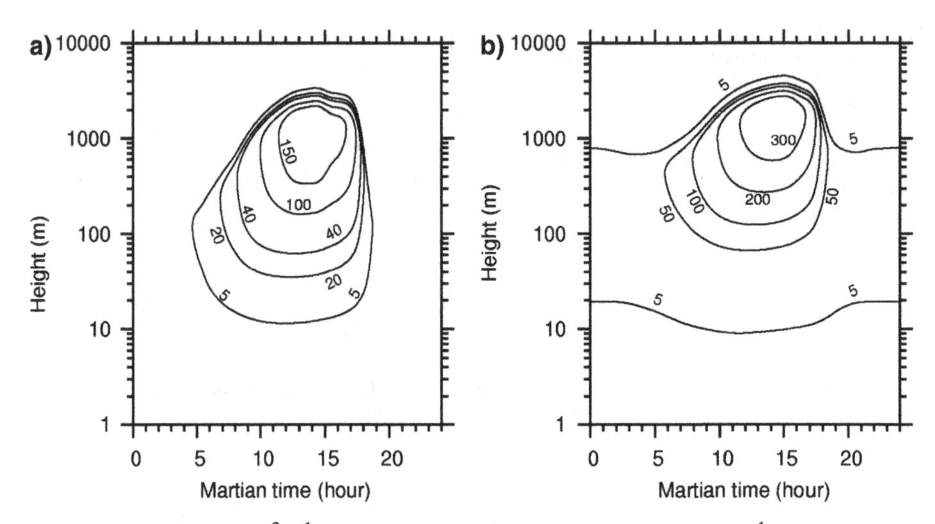

Fig. 4 Eddy diffusivity (m^2s^{-1}) during one sol: (**a**) geostrophic wind = 20 m s^{-1}; (**b**) geostrophic wind = 35 m s^{-1}

height near the surface we can apply the same condition at a model lower boundary, z_{lb}, a little way above. We will use $z_{lb} = 1$ m in the results presented here. This allows us to avoid issues related to the appropriate roughness length for diffusion of particles and its relationship with the momentum roughness length, z_0, which we have taken as 0.01 m.

While in principle we could allow the deposition velocity w_d to be different from the particle settling velocity w_p, in practice we will set them to be equal in the results to be presented here. In Eq. (13a, b), α_g (with units of # m^{-2} s^{-1}) is a factor in the dust source function, and u_{*t} is the threshold friction velocity. An explicit form for u_{*t} will be given below. Note that α_g is a flux per unit area. For particles with a mix of sizes, we can use N(r, z, t), or N(r) if the height and time dependence is not flagged explicitly, to represent a particle size distribution (# m^{-3} m^{-1}) and in these cases $F_0(r)$ will be the flux per unit increment of particle radius. For this case references available to help fix an expression for $\alpha_g(r)$ on Mars are scarce. Based on the findings in Tomasko et al. (1999), that the column integrated size distribution of Martian dust can be approximated by a modified gamma distribution, we have used that form for α_g in our present study. Noting that we will assume $u_{*t}(r)$ to be constant over most of the range of r that we consider, we use $u_{*t}(r)$ as a factor to give the correct dimensions for $\alpha_g(r)$ and further assume

$$\alpha_g(r) = u_{*t}(r)C(r/a)^{(1-3b)/b}\exp(-r/ab). \qquad (14)$$

The constant C has the same dimensions as N(r) and will be used as a normalising factor, which can in principle be adjusted to provide results appropriate to a specified optical depth if we make appropriate assumptions about the optical properties of the particles as in Sect. 6 below. In principle we could modify Eq. (14) in order to produce a desired size distribution in the air, at a fixed height or column integrated, but for these initial studies we specify the surface flux. Also note, that in Eq. (14) a is the cross-section-weighted average radius and a^2b is the cross-section-weighted variance.

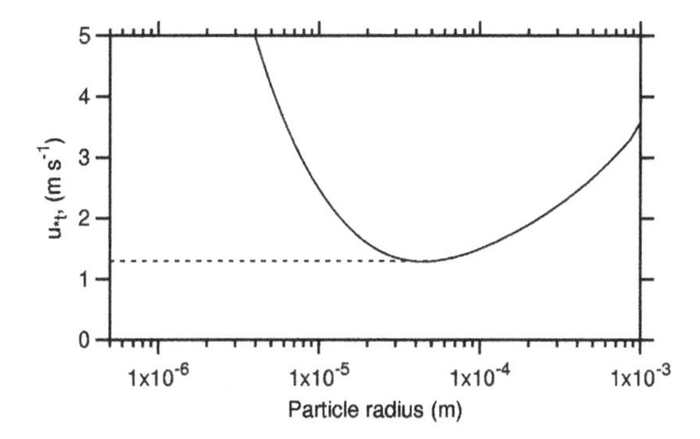

Fig. 5 Threshold friction velocity $u_{*t}(r)$. Solid line is obtained from Eq. (15). For particles of radius less than or equal to 50 μm, u_{*t} is set as a constant in this study, as shown by the dashed line

We also need to determine the threshold friction velocity. Based on results from Bagnold (1941) and Greeley and Iversen (1985), a semi-empirical form for u_{*t} can be determined as follows:

$$u_{*t} = A(2\,gr\rho_d/\rho_a)^{1/2}, \tag{15}$$

where

$$A = \begin{cases} 0.1291[1 + I_p/(\rho_d g(2r)^{2.5})]^{0.5}(1.928R_{*t}^{0.0922} - 1)^{-0.5} & \text{if } 0.03 \le R_{*t} \le 10 \\ 0.120[1 + I_p/(\rho_d g(2r)^{2.5})]^{0.5} \\ \times(1 - 0.0858\exp(-0.0617(R_{*t} - 10))) & \text{if } 10 < R_{*t}. \end{cases}$$

Here $I_p (= 3 \times 10^{-7}\ \mathrm{N\,m^{-1/2}})$ is the inter-particle cohesion parameter, and $R_{*t} = u_{*t}(2r)/\nu$ is the threshold friction Reynolds number. A plot of u_{*t} for a range of particle radius is shown in Fig. 5. The solid line in Fig. 5 is obtained by iteratively solving Eq. (15), and indicates that particles having the smallest u_{*t} are of radius $\approx 50\,\mu$m. The corresponding value of u_{*t} we denote by u_{*t_min} and is about 1.3 m s^{-1}. Figure 5 also shows that particles of size smaller than 50 μm would require an exceptionally high surface wind to lift them from the surface. In our analysis, however, we keep u_{*t} constant for particles of radius <50 μm as represented by the dashed line in Fig. 5. The rationale is based on the experimental study by Greeley et al. (1994), who showed that dust of radius a few microns in a layer of 0.002 m thick on top of small pebbles could be raised by wind with threshold friction velocities as low as 1.3 m s^{-1}. In addition, Shao (2000, Chapter 7) and others have argued that once sand size particles are saltating, their impacts (saltation bombardment or sandblasting) can overcome the cohesive forces that would otherwise hold the smaller dust particles in place on the ground. Shao and Lu (2000) and McKenna Neuman (2003) have presented alternatives to the Greeley and Iversen (1985) equations. McKenna Neuman (2003) includes results for particles of diameter 210 μm and above, and with appropriate scaling for Mars air density and gravity her results ($u_{*t} = 1.5$ m s^{-1} for r = 105 μm) are consistent with our assumptions for threshold friction velocity.

5 Model scenarios

A number of different scenarios have been used with the model, using a range of steady geostrophic wind speeds but all based on the Phoenix lander location and season. Two situations will be used for illustration of results with $U_g = 20\,m\,s^{-1}$ and $35\,m\,s^{-1}$.

As noted above the planetary boundary-layer dynamics model is first run to produce appropriate diurnal $K_s(z,t)$ fields. In principle we could couple the PBL and dust models and include the feedback of the dust on the radiation fields but that requires a series of additional assumptions and would lead to more specific results. We have preferred a simpler approach for now. In all cases we compute results using Eq. (5) for a broad range of individual particle sizes and then combine them to produce results for particle size distributions. The top and bottom boundaries are set at $z_{ub} = 10\,km$ and $z_{lb} = 1\,m$ respectively. Surface pressure is assumed to be constant at 675 Pa and the atmosphere is considered as isothermal with $T = 200\,K$ for the settling velocity calculations, though not in the dynamics model.

5.1 Dust advected from a remote source region—Scenario A

In this scenario dust is initially mixed uniformly to 10 km but then slowly settles out while being mixed through the boundary layer by turbulence. The initial size distribution of dust particles is assumed to satisfy the gamma distribution

$$N(r,z)/C = (r/a)^{(1-3b)/b} \exp(-r/ab), \tag{16}$$

where, consistent with the values obtained by Tomasko et al. (1999), we take $a = 1.6\,\mu m$ and $b = 0.2$. The geostrophic wind is assumed to be $20\,m\,s^{-1}$. The maximum boundary-layer height is diagnosed at about $H = 2.5\,km$ and the corresponding eddy diffusivity in one sol is shown in Fig. 4a. This shows that the maximum eddy diffusivity occurs in the late afternoon of each sol and that our formulation of turbulent velocity scale (Eq. (11)) does allow some diffusion above H, to a little over 3 km in this case. The minimum (over all particle sizes) threshold friction velocity for lifting dust from the ground is set at $1.3\,m\,s^{-1}$ and is never exceeded at this value of U_g. The surface is thus always a sink for the dust particles, which are present by virtue of the initial conditions. At the top boundary we have set $N = 0$; results with zero flux, i.e., $\partial N/\partial z = 0$ were essentially the same. At the lower boundary, as discussed above, we use Eq. (13) with $w_d = w_p$ in these simulations, so that $\partial N/\partial z = 0$ at z_{lb}.

Figure 6 shows the evolution of concentration of spherical particles of radius $1.6\,\mu m$ and $10\,\mu m$ over 30 sols and their diurnal variation on that 30th sol. Note that we will refer to particles of radius $1.6\,\mu m$ as "$1.6\,\mu m$ particles" in the text below. The concentrations are normalised by the initial concentration at $t = 0$. Figure 6a shows that due to settling in the absence of turbulent mixing above the maximum afternoon boundary-layer height, all particles are removed from the layer above about 3 km after about 20 sols. This roughly matches the particle settling velocity, which is of order $0.0024\,m\,s^{-1}$ at the surface but is higher at 10 km consistent with a settling of about 7 km over 20 sols ($0.004\,m\,s^{-1}$) over that period. Figure 6a shows that the effect of the settling velocity for particles of radius $1.6\,\mu m$ is overcome by the eddy diffusivity within the boundary layer, so that the particle concentration remains close to being well-mixed between the surface and the maximum (late afternoon) height, of the boundary layer. Note that, due to the increase in w_p with height, there is a convergence of the downward flux and normalised particle concentrations are increased to about 1.6 above the initial value (1.0) by sol 18. There is some diurnal variation, apparent in the sol 30 plot, for these $1.6\,\mu m$ particles but it is relatively weak. Near-surface (10 m) normalised dust concentrations show a general decline over sol 30, but there

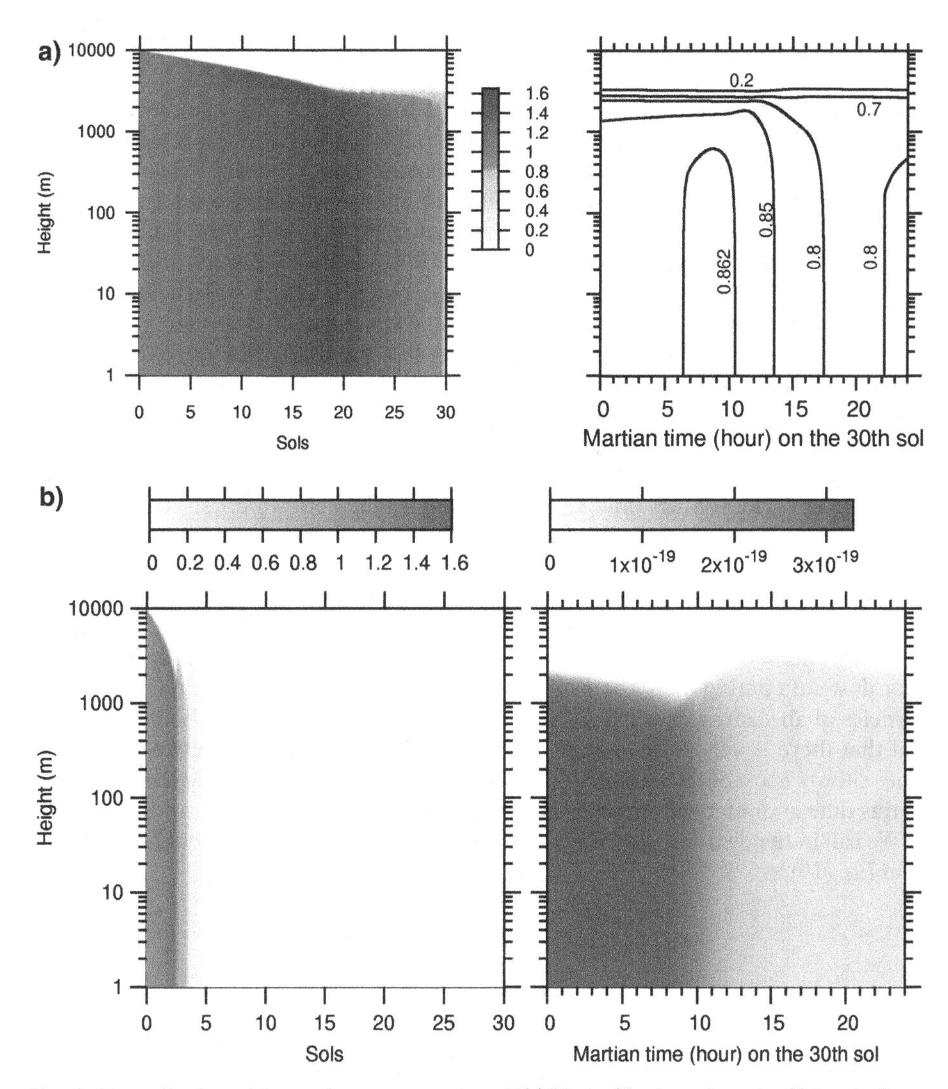

Fig. 6 Normalised particle number concentrations $N(z)/N_o$ in 30 sols and on the 30th sol of scenario A. (**a**) $r = 1.6\ \mu m$, (**b**) $r = 10\ \mu m$. N_o represents the initial concentration of particles of the size concerned

is a local maximum occurring near 0800, again due to downward flux convergence, and a local minimum in late afternoon as material is mixed back up to the top of the boundary layer. Deposition to the surface causes a net decrease of 6.7% over the sol. Once the initial material has sunk below the diurnal maximum boundary-layer height (H_m), if we simply assume a well-mixed concentration of C_M at that point in time, we would subsequently anticipate an exponential decay in concentration of the form $C = C_M \exp(-w_d t/H)$ with time scale, $H_m/w_d \approx 3\,km/0.0024\,m\ s^{-1} = 14$ sols. The corresponding decay rate is 7.1% per sol, consistent with the rate above.

During this phase, and for $1.6\ \mu m$ particles, settling is too slow ($210\,m\ sol^{-1}$) to allow the upper dust boundary to follow the top of the boundary layer as it descends in early evening (Fig. 3) and all that would be "marked" by these small particles is the diurnal maximum boundary-layer height.

Particles of larger radius have larger settling velocity and the relative concentration of particles of radius of 10 μm is significantly reduced after a few sols, as shown in Fig. 6b. It should however be noted that there is an initial increase due to faster settling from above as for the 1.6 μm particles. These 10 μm particles have a settling velocity of order 110 m h^{-1} but will still not sink fast enough to follow the late afternoon collapse of the boundary layer (about 1 km h^{-1}). However these concentration plots do show a clear response to the diurnal cycle of the boundary layer with the upper dust boundary descending from a maximum of about 3 km in mid afternoon to a minimum of about 1 km at 0900 the next morning. In this case the time scale for concentration decay is reduced to about 27 Mars hours.

Combining results for a range of single size computations, and taking account of the initial size distribution (Eq. (16)), allows us to produce size distribution plots. Figure 7 shows the size distributions (plotted as (r/a)N(r, z)/C vs log r) of particles at 0200 and 1400 (local Mars time, in Mars hours) on the 30th sol for selected heights. The times are approximately those corresponding to minimum and maximum values of diffusivity. The "extra" (r/a) factor arises because the r axis is logarithmic so the spectra represent contributions proportional to a unit increase in log r. The heights chosen are the lower boundary (1 m) plus three levels (2, 3 and 5 km) near and above H_m. Note that 0–5 km will be the height range in which we anticipate collecting most lidar dust data during the Phoenix mission. The figure indicates that, on the 30th sol, particles of all sizes are essentially well-mixed with height within the boundary layer and that there is very little change in the overall size distribution between the two times shown. Most of the particles of radius >2 μm have been deposited to the surface by this time and removal rates for the small particles are low.

We can fit the distributions N(r, z) well by a modified gamma distribution, defined, as in Eq. (16) as:

$$G(r,z) = c'(r/a')^{(1-3b')/b'} \exp(-r/a'b'), \tag{17}$$

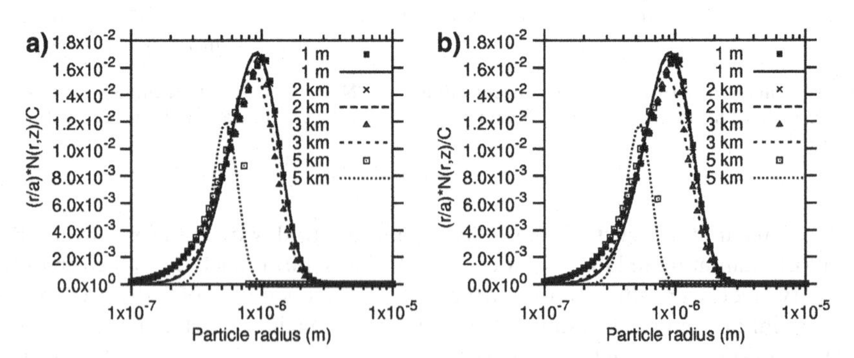

Fig. 7 Normalised particle size distributions, N(r, z), at different altitudes on the 30th sol of scenario A. (**a**) at 0600; (**b**) at 1800 (Mars hours). Symbols: results obtained from model predictions; lines: Gamma distribution fits computed from cross-section-weighted radius a′(z), cross-section-weighted variance (normalised by a'^2) b′(z) and best-fit values c′(z). Normalisation is based on a and C from the initial size distribution. Factor (r/a) is included to give contributions to the spectrum per unit increment in log r

where primed quantities $a'(z)$ and $b'(z)$ are the cross-section-weighted effective radius (denoted as r_{eff} in the recent review by Korablev et al. 2005) and cross-section-weighted variance (normalised by a'^2) respectively, and $c'(z)$ is a least squares fit parameter. Gamma distribution fits are shown in Fig. 7 together with the original model results. Overall, the results support the idea that a gamma distribution can continue to approximately represent particle size distributions within the boundary layer during the settling and mixing phase for these small particles. This applies especially for particles with r close to a′, but there is a tendency to underestimate the number density of small particles. Above the boundary layer (z = 5 km) where settling dominates and larger particles are removed, the modified gamma distribution is not a good fit. Note that a′ and b′ will differ from the initial (a, b) values. Values at z = 1 m for 1400 in this case are $a' = 1.24\,\mu m$, $b' = 0.13$ showing that the mean radius has decreased slightly and the size distribution has narrowed, relative to $a = 1.6\,\mu m$ and $b = 0.2$.

For the lidar it is the particle cross-section and its contribution to the back scatter that matters rather than particle number. The area-weighted (r^3N) and volume-weighted (r^4N) size spectra for 1800 h (Mars hours) are shown in Fig. 8 a,b. Again there are "extra" r factors because the r axis is logarithmic. As expected the peaks of these distributions occur at higher radii than in Fig. 7. The area-weighted size distribution of the dust initially present in the air is included in Fig. 8a. Many of the larger particles ($r > 2\,\mu m$) have been deposited on the ground by sol 30 while concentrations of particles with $r < 1.4\,\mu m$ have increased because of flux convergence associated with increases of the settling velocity with height, as discussed in relation to Fig. 6a.

An important quantity for the lidar is the total cross-section per unit area as a function of height, as shown in Fig. 8c. This confirms the result, anticipated from the 1.6 μm particle calculations, that the daily maximum boundary-layer height will be clearly defined. In this case we see a rapid decrease with height of the total particle cross-sectional area and we should be able to observe this with the Phoenix lidar system. It is informative to consider the diurnal cycles involved and the role of turbulence in modifying the evolution of the dust profile and concentration.

In the absence of any turbulence, particles would simply settle out and be deposited on the surface. In scenario A the initial concentration profile has particles mixed uniformly to a height of 10 km. Above the boundary layer with minimal mixing, particles simply fall and all but the finest dust has descended into the boundary layer after 30 sols. Daytime turbulence maintains a relatively uniform concentration within the boundary layer. In Fig. 8c we can observe a decrease in surface concentration over the sol 29–30 and can see that at 1400, when mixing is taking place, the profile is not yet well mixed. Instead it reflects upward mixing of dust from the lower parts of the boundary layer into the top 200 m or so to replace the dust that has been removed from this layer by settling during the previous night. Because of this mixing, surface concentrations and the rate of removal at the surface are reduced, thus extending the period during which there is dust in suspension.

5.2 Surface as the source of dust—Scenarios B1, B2

In this case, although the geostrophic wind is held constant at $35\,m\,s^{-1}$, the surface friction velocity, as shown in Fig. 3, exceeds the threshold value ($1.3\,m\,s^{-1}$) around the middle of the sol (0930–1400), so that the surface then acts as a source of dust particles. During the rest of the sol however, with reduced u_* there is deposition only. A deposition velocity ($w_d = w_p$) is still applied as in the previous scenario. The

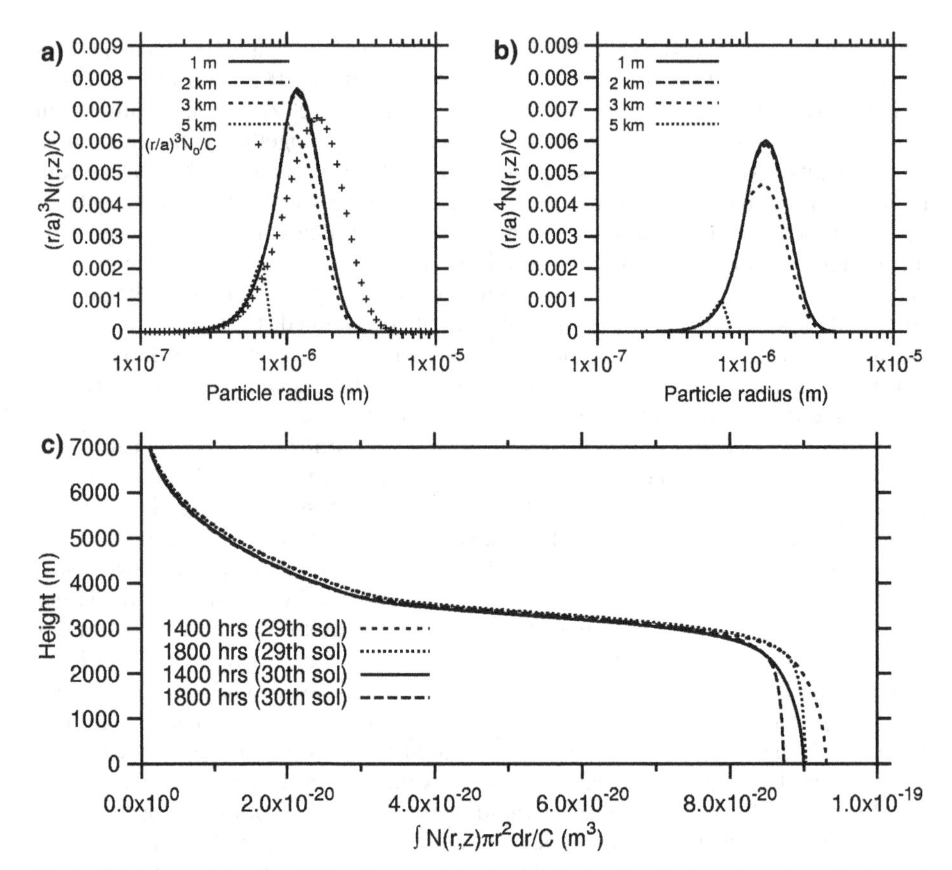

Fig. 8 (**a**) Area-weighted and (**b**) volume-weighted particle size distributions for selected heights at 6 pm on sol 30 for scenario A. Additional (r/a) factors are because of the logarithmic radius scale. (**c**) total cross-sectional area per unit volume at 4 times on sol 30

variation of eddy diffusivity in one sol is shown in Fig. 4b. The maximum analysed boundary-layer height is now at about 3 km, but our specification of eddy diffusivity allows diffusion above that level, to about 4 km. As in the previous case, we set $N = 0$ at the top boundary, and in this case we also set $N = 0$ as the initial condition.

The time evolution of particle concentrations (Nu_{*t}/α_g) for three specific sizes is shown in Fig. 9 assuming a source as specified by Eqs. (13a, b). Figure 9a shows that, because of the injection of particles from the surface, efficient mixing through the boundary layer and the low settling velocity, the concentration of particles of radius 1 μm increases over time and has not quite reached an equilibrium (i.e., a repeated diurnal cycle) after 15 sols. For particles of radius 10μm and 50 μm the daily averaged deposition is able to balance the source term after about 15 sols resulting in a periodic diurnal variation of particle concentration, as demonstrated in Figs. 9b and c. Note that 10 μm particles ($w_p \approx 0.03\,\mathrm{m\,s^{-1}}$) are mixed throughout the boundary layer, while 50 μm particles, with $w_p \approx 0.56\,\mathrm{m\,s^{-1}}$ are confined to the lowest 10–20 m. During each sol the maximum concentrations occur around midday, corresponding to the time when u_* and the flux are maximum. At this point an upper bound for the concentration can be determined if we balance the source and sink terms at the surface, i.e., set $F_0 = 0$ in Eq. (13) so that,

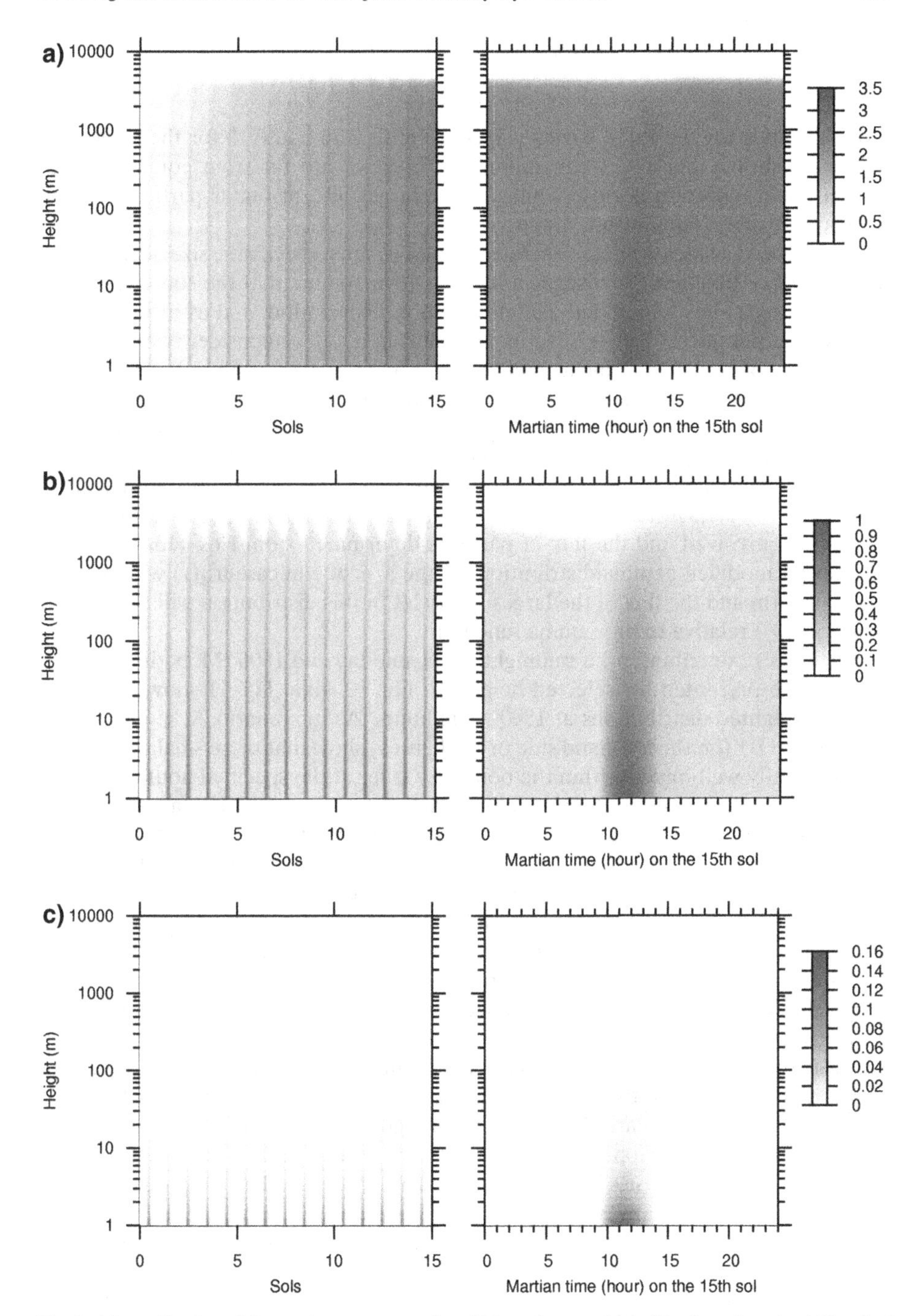

Fig. 9 Normalised particle number concentrations $N(r, z, t)u_{*t}/\alpha_g(r)$ in 15 sols and on the 15th sol of scenario B. (**a**) $r = 1\ \mu m$, (**b**) $r = 10\ \mu m$, (**c**) $r = 50\ \mu m$

$$Nu_{*t}/\alpha_g = (u_{*t}/w_d)(u_*/u_{*t})^3(u_*/u_{*t} - 1). \tag{18}$$

Upper bounds for the three particle sizes are 60.1, 2.76 and 0.15 for the three particle sizes shown. For the 1 μm radius particles we are far from equilibrium, at 10 μm the equilibrium is close while for the larger 50 μm radius particles we are in an approximate equilibrium situation.

In order to consider particle size distributions we must specify the size distributions for the source function. As discussed in Sect. 4 we assume that the source function coefficient $\alpha_g(r)$ has a modified gamma function distribution, and as in other cases we take the parameter $b = 0.2$. As mean radius in $\alpha_g(r)$ we consider two scenarios. In scenario B1 we set the mean radius $a = 1.6\,\mu$m, matching the observations of Tomasko et al. (1999) although, as noted in the review by Korablev et al. (2005), there is considerable uncertainty in the size distributions, especially for small particles. As a significantly different scenario we also consider B2 with a much coarser dust/sand distribution with $a = 50\,\mu$m. Both cases also involve $u_{*t}(r)$ to actually determine the flux. For the $a = 1.6\,\mu$m case there is essentially no variation in $u_{*t}(r)$ over the range of particles involved and the flux of particles determined from Eqs. (13a) and (14) will have a modified gamma distribution. In the $a = 50\,\mu$m case $u_{*t}(r)$ will increase for $r > 50\,\mu$m and the flux of the largest particles in this distribution will be reduced (via Eq. (13)) relative to the gamma function.

Particle size distributions at midnight, 0600, midday and 1800 (Mars time) on the 15th sol are presented for selected heights in Fig. 10 while Fig. 11 shows area and volume weighted distributions at 1800 Mars hours. As in scenario A, model results for scenario B1 (on the left-hand side of the figures) show that these smaller particles are essentially well-mixed within the boundary layer at the specified hours except at midday (Fig. 10c) when the surface source is active and the 1-m values are noticeably higher. At 4 km, just above the boundary layer, the model still has some late afternoon mixing and some small particles reach these levels. Size distribution results for scenario B2 with larger particles emitted show much more variation with height and time, and lower normalised concentration numbers. Note the different scales used for the different values of a. The larger particles cannot rise to any significant height under the wind and turbulence conditions represented here and the size distribution at each height shifts to lower radii as time advances from midday to 1800 and so on. As in the previous scenario, we have used the gamma distribution to approximate the particle size distributions at each height for the four specified hours on the 15th sol. Figure 10 shows that the approximations are in general good within the boundary layer for scenario B1, but for scenario B2 they are not a good fit and suggest that alternative functions may be more appropriate. Figure 11 shows area and volume-weighted distributions. These emphasise the differences between the smaller and larger particle size cases and clearly show much lower normalised particle cross-section values and their reduction with height in scenario B2. A shift in mean radius with height above the ground is also evident in B2.

Figure 12 shows that for scenario B1 the cross-section-weighted mean radius (a') and the total cross-sectional area per unit volume, $A(z)$, vary slowly with height within most of the boundary layer, while above they both decrease rapidly. Observations are not available for direct comparison with our model results. However, in Chassefière et al. (1995), vertical profiles of the effective radius were obtained by extrapolating observations made by Phobos 2 at 15 km to 25 km range down to the ground, which

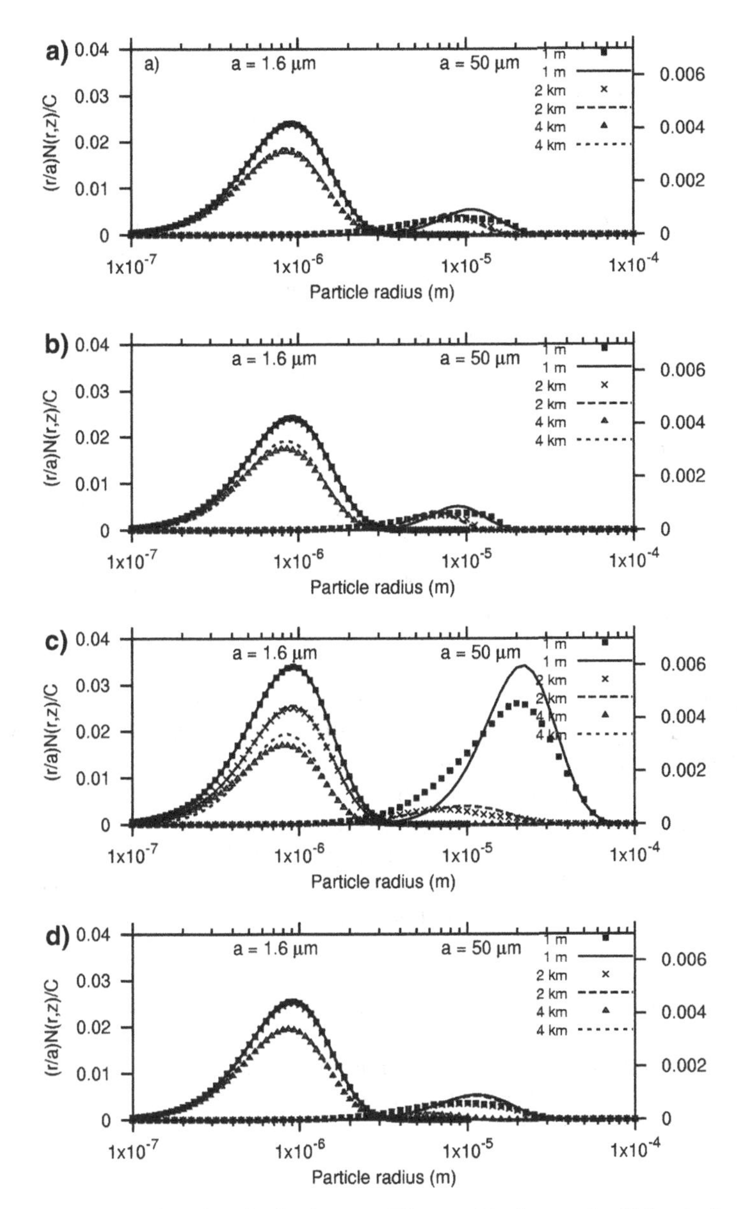

Fig. 10 Normalised particle size distributions at different altitudes on the 15th sol of scenario B. Symbols: N(r,z)/C obtained from model predictions; lines: Gamma distributions computed from cross-section-weighted radius $a'(z)$, cross-section-weighted variance (normalised by a'^2) $b'(z)$ and good-fitted values $c'(z)$. (**a**) at 0000; (**b**) at 0600; (**c**) at 1200; (**d**) at 1800 (Mars hours)

agree both qualitatively as well as quantitatively with the values of $a'(z)$ obtained in our scenario B1.

Scenario B2 is much different showing strong variation of a', and total cross-sectional area (Fig. 12c), with height and time. Direct interpretation of the relative numerical values between Figs. 12b and c is difficult because of the scaling of the

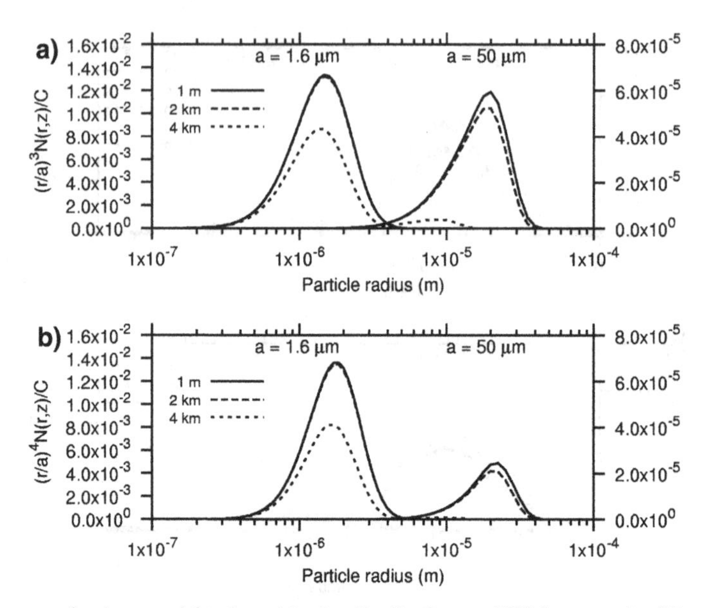

Fig. 11 Area- and volume-weighted particle size distributions at 1800 for scenarios B1 and B2

source strengths. What we can infer is that in scenario B1 the height of the afternoon maximum boundary layer is well represented by the decay with height of the cross-sectional area per unit volume in the height range 4–5 km at all times. For scenario B2 the boundary-layer height is reasonably well defined at 1200 or 1800. It is not so well defined at 0000 and at 0600 there is a shoulder in the profile at about 1 km, roughly equivalent to the top of the nocturnal boundary layer. Further analysis of this and other scenarios will be undertaken in conjunction with an evaluation of the lidar backscatter data collected during the Phoenix mission.

6 Optical properties, attenuation coefficients and determination of "C"

Dust scattering is the most important attenuation process in the Martian atmosphere. The extinction coefficient at optical wavelength λ due to aerosol (primarily dust), σ_λ (m^{-1}) can be modelled as,

$$\sigma_\lambda = \int_0^\infty N(r,z) b_{r,\lambda} \mathrm{d}r', \tag{19}$$

where $b_{r,\lambda}$ is the effective scattering cross-section of a single aerosol of radius r. For spherical dust particles, the effective scattering cross-section for a single particle is given by

$$b_{r,\lambda} = \pi r^2 Q_{r,\lambda}, \tag{20}$$

where πr^2 is the actual dust particle cross-sectional area, and $Q_{r,\lambda}$ (dimensionless) is the single particle scattering efficiency. The scattering efficiency is a function of a complex index of refraction, wavelength of the light and a size parameter.

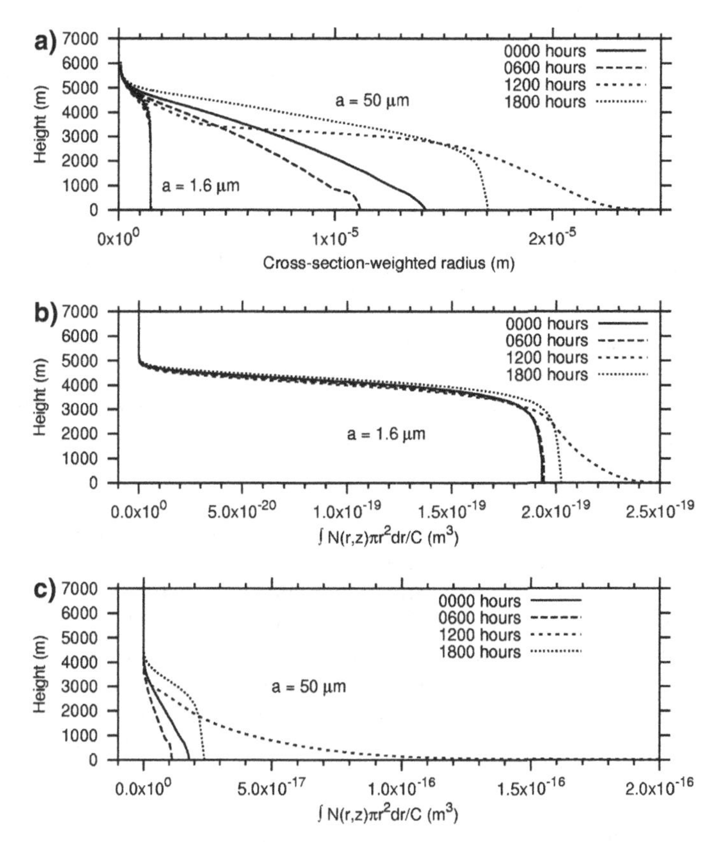

Fig. 12 Profiles of, (**a**) cross-section-weighted mean particle radius, and (b, c) total cross-sectional area per unit volume for scenarios B1 and B2 at indicated times (Mars hours). Curves for Scenario B1 are to the left of those for B2 in part (**a**)

Ockert-Bell et al. (1997) give a table of extinction efficiencies $Q_{ext} = Q_{r,\lambda}$ for Martian dust, indicating variation from about 2.8 at 500 nm to 3.3 at 1,000 nm. No dependence on particle size is indicated, and they assume that the dust has a size distribution and other properties presented by Pollack et al. (1995). Optical depth (dimensionless) quantifies the scattering and absorption that occurs between the top of the atmosphere and a given altitude, z_{lb}, and is given by

$$\tau_\lambda = \int_{z_{lb}}^{\infty} \sigma_\lambda dz. \tag{21}$$

For $Q_{r,\lambda} = Q_{ext}(\lambda)$, independent of r, and thus of height, z, we can write

$$\tau_\lambda = Q_{ext} \int_{z_{lb}}^{\infty} A(z)dz, \tag{22}$$

where A(z) is the cross-sectional area per unit volume,

$$A(z) = \int_0^\infty \pi r^2 N(r, z) \mathrm{d}r, \tag{23}$$

shown (scaled by C) in Figs. 8c and 12b, c. Performing the integrations indicated by Equation (22) for scenarios A, B1 and B2 gives the values for $\tau_\lambda C/Q_{ext}$ given in Table 1.

Taking representative values of Q_{ext} for the wavelengths and particle sizes considered then allows the determination of the scaling factor C needed to produce a desired optical depth. For example with $Q_{ext} = 2.8$ and optical depth $\tau_\lambda = 0.5$ at 0000 on sol 15 under scenario B1 we would have $C = \tau_\lambda/(7.83 \times 10^{-16}\ Q_{ext}) = 2.28 \times 10^{14}$ m^{-4}. Values of A(z) corresponding to this value of C within the well-mixed layer are of order $4.4 \times 10^{-5}\ m^{-1}$.

If we were to consider a uniform distribution of 1 μm particles then a representative concentration, again for Scenario B1 at 0000, would be $1.4 \times 10^7\ m^{-3}$ or 14 particles per cm^3. Column-integrated number densities would be of order $4{,}000 \times 1.4 \times 10^7\ m^{-2}$ or approximately $5.6 \times 10^{10}\ m^{-2}$ (5.6×10^6 per cm^2). With the actual size distribution we obtain the values in Table 2 and using $C = 2.28 \times 10^{14}\ m^{-4}$ we obtain approximately $5.4 \times 10^{10}\ m^{-2}$ as the column-integrated number density for Scenario B1 at 0000. These are in general agreement with the column particle integrated number densities used in Tomasko et al. (1999) of 4.6×10^6 per cm^2, which gave optical depths of order 0.5.

The vertically integrated number densities in Table 2 also reflect the evolution of the particle numbers. In Scenario A these decay slowly with time while in Scenario B they reflect the midday injection of new particles, offset by the continuous deposition to the surface.

Table 1 Values of $\int A(z)dz/C$ from $z_{lb} = 1$ m to infinity (m^4) corresponding to the scenarios presented. Note that C has dimensions m^{-4}

Time\Scenario	A: 30th sol	B1: 15th sol	B2: 15th sol
0000	3.39 E-16	7.83 E-16	3.46 E-14
0600	3.35 E-16	7.73 E-16	1.94 E-14
1200	3.31 E-16	8.17 E-16	11.35 E-14
1800	3.28 E-16	8.32 E-16	7.00 E-14

Table 2 Values of $\int\int N dz dr/C$ over all r and from $z_{lb} = 1$ m to infinity (m^2) corresponding to the scenarios presented

Time\Scenario	A: 30th sol	B1: 15th sol	B2: 15th sol
0000	1.53 E-4	2.37 E-4	1.51 E-4
0600	1.52 E-4	2.35 E-4	1.21 E-4
1200	1.51 E-4	2.47 E-4	2.12 E-4
1800	1.50 E-4	2.52 E-4	1.94 E-4

7 Conclusions

In this study, we present results from a one-dimensional, time dependent, eddy diffusion based model to study the size and vertical distributions of dust in the Martian boundary layer. We have applied our model to three scenarios. In the first case, the whole computational domain has been filled with particles to simulate dust advection from a remote source region. In the second and third cases, the atmosphere is initially dustless, and the surface is considered as a dust source. The only dust lifting mechanism considered in this case is surface wind stress. In all cases, a deposition velocity is applied at the surface. Model results from the two scenarios with fine dust (A and B1) show that the particle size distribution varies slowly with time and height within the boundary layer. In Scenario B2 with larger particles being available for lifting from the surface, the results are noticeably different with a much stronger diurnal variation.

Model results also show that if a modified gamma distribution is used to represent the initial dust concentration or the dust source function with $a = 1.6\,\mu m$ then the size distribution evolves so that it continues to be well approximated by a gamma function, although a and b will vary with time and height. For a source function with larger particles this was not the case, although in this case an extra factor in the source function may have affected this result.

Further work is needed to couple the boundary-layer and dust models and to make quantitative comparisons with data to be collected during the Phoenix mission.

Acknowledgements We are indebted to Allan Carswell of Optech Incorporated for the opportunity to become involved in the Canadian component of the NASA Phoenix Scout mission, and to Valerie Ussyshkin of Optech and Cameron Dickinson of Dalhousie University for prodding us to produce appropriate results. The research has been partially funded through contracts with the Canadian Space Agency and partially through a grant from the Natural Science and Engineering Research Council of Canada.

References

Bagnold RA (1941) (revised 1954) The physics of blown sand and desert dunes. Methuen, London, 265 pp

Blackadar AK (1962) The vertical distribution of wind and turbulent exchange in a neutral atmosphere. J Geophys Res. 67:3095–3102

Chapman S, Cowling TG (1964) The mathematical theory of non-uniform gases: an account of the kinetic theory of viscosity, thermal conduction, and diffusion in gases. Cambridge University Press, U.K. 431 pp

Chassefière E, Drossart P, Korablev O (1995) Post-Phobos model for the altitude and size distribution of dust in the low Martian atmosphere. J Geophys Res 100:5525–5539

Garratt JR (1994) The atmospheric boundary layer. Cambridge University Press, Cambridge, 316 pp

Greeley R, Iversen JD (1985) Wind as a geological process. No. 4 in Cambridge Planetary Science Series. Cambridge Univ. Press, New York, NY, 33 pp

Greeley R, Lacchia M, White B, Leach R, Trilling D, Pollack J (1994) Dust on Mars: new values for wind threshold. In XXV Lunar Planetary science Conference, pp 467–468

Hong CS, Lee KH, Kim YJ, Iwasaka Y (2004) LIDAR measurements of the vertical aerosol profile and optical depth during the ACE-Asia 2001 IOP at Gosan, Jeju Island, Korea. Environ Monitor Assess 92:43–57

Korablev O, Moroz VI, Petrova EV, Rodin AV (2005) Optical properties of dust and the opacity of the Martian atmosphere. Adv Space Res 35:21–30

Larsen SE, Jørgensen HE, Landberg L, Tillman JE (2002) Aspects of the atmospheric surface layers on mars and earth. Boundary-Layer Meteorol 105:451–470

McKenna Neuman C (2003) Effects of temperature and humidity upon the entrainment of sedimentary particles by the wind. Boundary-Layer Meteorol 108:61–89

Michelangeli DV, Toon OB, Haberle RM, Pollack JB (1993) Numerical simulations of the formation and evolution of water ice clouds in the Martian atmosphere. Icarus 102:261–285

Moudden Y, McConnell JC (2005) A new model for multiscale modeling of the Martian atmosphere, GM3. J Geophys Res 110, E04001, doi:10.1029/2004JE002354

Murphy JR, Toon OB, Haberle RM, Pollack JB (1990) Numerical simulations of the decay of Martian dust storms. J Geophys Res. 95:14629–14648

Newman CE, Lewis SRP, Read L, Forget F (2002a) Modeling the Martian dust cycle, 1, Representations of dust transport processes. J Geophys Res 107:E12, 5123

Newman CE, Lewis SR, Read PL, Forget F (2002b) Modeling the Martian dust cycle, 2, Multiannual radiatively active dust transport simulations. J Geophys Res 107:E12, 5124

Ockert-Bell ME, Bell JF, III, Pollack JB, McKay CP, Forget F (1997) Absorption and scattering properties of the Martian dust in the solar wavelengths. J Geophys Res 102:9039–9050

Pankine AA, Ingersoll AP (2004) Interannual variability of Mars global dust storms: an example of self-organized criticality? Icarus 170:514–518

Pollack JB, Ockert-Bell ME, Shepard MK (1995) Viking Lander image analysis of Martian atmospheric dust. J Geophys Res 100:5235–5250

Pruppacher HR, Klett JD (1997) Microphysics of clouds and precipitation, Kluwer Academic Publishers, Dordrecht, 976 pp

Rennó NO, Nash AA, Lunine J, Murphy J (2000) Martian and terrestrial dust devils: test of a scaling theory using Pathfinder data. J Geophys Res 105:1859–1865

Savijarvi H (1999) A model study of the atmospheric boundary layer in the Mars Pathfinder lander conditions. Quart J Roy Meteorol Soc 125:483–493

Seinfeld JH, Pandis SN (1998) Atmospheric chemistry and physics: from air pollution to climate change. Wiley, New York, 1326 pp

Shao Y (2000) Physics and modelling of wind erosion. Kluwer Academic Publishers, Dordrecht, 408 pp

Shao Y, Lu H (2000) A simple expression for wind erosion threshold friction velocity. J Geophys Res 105:22437–22443

Stull RB (1988) An introduction to boundary layer meteorology. Kluwer Academic Publishers, Dordrecht, 666 pp

Taylor PA, Li P-Y, Wilson JD (2002) Lagrangian simulation of suspended particles in the neutrally stratified atmospheric boundary layer. J Geophys Res 10.1029/2001 JD002049

Toigo AD, Richardson MI, Ewald SP, Gierasch PJ (2003) Numerical simulation of Martian dust devils. J Geophys Res 108(E6):5047, doi:10.1029/2002 JE002002

Tomasko MG, Doose LR, Lemmon M, Smith PH, Wegryn E (1999) Properties of dust in the Martian atmosphere from the Imager on Mars Pathfinder. J Geophys Res 104:8987–9007

Washington R, Todd MC, Engelstaedter S, Mbainayel S, Mitchell F (2006) Dust and the low-level circulation over the Bodélé Depression, Chad: Observations from BoDEx 2005. J Geophys Res 111, D03201, doi:10.1029/2005JD006502

Xiao J, Taylor PA (2002) On equilibrium profiles of suspended particles. Boundary-layer Meteorol. 105:471–482

On the turbulent Prandtl number in the stable atmospheric boundary layer

Andrey A. Grachev · Edgar L Andreas · Christopher W. Fairall · Peter S. Guest · P. Ola G. Persson

Abstract This study focuses on the behaviour of the turbulent Prandtl number, Pr_t, in the stable atmospheric boundary layer (SBL) based on measurements made during the Surface Heat Budget of the Arctic Ocean experiment (SHEBA). It is found that Pr_t increases with increasing stability if Pr_t is plotted vs. gradient Richardson number, Ri; but at the same time, Pr_t decreases with increasing stability if Pr_t is plotted vs. flux Richardson number, Rf, or vs. $\zeta = z/L$. This paradoxical behaviour of the turbulent Prandtl number in the SBL derives from the fact that plots of Pr_t vs. Ri (as well as vs. Rf and ζ) for individual 1-h observations and conventional bin-averaged values of the individual quantities have built-in correlation (or self-correlation) because of the shared variables. For independent estimates of how Pr_t behaves in very stable stratification, Pr_t is plotted against the bulk Richardson number; such plots have no built-in correlation. These plots based on the SHEBA data show that, on the average, Pr_t decreases with increasing stability and $\mathrm{Pr}_t < 1$ in the very stable case. For specific heights and stabilities, though, the turbulent Prandtl number has more complicated behaviour in the SBL.

A. A. Grachev · P. O. G. Persson
Cooperative Institute for Research in Environmental Sciences,
University of Colorado,
Boulder, CO, USA

A. A. Grachev (✉) · C. W. Fairall · P. O. G. Persson
NOAA Earth System Research Laboratory,
Boulder, CO, USA
e-mail: Andrey.Grachev@noaa.gov

E. L Andreas
NorthWest Research Associates, Inc. (Bellevue Division),
Lebanon, NH, USA

P. S. Guest
Naval Postgraduate School,
Monterey, CA, USA

Atmospheric Boundary Layers. A. Baklanov & B. Grisogono (eds.),
doi: 10.1007/978-0-387-74321-9_12,

Keywords Richardson number · SHEBA · Stable boundary layer · Turbulent Prandtl number

1 Introduction

One major uncertainty in the atmospheric stable boundary layer (SBL) is associated with the stability dependence of the turbulent Prandtl number defined by

$$\mathrm{Pr}_t = \frac{k_m}{k_h} = \frac{\langle u'w'\rangle \frac{\mathrm{d}\theta}{\mathrm{d}z}}{\langle w'T'\rangle \frac{\mathrm{d}U}{\mathrm{d}z}} \equiv \frac{\varphi_h}{\varphi_m}, \tag{1}$$

where $k_m = -\frac{\langle u'w'\rangle}{\mathrm{d}U/\mathrm{d}z}$ is the turbulent viscosity, and $k_h = -\frac{\langle w'T'\rangle}{\mathrm{d}\theta/\mathrm{d}z}$ is the turbulent thermal diffusivity ($-\langle u'w'\rangle$ is the downwind stress component and $\langle w'T'\rangle$ is the temperature flux). The turbulent Prandtl number (1) describes the difference in turbulent transfer between momentum and sensible heat. Turbulent momentum transfer is more efficient than turbulent heat transfer when $\mathrm{Pr}_t > 1$ and vice versa. If the turbulent Prandtl number is not unity, Eq. 1 also demonstrates a difference between the stability profile functions of momentum, φ_m, and sensible heat, φ_h. These functions are defined as non-dimensional vertical gradients of mean wind speed, U, and potential temperature, θ:

$$\varphi_m = \frac{\kappa z}{u_*}\frac{\mathrm{d}U}{\mathrm{d}z}, \tag{2a}$$

$$\varphi_h = \frac{\kappa z}{T_*}\frac{\mathrm{d}\theta}{\mathrm{d}z}, \tag{2b}$$

where $u_* = \sqrt{-\langle u'w'\rangle}$ is the friction velocity, $T_* = -\langle w'T'\rangle/u_*$ is the temperature scale, and κ is the von Kármán constant. The turbulent Prandtl number is an important characteristic of momentum and heat turbulent mixing for calibrating turbulence models and other applications (e.g., Sukoriansky et al. 2006).

In spite of progress in understanding SBL physics, a unified picture on the stability dependence of Pr_t does not exist. First, we survey the experimental results. Kondo et al. (1978), Ueda et al. (1981), Kim and Mahrt (1992), Ohya (2001), Strang and Fernando (2001), and Monti et al. (2002) found that the turbulent Prandtl number increases with increasing stability on plotting Pr_t (or $1/\mathrm{Pr}_t$) vs. the gradient Richardson number, Ri. In plots of Pr_t vs. the stability parameter $\zeta = z/L$ (L is the Obukhov length and z is the measurement height), on the other hand, Howell and Sun (1999) found that Pr_t estimates are generally scattered around unity and do not strongly depend on stability. However, for the very stable regime ($\zeta > 1$), their estimates of the turbulent Prandtl number tended to be less than unity (their Fig. 9). Yagüe et al. (2001) reported a mixed result. According to their Fig. 10a, Pr_t increases (or $1/\mathrm{Pr}_t$ decreases) with increasing Ri; but they found no clear dependence of Pr_t on ζ (their Fig. 10b). Grachev et al. (2003, 2007) found Pr_t to decrease with increasing ζ. This finding is directly related to the different behaviours of φ_m and φ_h, Eqs. 2a and 2b, in the limit of very strong stability. According to Grachev et al. (2005, 2007), φ_m increases with increasing stability, whereas φ_h initially increases with increasing ζ, reaches a maximum at $\zeta \approx 10$, and then tends to level off with further increasing ζ.

Theoretical studies by Zilitinkevich and Calanca (2000), Zilitinkevich (2002), and Sukoriansky et al. (2006) and Beljaars and Holtslag's (1991) parameterization argue in favour of an increasing turbulent Prandtl number with increasing stability. Likewise, when Andreas (2002) reviewed seven different formulations for $\varphi_m - \varphi_h$ pairs in stable stratification, he found five predicted $\mathrm{Pr}_t = 1$ in very stable stratification and two predicted that Pr_t increased with increasing ζ. None predicted that Pr_t decreased. The result $\mathrm{Pr}_t > 1$ is usually associated with the presence of the internal gravity waves in the SBL. They are presumed to enhance the momentum transfer through pressure terms in the Navier–Stokes equations, whereas gravity waves do not affect the sensible heat flux (e.g., Monin and Yaglom 1971). However, the result $\mathrm{Pr}_t < 1$ was obtained in recent large-eddy simulation (LES) studies of the SBL. Beare et al. (2006) plotted the turbulent viscosity and the turbulent thermal diffusivity as a function of z. According to their Fig. 14, k_m is clearly less than k_h, corresponding to $\mathrm{Pr}_t < 1$ (note different scales on the horizontal axes for k_m and k_h in their Fig. 14). Also Basu and Porté-Agel (2006) reported $\mathrm{Pr}_t < 1$ from their LES simulations (p. 2082).

The purpose of this study is to examine how the turbulent Prandtl number depends on different stability parameters to shed light on the behaviour of Pr_t in the SBL.

2 Turbulent Prandtl number vs. different stability parameters

One may notice that authors who found the turbulent Prandtl number to increase with stability, i.e., $\mathrm{Pr}_t > 1$, (Kondo et al. 1978; Ueda et al. 1981; Kim and Mahrt 1992; Ohya 2001; Strang and Fernando 2001; Yagüe et al. 2001; Monti et al. 2002) plotted Pr_t (or $1/\mathrm{Pr}_t$) solely vs. the gradient Richardson number, defined by

$$\mathrm{Ri} = \frac{g}{\theta_v} \frac{\mathrm{d}\theta_v/\mathrm{d}z}{(\mathrm{d}U/\mathrm{d}z)^2} = \frac{\zeta \varphi_h}{\varphi_m^2}, \tag{3}$$

where θ_v is the virtual potential temperature of air and g is the acceleration due to gravity. On the other hand, those who reported $\mathrm{Pr}_t < 1$ (Howell and Sun 1999; Grachev et al. 2003, 2007) plotted Pr_t against the Monin-Obukhov stability parameter ζ defined by

$$\zeta = \frac{z}{L} = -\frac{z\,\kappa g \langle w' T_v' \rangle}{u_*^3\,\theta_v}. \tag{4}$$

In this context, the result obtained by Yagüe et al. (2001) is remarkable. As mentioned above, they found that Pr_t increased with increasing stability if Eq. 1 was plotted against Eq. 3; but the same data showed no clear stability dependence in Pr_t when Eq. 1 was plotted against Eq. 4. In the light of this result, it makes sense to use the same dataset to explore in detail how Pr_t depends on different indicators of stability.

Another widely used stability parameter, along with Eqs. 3 and 4, is the flux Richardson number defined by

$$\mathrm{Rf} = \frac{g}{\theta} \frac{\langle w' T_v' \rangle}{\text{[illegible]}\,(\mathrm{d}U/\mathrm{d}z)} \equiv \frac{\zeta}{\varphi_m}. \tag{5}$$

Near the surface, when surface temperature is available, it is convenient also to use a bulk Richardson number:

$$\mathrm{Ri}_B = -\frac{gz}{\theta_v}\frac{(\Delta\theta + 0.61\theta_v\Delta q)}{U^2}, \tag{6}$$

where $\Delta\theta$ and Δq are differences in the potential temperature and the specific humidity, respectively, between the *surface* and reference level z.

In this study, measurements of atmospheric turbulence made during the Surface Heat Budget of the Arctic Ocean experiment (SHEBA) are used to examine the turbulent Prandtl number, Eq. 1, as a function of four different stability parameters, given by Eqs. 3–6. Turbulent fluxes and mean meteorological data were continuously measured at five levels, nominally 2.2, 3.2, 5.1, 8.9 and 18.2 m (or 14 m during most of the winter), on the 20-m main SHEBA tower. This stood on sea ice in the Beaufort Gyre from October 1997 through September 1998 and yielded 11 months of data.

Each level on the main tower had a Väisälä HMP-235 temperature and relative humidity probe and an Applied Technologies, Inc. (ATI) three-axis sonic anemometer/thermometer, which sampled at 10 Hz. The surface temperature (necessary for computing Ri_B) was obtained from down-looking and up-looking Eppley broadband, hemispherical radiometers (model PIR). Turbulent covariance values and appropriate variances at each level are based on 1-h averaging and derived through frequency integration of the cospectra and spectra. Observations with a temperature difference between the air and the snow surface less than 0.5°C and wind speed smaller than $1\mathrm{ms}^{-1}$ have been excluded from our analysis to avoid the large uncertainty in determining the turbulent fluxes. Other relevant information on flux and profile measurements and calculations including quality-control criteria, can be found in Persson et al. (2002), Grachev et al. (2005, 2007), and Andreas et al. (2006).

Figure 1 shows the turbulent Prandtl number as a function of the stability parameters (3)–(5) for the same SHEBA dataset. The averaged points in Fig. 1 based on the conventional bin-averaging of the individual 1-h data for Pr_t and proper stability parameters are indicated by different symbols for each measurement level. The individual 1-h-averaged data based on the median fluxes and other medians (heights, temperatures, etc.) for the five levels are also shown in Fig. 1 as background x-symbols. These points give an estimate of the available data at all levels and the typical scatter of the data. In the case when data at all five levels are available, the medians represent the level 3 data.

The results presented in Fig. 1, though, are contradictory. The turbulent Prandtl number increases with increasing stability if Pr_t is plotted vs. Ri (Fig. 1a); but at the same time, Pr_t decreases with increasing stability if Pr_t is plotted vs. Rf (Fig. 1b) or vs. ζ (Fig. 1c).

Vertical gradients of the mean wind speed, potential temperature and specific humidity appeared in Pr_t, Ri, and Rf (Eqs. 1, 3, and 5) in Fig. 1 and were obtained by fitting a second-order polynomial through the 1-h profiles followed by evaluating the derivative with respect to z for levels 1–5 (Grachev et al. 2005, their Eq. 8). In Fig. 2, the vertical gradients at levels $n = 2$–4 that appear in Pr_t, Ri, and Rf are based on linear interpolations of mean wind speed and potential temperature derived from the two adjoining levels $n - 1$ and $n + 1$. Figure 2 confirms the results in Fig. 1 and, therefore, shows that the results in Fig. 1 are not sensitive to how we evaluated the wind speed and temperature gradients.

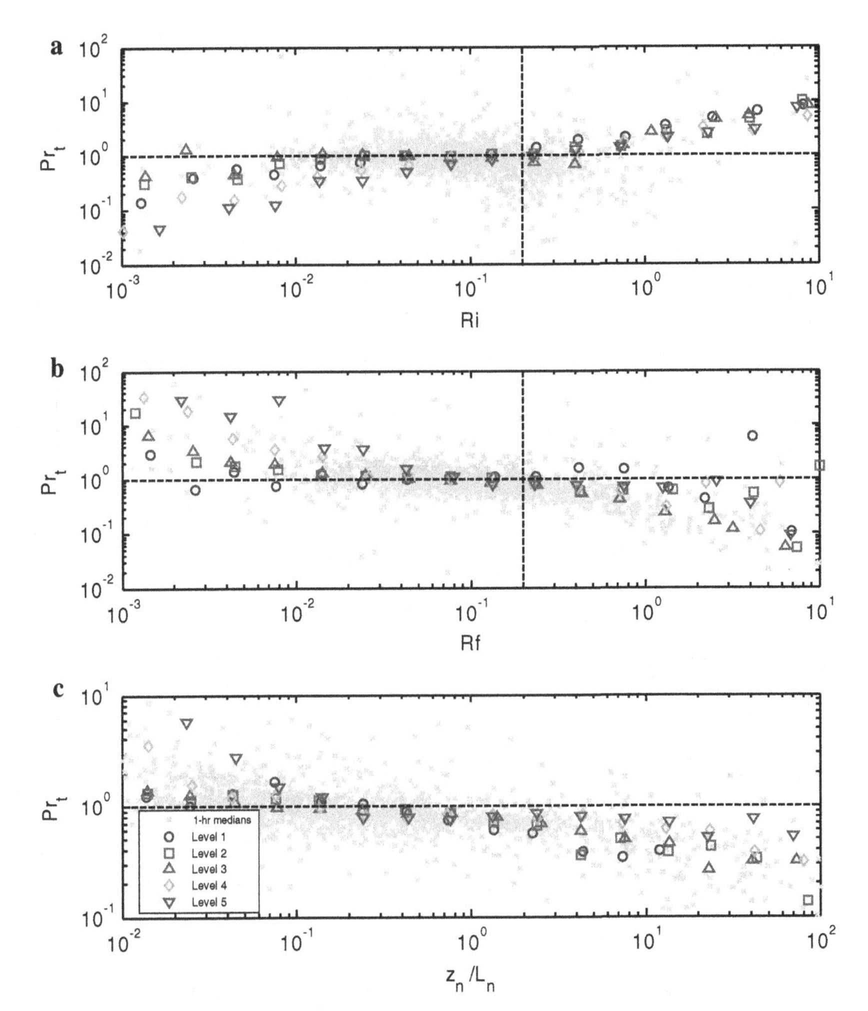

Fig. 1 Plots of the bin-averaged turbulent Prandtl number (bin medians) as functions of (**a**) Ri, (**b**) Rf, and (**c**) z_n/L_n (bin means) during the 11 months of the SHEBA measurements. The vertical dashed lines correspond to the critical Richardson number 0.2. Individual 1-h averaged data based on the median fluxes for the five levels are shown as background crosses

Thus we have a contradictory picture. On the one hand, according to Figs. 1a and 2a, the SHEBA data suggest that $\mathrm{Pr}_t > 1$ in the SBL, and this result agrees with findings reported by Kondo et al. (1978), Ueda et al. (1981), Kim and Mahrt (1992), Ohya (2001), Strang and Fernando (2001), and Monti et al. (2002). On the other hand, the SHEBA data also support the opposite opinion, $\mathrm{Pr}_t < 1$ (Figs. 1b, c, 2b, c).

3 Self-correlation

The above contradiction is likely associated with self-correlation. The problem is that the two quantities–for example, φ_m and ζ or Pr_t and Ri—between which a functional

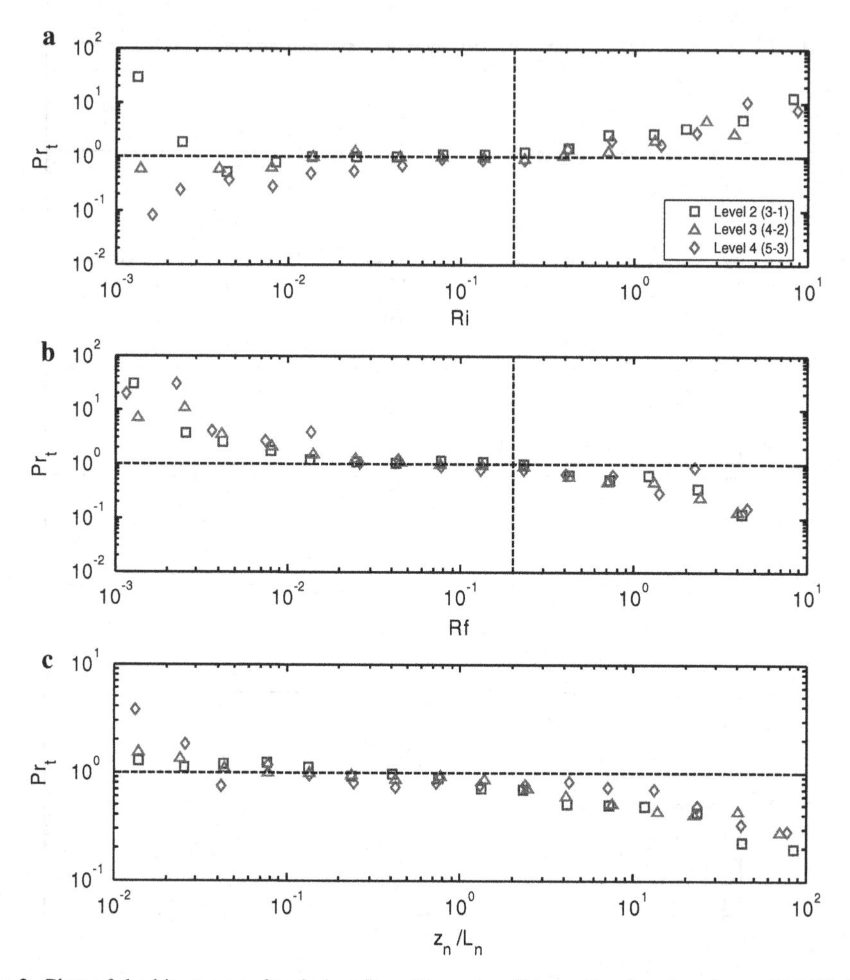

Fig. 2 Plots of the bin-averaged turbulent Prandtl number (bin medians) as functions of (**a**) Ri, (**b**) Rf, and (**c**) z_n/L_n (bin means) during the 11 months of the SHEBA measurements for level $n = 2$–4. Vertical gradients that appear in Pr_t, Ri, and Rf are based on the linear interpolation of mean wind speed and potential temperature derived from the two adjoining levels $n-1$ and $n+1$. For example, the gradients at level 2 are based on the temperature and wind speed differences between levels 3 and 1

relationship is sought have built-in correlation because of their shared variables (e.g., Hicks 1978; Mahrt et al. 1998; Andreas and Hicks 2002; Klipp and Mahrt 2004; Lange et al. 2004; Baas et al. 2006 and references therein). Self-correlation is also referred to as artificial, fictitious or spurious correlation. As an illustration, note that the turbulent Prandtl number is correlated to the gradient and flux Richardson numbers because $\mathrm{Pr}_t = \mathrm{Ri}/\mathrm{Rf}$ (see Eqs. 1, 3, and 5); that is, Pr_t varies proportionally with Ri but inversely with Rf (cf. Fig. 1a, b).

Usually, the self-correlation problem is discussed for plots of φ_m, φ_h, and variances vs. ζ primarily because of the shared friction velocity (e.g., Andreas and Hicks 2002; Baas et al. 2006). But self-correlation arises in several other atmospheric surface-layer applications (e.g., Andreas et al. 2006). In our particular application, self-correlation occurs in plots of Pr_t vs. Ri (Figs. 1a, 2a) because of the shared vertical gradients of mean wind speed and potential temperature. Furthermore, random variability in both shared variables leads to increasing Pr_t with increasing Ri. According to Eqs. 1 and 3, increasing $d\theta/dz$ leads to increases in both Pr_t and Ri; and increasing dU/dz leads to decreases in both Pr_t and Ri. Therefore, the self-correlation associated with variations in both $d\theta/dz$ and dU/dz leads to the tendency for Pr_t to increase with increasing Ri, as demonstrated in Figs. 1a and 2a.

Similarly, in plots of Pr_t vs. ζ, random variability in the shared variables u_* and $\langle w'T'\rangle$ leads to decreasing Pr_t with increasing ζ, and vice versa (see Eqs. 1 and 3). Plots of Pr_t vs. Rf have three shared variables: $\langle u'w'\rangle$, $\langle w'T'\rangle$, and dU/dz (see Eqs. 1 and 5). Random variability in two of them, $\langle u'w'\rangle$ and $\langle w'T'\rangle$, leads to decreasing Pr_t with increasing Rf, and vice versa; while variability in dU/dz gives the opposite result: that is, increasing Pr_t with increasing Rf, and vice versa.

However, not all self-correlations are serious. The degree of self-correlation is related to the variation in the shared variables compared to those of the other (non-shared) variables, and it is described by the coefficient of variation $V_x = \sigma_x/\bar{X}$ (e.g., Klipp and Mahrt 2004), where the standard deviation, σ_x, and the mean value, $\bar{X}$, are the statistics for the whole dataset. According to the SHEBA data, a typical coefficient V_x (computed for median values) may be as much as $V_x \approx 1.2$ for $\langle u'w'\rangle$, $V_x \approx 0.8$ for $\langle w'T'\rangle$, $V_x \approx 0.3$ for dU/dz, and $V_x \approx 1.1$ for $d\theta/dz$. Thus, more serious self-correlation for SHEBA data is associated with variations in both $\langle u'w'\rangle$ and $d\theta/dz$.

Another sign of the self-correlation in Figs. 1 and 2 is associated with the behaviour of Pr_t for weakly stable conditions. According to Figs. 1 and 2, Pr_t decreases as $\mathrm{Ri} \to 0$ and $\mathrm{Pr}_t \approx 0.1$ at $\mathrm{Ri} \approx 0.001$ (Figs. 1a and 2a); while Pr_t increases as $\mathrm{Rf} \to 0$ and $\mathrm{Pr}_t \approx 10$ at $\mathrm{Rf} \approx 0.001$ (Figs. 1b and 2b). Yagüe et al. (2001, their Fig. 10a) found a similar discrepancy for small Ri in plots of $1/\mathrm{Pr}_t$ vs. Ri. Obviously, this experimental result contradicts the canonical limit that $\mathrm{Pr}_t \approx 1$ for neutral conditions. Thus, self-correlation severely influences functional dependencies between Pr_t and different stability parameters in Figs. 1 and 2.

To obtain a more reliable and independent picture of how the turbulent Prandtl number behaves over a wide range of stable conditions, we plot Pr_t vs. Ri_B, Eq. 6, in Fig. 3. Obviously, these plots have no built-in correlation. Vertical gradients that appear in Pr_t in Fig. 3 are based on fitting a second-order polynomial through the 1-h profiles similarly to Fig. 1. The bulk Richardson number in the upper panel (Fig. 3a) is based on the wind speed at reference level z_n ($n = 1 - 5$) and differences in the potential temperature and the specific humidity between the surface and the level z_n.

Whereas, the bulk Richardson number in the bottom panel (Fig. 3b) is based on the wind speed at median level z_m and differences between the surface and the level z_m, i.e., Pr_t at a level z_n ($n = 1-5$) is plotted vs. the bulk Richardson number with fixed z. Thus plotting Pr_t vs. $\mathrm{Ri}_{B\,m}$ in Fig. 3b provides information on the height dependence of the turbulent Prandtl number. According to the SHEBA data in Fig. 3, the Pr_t data are scattered around 1 for weakly stable conditions (around 0.01). The greater scatter of points in Fig. 3 for $\mathrm{Ri}_B < 0.01$ results from the relatively small sensible heat flux and unreliable temperature gradient measurements in near-neutral conditions.

With increasing stability, Pr_t decreases *on the average*, although Pr_t at different levels behaves variously. Figure 3b shows that in the range $0.01 < \mathrm{Ri}_{B\,m} < 0.03$, the turbulent Prandtl number decreases with increasing height for fixed $\mathrm{Ri}_{B\,m}$, i.e., $d\,\mathrm{Pr}_t\,/dz < 0$. Furthermore Pr_t at two lower levels is above 1: $\mathrm{Pr}_t(z_5) < \mathrm{Pr}_t(z_4) < \mathrm{Pr}_t(z_3) \approx 1 < \mathrm{Pr}_t(z_2) < \mathrm{Pr}_t(z_1)$. However, with further increasing stability, Pr_t at all levels tends to

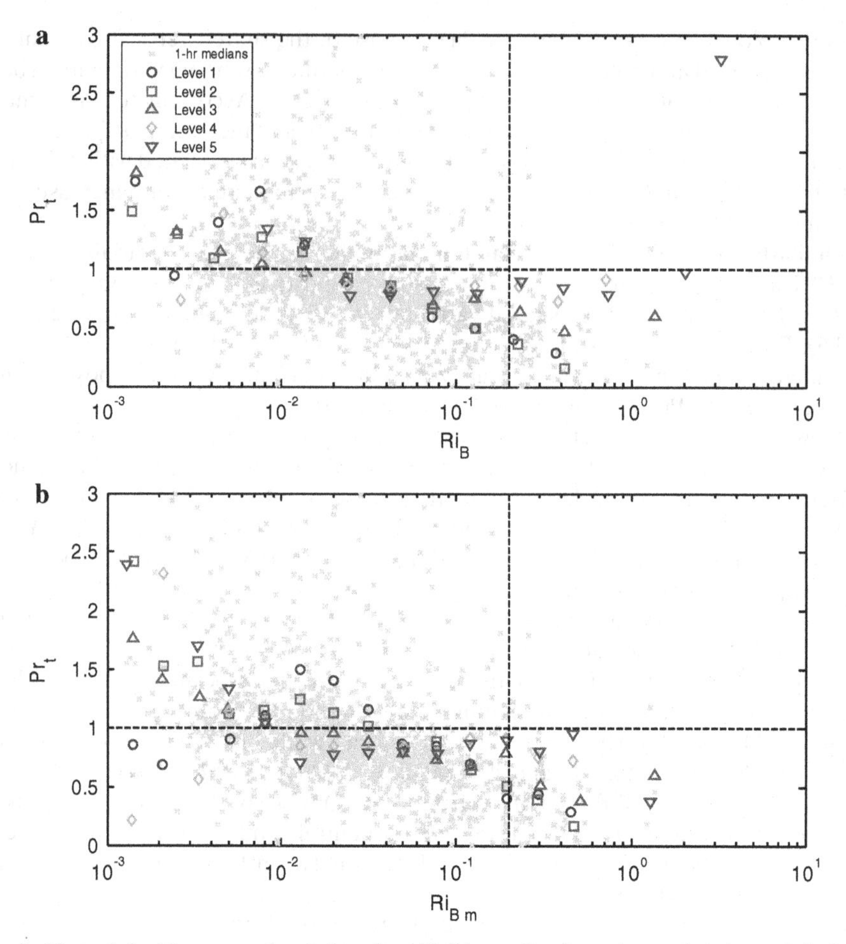

Fig. 3 Plots of the bin-averaged turbulent Prandtl (bin medians) number as functions of the bulk Richardson number (bin means) which is based on the (**a**) differences between the surface and reference level z_n (Ri_B) and (**b**) differences between the surface and median level z_m ($\mathrm{Ri}_{B\,m}$) during the 11 months of the SHEBA measurements. The vertical dashed lines correspond to the critical Richardson number, $\mathrm{Ri}_B = 0.2$. Individual 1-h averaged data based on the median fluxes for the five levels are shown as background crosses

be less than 1 in the subcritical regime, $\mathrm{Ri}_B < 0.2$ (see more discussion in the next Section). Thus Fig. 3 supports the conclusion that Pr_t on the average decreases with increasing stability for the SHEBA data. It seems that the trend in the Pr_t–Ri data in the SHEBA set (Figs. 1a and 2a) results strictly from self-correlation.

We suspect that self-correlation also influenced the conclusion, reported by others, that Pr_t increases with increasing Ri. Another notable example of self-correlation is the suggestion that the von Kármán constant depends on the roughness Reynolds number. Andreas et al. (2006) found recently that artificial correlation seems to explain the tendency for the von Kármán constant to decrease with increasing roughness Reynolds number in the atmospheric surface layer (i.e., Frenzen and Vogel 1995a, b; Oncley et al. 1996). According to Andreas et al. (2006) the von Kármán constant is, indeed, constant at 0.38–0.39.

In any event, analyzing self-correlation should be central for estimating how Pr_t behaves in the stable boundary layer because built-in correlation is unavoidable in the relations between Pr_t and the stability parameters Ri, Rf, and ζ. Ultimately, we must separate the effects of self-correlation and the physics on the dependency between Pr_t and the stability parameters.

4 Case Study

Although plots of averaged turbulent Prandtl number vs. different stability parameters are useful for qualitative analyses, additional detailed information can be obtained from time series of the turbulent Prandtl number and other relevant variables plotted for different conditions. Note that such plots by definition contain no built-in correlation. Typical time series of hourly averaged Pr_t, z_n/L_n, and Ri_B for moderately and very stable conditions during the dark period at SHEBA are shown in Figs. 4 and 5, respectively. Note that for data presented in Fig. 4, wind speed, wind direction, air temperature, and turbulent fluxes at each level (not shown) are approximately constant during 1997 YD (Year Day) 344.1–345.3 which lasts longer than 1 day. Therefore, z_n/L_n, and Ri_B (Fig. 4b, c) are also approximately constant for this period. At high latitudes, especially during the polar night, stable conditions are long lasting and can reach quasi-stationary states (e.g., Fig. 4) compared to measurements in the traditional nocturnal boundary layer in mid-latitudes. Such long-lived SBLs eventually can reach very stable states (e.g., Fig. 5).

Three-day's evolution of the turbulent Prandtl number, z_n/L_n, and the bulk Richardson for moderately stable conditions is shown in Fig. 4. It is particularly remarkable that the turbulent Prandtl number decreases with increasing height (Fig. 4a). In addition, Pr_t at levels 4 and 5 is systematically less than 1, whereas Pr_t at the two lower levels tends to be above 1 (Pr_t at level 3 is basically scattered around 1). Similar behaviour of Pr_t during SHEBA also has been observed at other times, for example, during 1998, YD 13–15, 52–53, 65–67, 181.5–182.5, 201–203.

Figure 4a supports the result $d\,\mathrm{Pr}_t/dz < 0$ for $0.01 < \mathrm{Ri}_{B\,m} < 0.03$ presented in Fig. 3b. This finding is also in good agreement with Howell and Sun (1999) measurements and LES simulations by Basu and Porté-Agel (2006). Howell and Sun (1999) found on average that Pr_t estimates at the 3-m level are higher than at the 10-m level. According to Basu and Porté-Agel's (2006) study, $\mathrm{Pr}_t \approx 0.7$ inside the boundary layer (up to 150 m), but values of Pr_t increase to ~ 1 in the surface layer (Ibid. p. 2082). The result $d\,\mathrm{Pr}_t/dz < 0$ indicates that, for the stability range $0.01 < \mathrm{Ri}_{B\,m} < 0.03$, turbulent momentum transfer is relatively more efficient near the surface. It should be mentioned that measurements at two lower levels may be influenced by a surface flux footprint effect or a blowing snow effect (see Fig. 6 and relevant discussion in Grachev et al. 2007).

Figure 5 shows a typical 1-day time series of Pr_t, z_n/L_n, and Ri_B for the very stable conditions observed during December 27–28, 1997 (YD 361–363). The data are based on 1-h averaging. Time series of the basic meteorological variables and turbulent fluxes for YD 361.8–363 (Fig. 5) can be found in Grachev et al. (2003, their Fig. 1). According to Fig. 5a, the turbulent Prandtl number for the very stable conditions tends mainly to be less than 1 at all levels. Similar time series of Pr_t for the very stable conditions observed during SHEBA also can be found on 1998 YD 56, 64–65, and

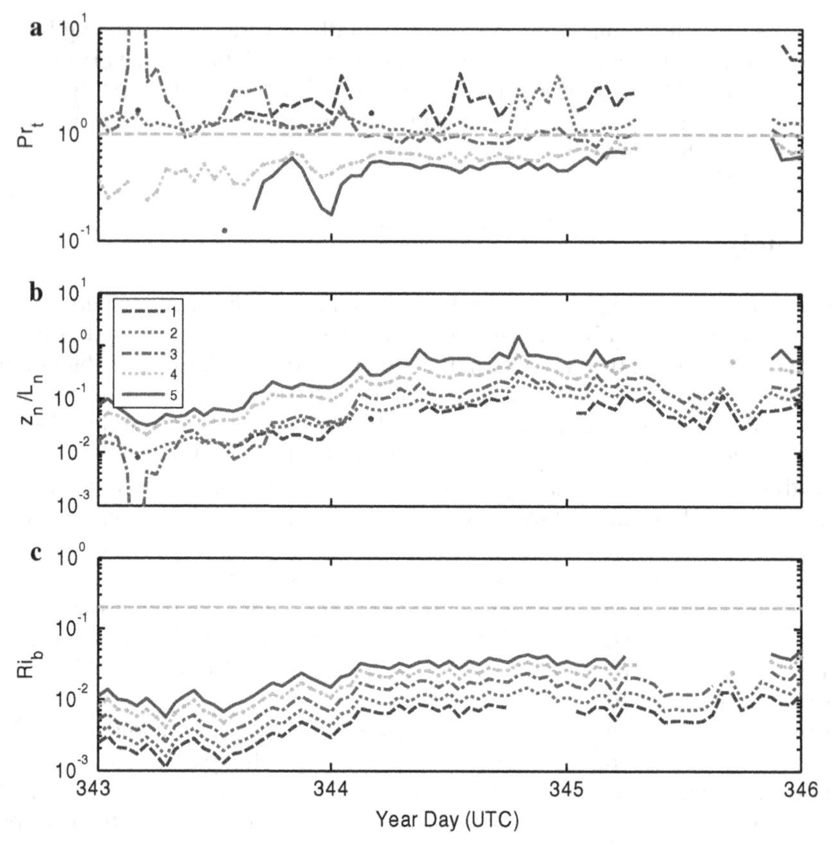

Fig. 4 Time series of the (**a**) the turbulent Prandtl number, (**b**) z_n/L_n, and (**c**) the bulk Richardson number measured at the five levels during moderately stable conditions, 1997 year days 343–346 (December 9–12, 1997 UTC). The data are based on 1-h averaging. The horizontal dashed line in the bottom panel correspond, to the critical Richardson number, $\mathrm{Ri}_B = 0.2$

142–143 among others [see also Fig. 2 in Grachev et al (2003) for YD 142–143 time series].

Although *on average* the turbulent Prandtl number decreases with increasing stability and $\mathrm{Pr}_t < 1$ in the very stable case (Fig. 3a), our study does not find that $\mathrm{Pr}_t < 1$ is a general result for the SBL. One may speculate that Pr_t generally does not have a universal behaviour in the stable atmospheric boundary layer in the framework of the Monin–Obukhov similarity. As mentioned above, Pr_t describes the difference in turbulent transfers of momentum and sensible heat. Similarity in the turbulent mixing of momentum and heat suggests that $\mathrm{Pr}_t \approx 1$. However, physical processes overlooked in Monin–Obukhov similarity theory (e.g., internal gravity waves, Kelvin–Helmholtz billows, an upside-down boundary layer, radiative flux divergence, etc.) may increase only the momentum flux ($\mathrm{Pr}_t > 1$), only the heat flux ($\mathrm{Pr}_t < 1$), or may produce a mixed effect and therefore violate similarity.

Eleven months of multi-level measurements during SHEBA cover a wide range of stability conditions and can shed light on the discrepancy of Pr_t measurements in the literature (see Sect. 1). In addition to the self-correlation problem discussed

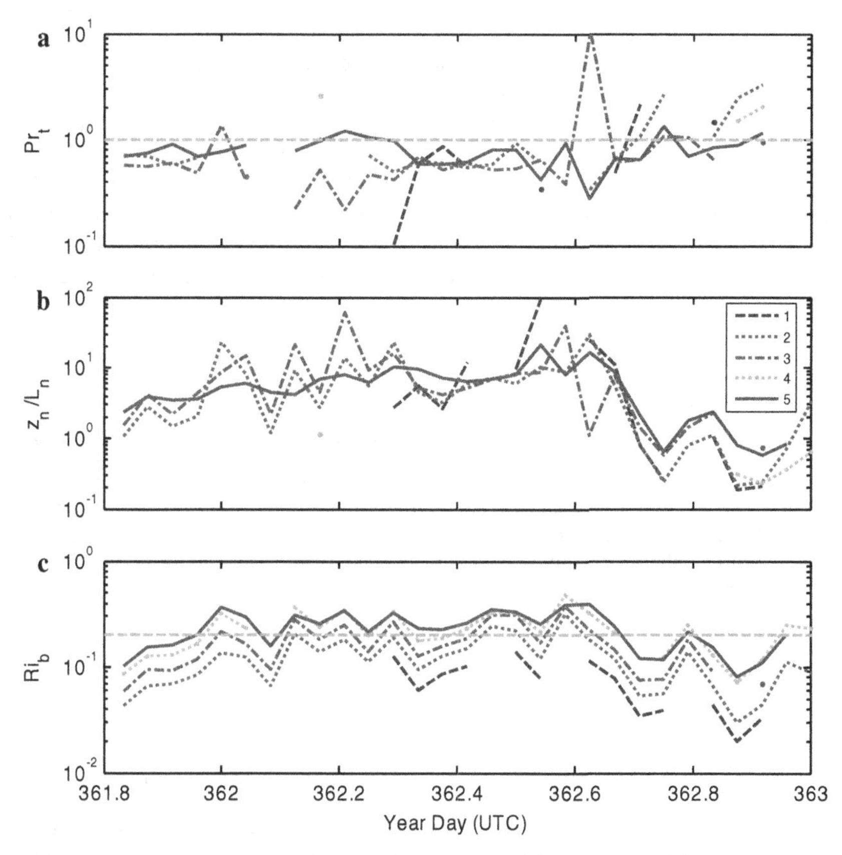

Fig. 5 Same as Fig. 4 but for data obtained during very stable conditions, 1997 year days 361–362 (December 27–28, 1997 UTC)

in Sect. 3, use of the limited datasets may be responsible for the discrepancy. For example, measurements only at heights less than 4 m (levels 1 and 2) for the limited stability range $Ri_{B\,m} < 0.03$ result in $Pr_t > 1$ (see Figs. 3b and 4). At the same time, Pr_t is systematically less than 1 for measurements at heights higher than 4 m (levels 3–5). As mentioned earlier, the *whole* SHEBA dataset *on the average* suggests that the turbulent Prandtl number decreases with increasing stability and $Pr_t < 1$ in the very stable regime for all levels.

5 Conclusions

The turbulent Prandtl number in the SBL is discussed based on measurements made during SHEBA. Plots of Pr_t vs. Ri (as well as vs. Rf and ζ) for individual 1-h observations and relevant conventional bin-averaged values of the individual quantities suffer severely from self-correlation because of the shared variables. As a result, such analyses conceal any real physical correlation. For example, plots of Pr_t vs. different stability parameters for the same dataset give conflicting dependencies (Figs. 1 and 2). The turbulent Prandtl number increases with increasing stability if Pr_t is plotted vs. Ri; but at the same time, Pr_t decreases with increasing stability if Pr_t is plotted vs. Rf or vs. $\zeta = z/L$. In addition, the data fail to agree with the canonical value $\mathrm{Pr}_t \approx 1$ for weakly stable conditions (Figs. 1a, b, 2a, b).

In contrast, plots of Pr_t vs. the bulk Richardson number, which have no built-in correlation, show that *on the average*, at least for the SHEBA data, Pr_t decreases with increasing stability and $\mathrm{Pr}_t < 1$ for all levels in the very stable cases (Fig. 3). However, the turbulent Prandtl number has more intricate behaviour for specific stability ranges and heights (Figs. 4 and 5). It is conceivable that the turbulent Prandtl number does not have a universal behaviour and $\mathrm{Pr}_t < 1$ is not a general result in the stable atmospheric boundary layer.

Acknowledgements The U.S. National Science Foundation supported this work with awards to the NOAA Environmental Technology Laboratory (now Earth System Research Laboratory) (OPP-97-01766), the Cooperative Institute for Research in Environmental Sciences (CIRES), University of Colorado (OPP-00-84322, OPP-00-84323), the U.S. Army Cold Regions Research and Engineering Laboratory (OPP-97-02025, OPP-00-84190), and the Naval Postgraduate School (OPP-97-01390, OPP-00-84279). The U.S. Department of the Army also supported ELA through Project 611102T2400. Thanks to all who participated in conversations which we held during the NATO Advanced Research Workshop (ARW) in Dubrovnik, Croatia, 18–22 April 2006, about the stability dependence of the turbulent Prandtl number in the atmospheric stable boundary layer. Special thanks go to Igor Esau for initiating that discussion.

References

Andreas EL (2002) Parameterizing scalar transfer over snow and ice: a review. J Hydormeterol 3:417–432

Andreas EL, Hicks BB (2002) Comments on 'Critical test of the validity of Monin-Obukhov similarity during convective conditions'. J Atmos Sci 59:2605–2607

Andreas EL, Claffey KJ, Jordan RE, Fairall CW, Guest PS, Persson POG, Grachev AA (2006) Evaluations of the von Kármán constant in the atmospheric surface layer. J Fluid Mech 559:117–149

Baas P, Steeneveld GJ, van de Wiel BJH, Holtslag AAM (2006) Exploring self-correlation in flux-gradient relationships for stably stratified conditions. J Atmos Sci 63(11):3045–3054

Basu S, Porté-Agel F (2006) Large-eddy simulation of stably stratified atmospheric boundary layer turbulence: a scale-dependent dynamic modeling approach. J Atmos Sci 63(8):2074–2091

Beare RJ, MacVean MK, Holtslag AAM, Cuxart J, Esau I, Golaz J-C, Jimenez MA, Khairoutdinov M, Kosovic B, Lewellen D, Lund TS, Lundquist JK, McCabe A, Moene AF, Noh Y, Raasch S, Sullivan P (2006) An intercomparison of large-eddy simulations of the stable boundary layer. Boundary-Layer Meteorol 118(2):247–272

Beljaars ACM, Holtslag AAM (1991) Flux parameterization over land surfaces for atmospheric models. J Appl Meteorol 30:327–341

Frenzen P, Vogel CA (1995a) On the magnitude and apparent range of variation of the von Kármán constant in the atmospheric surface layer. Boundary-Layer Meteorol 72:371–392

Frenzen P, Vogel CA (1995b) A further note 'On the magnitude and apparent range of variation of the von Kármán constant'. Boundary-Layer Meteorol 73:315–317

Grachev AA, Fairall CW, Persson, POG, Andreas EL, Guest PS, Jordan RE (2003) Turbulence decay in the stable arctic boundary layer. In: seventh conference on polar meteorology and oceanography and joint symposium on high-latitude climate variations. Amer. Meteorol. Soc., Hyannis, Massachusetts, Preprint CD-ROM

Grachev AA, Fairall CW, Persson POG, Andreas EL, Guest PS (2005) Stable boundary-layer scaling regimes: the SHEBA data. Boundary-Layer Meteorol 116(2):201–235

Grachev AA, Andreas EL, Fairall CW, Guest PS, Persson POG (2007) SHEBA flux-profile relationships in the stable atmospheric boundary layer. Boundary-Layer Meteorol (in press) DOI: 10.1007/s10546-007-9177-6

Hicks BB (1978) Comments on 'The characteristics of turbulent velocity components in the surface layer under convective conditions by H A Panofsky, et al. Boundary-Layer Meteorol 15(2):255–258

Howell JF, Sun J (1999) Surface-layer fluxes in stable conditions. Boundary-Layer Meteorol 90(3):495–520
Kim J, Mahrt L (1992) Simple formulation of turbulent mixing in the stable free atmosphere and nocturnal boundary layer. Tellus 44(5):381–394
Klipp CL, Mahrt L (2004) Flux-gradient relationship, self-correlation and intermittency in the stable boundary layer. Quart J Roy Meteorol Soc 130(601):2087–2103
Kondo J, Kanechika O, Yasuda N (1978) Heat and momentum transfers under strong stability in the atmospheric surface layer. J Atmos Sci 35:1012–1021
Lange B, Johnson HK, Larsen S, Højstrup J, Kofoed-Hansen H, Yelland MJ (2004) On detection of a wave age dependency for the sea surface roughness. J Phys Oceanogr 34(6):1441–1458
Mahrt L, Sun J, Blumen W, Delany T, Oncley S (1998) Nocturnal boundary-layer regimes. Boundary-Layer Meteorol 88:255–278
Monin AS, Yaglom AM (1971) Statistical fluid mechanics: mechanics of turbulence, vol. 1. MIT Press, Cambridge, MA pp 769
Monti P, Fernando HJS, Princevac M, Chen WC, Kowalewski TA, Pardyjak ER (2002) Observation of flow and turbulence in the nocturnal boundary layer over a slope. J Atmos Sci 59(17):2513–2534
Ohya Y (2001) Wind tunnel study of atmospheric stable boundary layers over a rough surface. Boundary-Layer Meteorol 98(1):57–82
Oncley SP, Friehe CA, Larue JC, Businger JA, Itswiere EC, Chang SS (1996) Surface-layer fluxes, profiles, and turbulence measurements over uniform terrain under near-neutral conditions. J Atmos Sci 53:1029–1044
Persson POG, Fairall CW, Andreas EL, Guest PS, Perovich DK (2002) Measurements near the atmospheric surface flux group tower at SHEBA: near-surface conditions and surface energy budget. J Geophys Res 107(C10):8045. DOI: 10.1029/2000JC000705
Strang EJ, Fernando HJS (2001) Vertical mixing and transports through a stratified shear layer. J Phys Oceanog 31(8):2026–2048
Sukoriansky S, Galperin B, Perov V (2006) A quasi-normal scale elimination model of turbulence and its application to stably stratified flows. Nonlinear Processes Geophys 13(1):9–22
Ueda H, Mitsumoto S, Komori S (1981) Buoyancy effects on the turbulent transport processes in the lower atmosphere. Quart J Roy Meteorol Soc 107(453):561–578
Yagüe C, Maqueda G, Rees JM (2001) Characteristics of turbulence in the lower atmosphere at Halley IV Station, Antarctica. Dyn Atmos Ocean 34:205–223
Zilitinkevich SS (2002) Third-order transport due to internal waves and non-local turbulence in the stably stratified surface layer. Quart J Roy Meteorol Soc 128:913–925
Zilitinkevich S, Calanca P (2000) An extended similarity-theory for the stably stratified atmospheric surface layer. Quart J Roy Meteorol Soc 126:1913–1923

Micrometeorological observations of a microburst in southern Finland

Leena Järvi · Ari-Juhani Punkka · David M. Schultz · Tuukka Petäjä · Harri Hohti · Janne Rinne · Toivo Pohja · Markku Kulmala · Pertti Hari · Timo Vesala

Abstract On the afternoon of 3 July 2004 in Hyytiälä (Juupajoki, Finland), convective cells produced a strong downburst causing forest damage. The SMEAR II field station, situated near the damage site, enabled a unique micrometeorological analysis of a microburst with differences above and inside the canopy. At the time of the event, a squall line associated with a cold front was crossing Hyytiälä with a reflectivity maximum in the middle of the squall line. A bow echo, rear-inflow notch, and probable mesovortex were observed in radar data. The bow echo moved west-north-west, and its apex travelled just north of Hyytiälä. The turbulence data were analysed at two locations above the forest canopy and at one location at sub-canopy. At 1412 EET (Eastern European Time, UTC+2), the horizontal and vertical wind speed increased and the wind veered, reflecting the arrival of a gust front. At the same time, the carbon dioxide concentration increased due to turbulent mixing, the temperature decreased due to cold air flow from aloft and aerosol particle concentration decreased due to rain scavenging. An increase in the number concentration of ultra-fine particles ($< 10\,\mathrm{nm}$) was detected, supporting the new particle formation either from cloud outflow or due to rain. Five minutes after the gust front (1417 EET), strong horizontal and downward vertical wind speed gusts occurred with maxima of 22 and $15\,\mathrm{m\,s^{-1}}$, respectively, reflecting the microburst. The turbulence spectra before, during and after the event were consistent with traditional turbulence spectral theory.

Keywords Aerosol particles · Convective storm · Microburst · Micrometeorology · Trace gases

---

L. Järvi (✉) · D. M. Schultz · T. Petäjä · J. Rinne · M. Kulmala · T. Vesala
Department of Physical Sciences, University of Helsinki, P.O. Box 64, Helsinki 00014, Finland
e-mail: leena.jarvi@helsinki.fi

A.-J. Punkka · D. M. Schultz · H. Hohti
Finnish Meteorological Institute, P.O. Box 503, Helsinki 00101, Finland

T. Pohja
Hyytiälä Forestry Field Station, University of Helsinki, Hyytiäläntie 124, Korkeakoski 35500, Finland

P. Hari
Department of Forest Ecology, University of Helsinki, P.O. Box 27, Helsinki 00014, Finland

Atmospheric Boundary Layers. A. Baklanov & B. Grisogono (eds.),
doi: 10.1007/978-0-387-74321-9_13,

1 Introduction

On the afternoon of 3 July 2004, an area of scattered thunderstorms occurred in southern Finland. One of the convective cells produced a small downburst, known as a microburst, that was detected at the SMEAR II *(Measuring Forest Ecosystem—Atmosphere Relations)* field station in Hyytiälä, where a large array of micrometeorological instrumentation is operated continuously (Vesala et al. 1998; Hari and Kulmala 2005). The main microburst damage track was around 10 m long and approximately 20 m wide, and trees fell mostly to the south-west, although areas outside the main damage track also experienced occasional forest damage. A microburst passing through a well-equipped observation site provides a unique opportunity to study high-resolution turbulence data, trace gas and total aerosol particle concentrations, and aerosol size spectra during an extreme weather event.

Severe thunderstorms mainly occur in Finland from June to August, and where lightning and heavy rain are sometimes accompanied by downbursts, and tornadoes are occasionally observed. The short lifetime and small size of microbursts and tornadoes make direct observations difficult. Because eyewitnesses are rare in sparsely populated areas of Finland, these phenomena are mostly detected only by their damage track on the earth's surface, supported by radar imagery (e.g. Punkka et al. 2006).

Although much previous research has been performed on severe convective storms and microbursts, we are not aware of any other study that has presented high frequency (>10 Hz) data from a microburst. To our knowledge, Sherman (1987) is the only known report of vertical velocity data from a microburst (at an instrumented tower in Brisbane, Australia), but no other micrometeorological analysis was made in that study. Our observations are novel in that we are able to link the meteorological observations to measurements of turbulence characteristics, trace gas (water vapour, carbon dioxide, and ozone) concentrations, and aerosol particle concentrations before, during, and after the microburst.

The aim of this study is to investigate the micrometeorological features of the 3 July 2004 microburst. Three-dimensional high-frequency wind measurements are analyzed, including comparisons between above-canopy and sub-canopy measurements. Trace gas concentrations, total aerosol particle concentrations and aerosol particle size spectra are shown, providing an example of their behaviour in extreme weather conditions. Turbulence spectra of horizontal and vertical wind speed components before, during and after the microburst event are studied, and dissipation rates and turbulent kinetic energy are calculated. Radar data are analyzed to investigate the larger mesoscale conditions associated with the microburst. These observations, along with the damage-track analysis, allowed us to classify the event as a microburst, instead of a tornado.

The measurement site, instrumentation and methodology applied in this study are introduced in Sect. 2. Section 3 presents the results and discussion, beginning with the synoptic and mesoscale overview in Sect. 3.1. Section 3.2 describes the local weather conditions on 3 July 2004 and Sect. 3.3 presents the micrometeorological analysis of the gust front and microburst and relates these observations to existing conceptual models of convective systems. Section 3.4 presents the trace gas and aerosol particle measurements and Sect. 3.5 presents the turbulence spectra, turbulent kinetic energy (TKE) and its dissipation calculations before, during and after the microburst event. Finally, Sect. 4 provides conclusions.

2 Measurements and methods

The microburst occurred in Hyytiälä, Juupajoki (southern Finland), 210 km north of Helsinki (Fig. 1a). The SMEAR II field station (61° 51′N, 24° 17′E, 181 m above sea level) is located close to the Hyytiälä Forestry Field Station (University of Helsinki). A large number of diverse meteorological measurements are collected at the SMEAR II station and more detailed description about the station and its measurements are presented in Vesala et al. (1998) and Hari and Kulmala (2005). The station is surrounded mainly by a 42-year-old Scots pine stand with an average tree height of 16 m in 2004, and the forest damage site is about 200 m south-west of the measuring station (Fig. 2). A standard SYNOP weather station (Finnish Meteorological Institute, Hyytiälä, WMO 05174) is located 450 m south-west from the SMEAR II station. The C-band weather radar used in the study (Finnish Meteorological Institute) is situated at Ikaalinen (61°46′N, 23°04′E, 154 m above sea level), 65 km west of Hyytiälä.

The SMEAR II station has three eddy-covariance systems recording turbulence data. Two systems above the canopy are at the same height of 23 m and are separated by a horizontal distance of 30 m. These are henceforth referred as locations A and B (Fig. 2). The third system is sub-canopy, below the foliage at 3 m, about 20 m south from location B. The above-canopy systems include 3-dimensional acoustic anemometers Solent 1012R2 and Solent HS1199 (Gill Instruments Ltd., Lymington UK) and the sub-canopy sonic anemometer is a Metek USA-1 (Metek GmbH, Germany). For all turbulence measurements, the measuring frequency is 10.4 Hz. The clocks of the two above-canopy instruments were synchronized but differences of some seconds may exist between these and the sub-canopy measurements.

At location A, wind speed and direction were measured at heights of 73.0 and 8.4 m with a 2-dimensional ultrasonic anemometer (Adolf Thies GmbH, Göttingen, Germany), whereas temperature was measured at heights of 67.2 m and 8.4 m with a Pt-100 resistance thermometer. Also at location A, pressure was measured at ground level with a digital pressure indicator (DPI260, Druck, Williston, VT), and dew-point temperature was measured at a height of 23 m (General Eastern Hygro E4, Billerica, MA, USA). At location B, rainfall was collected above the canopy at a height of 18 m with a rain collector (ARG-100). In addition, precipitation recorded at the local SYNOP weather station (Finnish Meteorological Institute, WMO 05174) was used.

Carbon dioxide and water vapour concentrations were measured with a fast-response gas analyser LI-6262 (Li-Cor Inc., Lincoln, NE, USA), and an ozone analyser TEI 49C (Thermo Electron Corporation, Waltham, MA, USA) and a condensation particle counter (CPC) (TSI 3010, Shoreview, MN, USA) were used to measure ozone and total particle concentrations, respectively. These measurements were made at location A at a height of 23 m. Aerosol particle size distribution was observed by means of twin differential mobility particle sizer (DMPS) systems (Aalto et al. 2001), and ion size distribution was measured with a balanced scanning mobility analyzer (BSMA, Airel Ltd., Estonia). The DMPS technique is based on the electrical classification of particles combined with the concentration measurement by the CPC. One system measured sizes from 3 to 10 nm (TSI 3025, Shoreview, MN, USA) and the other from 10 to 500 nm (TSI 3010, Shoreview, MN, USA). Both of the size distribution measurements were situated in a cabin north of location A (Fig. 2), with the air drawn into the cabin from an outside height of 2 m.

The 10-min average data are used for pressure, wind speed and temperature for the whole of 3 July 2004 to obtain information on the general weather conditions. A more detailed meteorological analysis was made for 1400–1430 EET (Eastern European Time, UTC + 2) when the event was observed. Mean or instantaneous values for wind speed, wind direction and vertical wind speed in all locations are used. For gas and total aerosol particle concentrations, 15-sec average values are analyzed, and 1-min average pressure at ground level, the 15-sec average temperature and the 1-min average dew point temperature are presented.

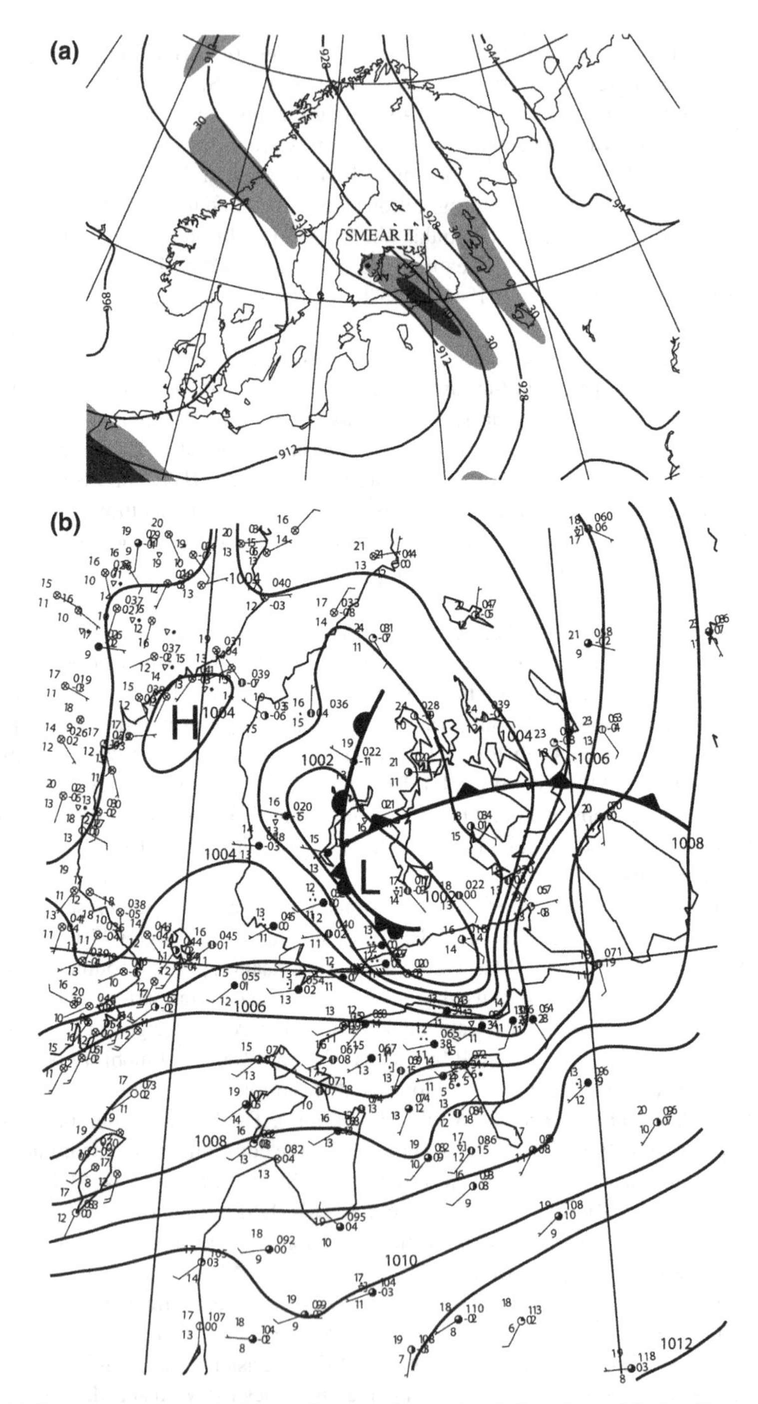

Fig. 1 (**a**) Constant pressure analysis from European Centre for Medium-Range Weather Forecasts model analysis at 300 hPa at 0200 EET 3 July 2004 (EET = UTC + 2): isohypses (solid lines every 80 dam) and isotachs (shaded every $10\,\mathrm{m\,s^{-1}}$ for speeds greater than $30\,\mathrm{m\,s^{-1}}$). (**b**) Surface SYNOP observation map at 1400 EET 3 July 2004: manually analyzed isobars every 1 hPa, with standard frontal analysis

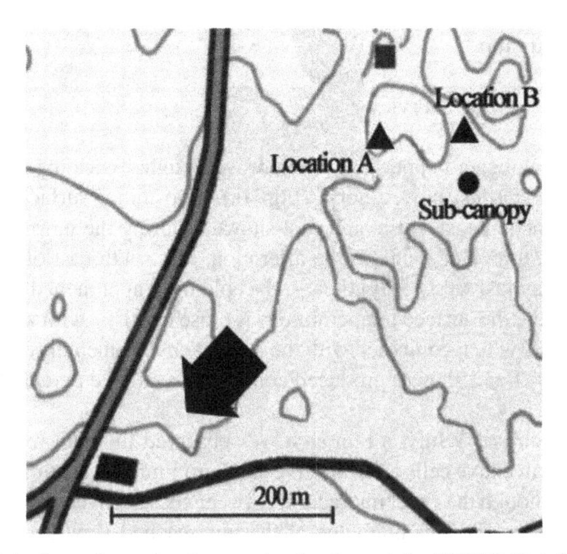

Fig. 2 Location of the forest damage and measuring locations at the SMEAR II station at Hyytiälä. The black arrow shows the location of the damage site and the direction of fallen trees. The triangle represents the above-canopy locations A and B and the circle stands for sub-canopy measurements. The square represents the instrument cabin, and the rectangle represents an un-instrumented barn. The grey lines represent schematic topography. Other lines represent roads

Finnish Meteorological Institute radar data, with SYNOP observation maps and European Centre for Medium-Range Weather Forecasts numerical model analyses, are used to analyze the synoptic and mesoscale environment of the microburst.

The atmospheric spectra were calculated using a fast Fourier transform for 5- to 10-min sections, after having been de-trended and Hamming windowed (Kaimal and Kristensen 1991). A two-dimensional coordinate rotation (i.e., u_{rot} is the wind component rotated to the direction of the mean wind and v_{rot} is the wind component directed perpendicular to the mean wind) was applied for the spectral calculations (Kaimal and Finnigan 1994). Turbulence dissipation rate ε was calculated according to the inertial dissipation technique

$$\varepsilon = \frac{2\pi}{\overline{U}} \sum_{j=1}^{n} \left[\frac{f_j^{5/3} S_{u_i}(f_j)}{\alpha_i} \right]^{3/2}, \tag{1}$$

$S_{ui}(f_j)$ is the frequency spectrum of wind component u_i at frequency f_j (n is the number of frequencies) and $\alpha_i = 0.53$ is the Kolmogorov constant. (Piper and Lundquist 2004). Turbulence dissipation was calculated only for u_{rot}. The turbulent kinetic energy, TKE, was calculated according to

$$\mathrm{TKE} = 0.5\left(\overline{u'^2} + \overline{v'^2} + \overline{w'^2}\right), \tag{2}$$

where u and v are horizontal wind components directed east and north, respectively, and w is the vertical wind component (e.g. Holton 1992). The overbar denotes a 15-sec time average and the prime refers to instantaneous turbulent fluctuations around the average.

3 Results and discussion

3.1 Synoptic and mesoscale overview

On 3 July 2004, an elongated upper-level short-wave trough, extending from southern Norway to the Baltic countries, moved north (Fig. 1a). Two minor surface depressions were situated over Estonia and east of Finland (not shown). During the morning, the depression over Estonia started to occlude and, in the afternoon, was south-east of Hyytiälä. At 1400 EET, a warm front lay just west of Hyytiälä, and a cold front approached the microburst area (Fig. 1b). At Hyytiälä, the surface temperature was close to 20°C, with a dew-point temperature of around 15°C, which combined with the nearest environmental sounding (Jyväskylä, 0800 EET 3 July 2004, not shown) produced convective available potential energy of about $500\,\mathrm{J\,kg^{-1}}$.

During the morning of 3 July, a rain area was observed in south-western Finland (not shown). A line of convective cells arrived at the southern Finnish coastline by 1200 EET and moved north-west, though the exact interaction between these convective storms and the cold front remains unknown. The leading edge of the line reached Hyytiälä around 1330 EET. Unlike other mesoscale convective systems (e.g., Parker and Johnson 2000), the maximum in radar reflectivity factor was located in the middle of the line (Fig. 3a–c). About 30 min before the microburst, the patterns in radar reflectivity factor showed the development of a bow echo, a rear-inflow jet and a rear-inflow notch. These radar signatures moved just north of Hyytiälä at 1415 EET (Fig. 3b), and are all considered important indicators of convective systems producing strong surface winds (e.g., Fujita 1979; Przybylinski and DeCaire 1985; Smull and Houze 1987; Przybylinski 1995; Wakimoto et al. 2006). Within a mesoscale convective system, a rear-inflow jet may advect dry mid-level air into the convective system, leading to evaporative cooling of precipitation, negative buoyancy, convective downdrafts, and the bowing of the convective line (e.g., Atkins and Wakimoto 1991). The rear-inflow notch, a channel of weak radar echoes pointing from the stratiform precipitation region toward the convective line, is the indication of the evaporation of the precipitation by the rear-inflow jet. Upon passing over Hyytiälä at 1412 EET, this convective line produced the microburst.

Recent numerical modelling (Weisman and Trapp 2003; Trapp and Weisman 2003) and observational studies (Atkins et al. 2004, 2005; Wheatley et al. 2006; Wakimoto et al. 2006) have shown that the most intense straight-line winds within bow echoes are frequently associated with low-level mesovortices. In the United States, where the majority of the bow echoes move to the east or south-east, the most severe wind damage is observed north or north-west (left) of the bow echo apex in close association with mesovortices (e.g., Wheatley et al. 2006). Interestingly, the 3 July 2004 microburst occurred about 10 km left of the apex. Examination of the Ikaalinen radar radial velocities at 1415 EET (Fig. 3d) showed a couplet of inbound and outbound velocities, with the northern flank of the possible vortex travelling over Hyytiälä at the time of the microburst. The difference between the inbound and outbound velocities was $15–20\,\mathrm{m\,s^{-1}}$, and the distance between the maximum and minimum radial velocity was about 5 km, both comparable to previous studies (e.g., Atkins et al. 2004, 2005). However, closer examination of the structure and evolution of the possible mesovortices and other radar signatures is beyond the scope of this study.

Downdrafts in thunderstorms are believed to be generated by evaporative cooling, precipitation loading, and precipitation drag (e.g., Wakimoto 2001; Choi 2004). Processes affecting the amount of negative buoyancy of the descending air parcels govern the intensity of the subsequent downdraft. Thus, melting, evaporation, and sublimation of hydrometeors lead to colder and denser downdraft air, which accelerates towards the surface. The cooling effects

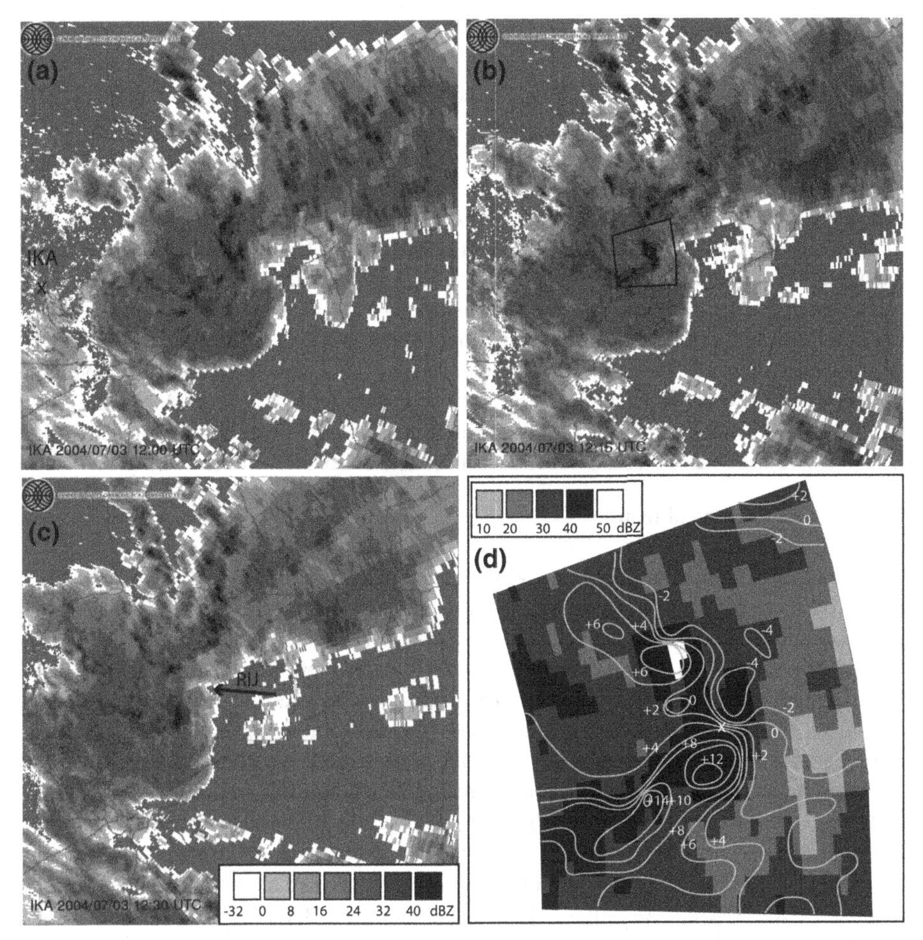

Fig. 3 Radar reflectivity factor (dBZ, shaded) from the Ikaalinen radar (labelled IKA in panel **a**) at (**a**) 1400 EET 3 July 2004, (**b**) 1415 EET with inset box, and (**c**) 1430 EET with location of rear-inflow jet (RIJ) shown by the arrow. (**d**) inset location from **b**) showing 1415 EET storm-relative radial velocity field (white solid lines every $2\,\mathrm{m\,s^{-1}}$, positive values away from radar) overlain on radar reflectivity factor (dBZ, shaded), and the location of the wind damage area marked with white X

associated with these phase changes usually offset the adiabatic warming of descending air. In addition, Ogura and Liou (1980) showed that convective downdrafts effectively transport horizontal momentum vertically. Therefore, convective cells developing in moderate to strong low- and mid-level flow should aid in the production of microbursts associated with strong winds.

These synoptic and mesoscale observations suggest that the convection in the Hyytiälä case occurred in an environment of modest convective available potential energy and moderate vertical shear of the horizontal wind. Such an environment is characteristic of convective systems formed under strong synoptic forcing. In such cases, the lack of abundant instability can be compensated for by greater wind shear to produce strong surface winds (e.g., Evans and Doswell 2001).

3.2 Local weather conditions on 3 July 2004

On 3 July 2004, the surface pressure decreased from 984 hPa to 979 hPa from 0000 to 2000 EET at location A at the SMEAR II station (not shown). Two clear peaks at approximately 0500 EET and 1415 EET were observed, with the latter appearing at the same time as the thunderstorm moved over Hyytiälä. The maximum air temperature was reached around 1200 EET, after which the temperature decreased sharply, 4.5°C within 1.5 h (not shown). The minimum temperatures of 13°C and 13.5°C at heights of 67.2 and 8.4 m, respectively, coincided with the maximum pressure (around 1415 EET). The decrease of the temperature and the pressure jump indicate the arrival of the microburst event. The mean wind speed was less than 3 $m\,s^{-1}$ at the lower level (8.4 m) throughout the day, but at the higher level (73 m), wind speed reached 13 $m\,s^{-1}$, coinciding with the pressure peak and local minimum temperature (not shown). Between 0800 EET 3 July 2004 and 0800 EET 4 July 2004, the measured rainfall was 30.3 mm. During the thunderstorm, the maximum precipitation intensity (with 15-min resolution) was 3.5 mm h^{-1} between 1400 and 1415 EET (not shown).

3.3 Micrometeorological features

More detailed examination of the period when the microburst occurred (1400–1430 EET) indicated a more complex structure. Before the event, the instantaneous wind speed above the forest was about 2 $m\,s^{-1}$ and near-zero at the sub-canopy, and the wind direction was from the north-north-west (Fig. 4). The vertical wind velocity was a few tens of $mm\,s^{-1}$ (Figs. 5 and 6), and temperature and dew-point temperature at location A were 18°C and 13.2°C, respectively (Fig. 7). At 1405 EET, the pressure started to increase from 980.2 hPa (Fig. 7) and one minute later, the wind slowly began turning to the north-east. The gust front reached the measurement site at 1412 EET, detected as an increase in the horizontal and vertical wind speeds at all three measurement locations (Figs. 4–6). Half a minute later, the pressure reached a maximum of 981.4 hPa, and both temperature and dew-point temperature experienced a small increase of 0.5°C (Fig. 7). Within two minutes at 1413 EET, both temperature and dew-point temperature fell by 2.5°C on arrival of the cold air (Fig. 7). Other gust-front studies (e.g., Charba 1974; Mueller and Carbone 1987) have observed this pressure peak and wind shift together with the temperature fall, consistent with our measurements. The pressure increase and wind shift prior to the gust front is believed to be caused by the dynamic effects of the temperature difference between the air masses (Figure 1 in Droegemeier and Wilhelmson 1987).

A windy and gusty period ensued between 1412 and 1418 EET (Figs. 4–6). Two peaks in horizontal wind speed were evident at location A, the first (19 $m\,s^{-1}$) after 1413 EET and the second (22 $m\,s^{-1}$) at 1417 EET (Fig. 4a). The peak wind speed values at location B of 21 $m\,s^{-1}$) and at sub-canopy of 9 $m\,s^{-1}$ occurred at the same time as the second wind speed maximum at location A (Fig. 4). During this 7-min period of gusty winds, the vertical wind speed fluctuated between -15 and $+10\,m\,s^{-1}$ at location A, between -9 and $+9\,m\,s^{-1}$ at location B and between -1.6 and $+1.8\,m\,s^{-1}$ in the sub-canopy (Fig. 5). Examining the vertical wind speed at location A with a higher temporal resolution revealed a brief episode of strong downward motion around 1417 EET (Fig. 6). This downward motion was also seen at location B and in the sub-canopy, although the strongest downward motion was measured a half a minute before at location B (not shown). Comparable descent in microbursts has been seen by others (e.g., Proctor 1988; Hjelmfelt et al. 1989). The horizontal wind speed maximum and the downward motion 5-min after the gust front at 1417 EET indicated the microburst arriving at the measuring station. The time lag of five minutes between the first

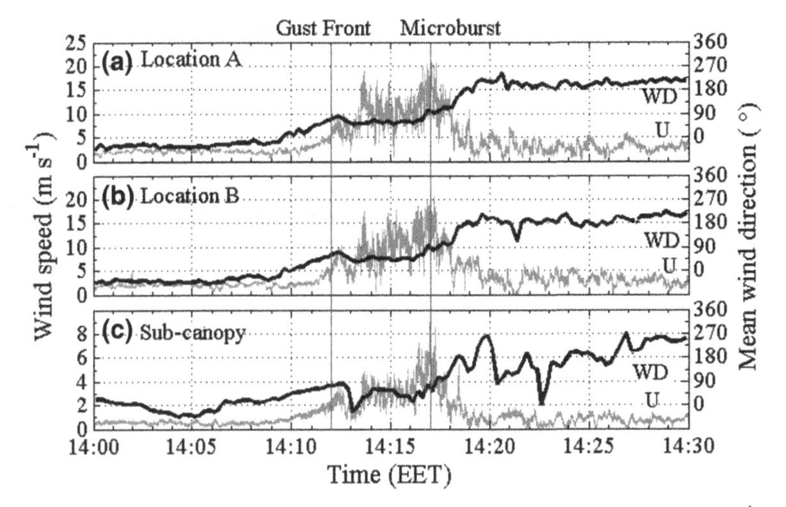

Fig. 4 Time series at SMEAR II station for 1400–1430 EET: instantaneous wind speed (m s^{-1}, thin grey lines) and 15-sec average wind direction (°, thick black lines) measured at (**a**) location A, (**b**) location B and (**c**) sub-canopy. Note the different velocity scale of the sub-canopy plot. The vertical lines represent the arrival of the gust front and microburst, respectively

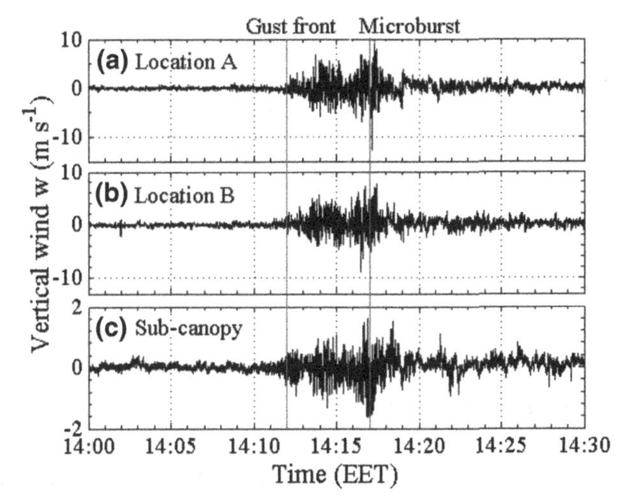

Fig. 5 Time series at SMEAR II station for 1400–1430 EET: instantaneous vertical wind speed (m s^{-1}) measured at (**a**) location A, (**b**) location B and (**c**) sub-canopy. Note the different velocity scale of the sub-canopy plot. The vertical lines represent the arrival of the gust front and microburst, respectively

gusty winds (gust front) and the maximum wind speed (microburst) has also been observed by Takayama et al. (1997).

Between the gust front and the microburst, the pressure experienced a minimum and started to increase, along with temperature and dew-point temperature, at 1416 EET at the same time as the strengthening microburst (Fig. 7). The wind veered toward the east preceding the gust front and during the microburst (Fig. 4), and a second pressure maximum occurred just after the highest wind speed values (Fig. 7). After the windy episode, both the horizontal and

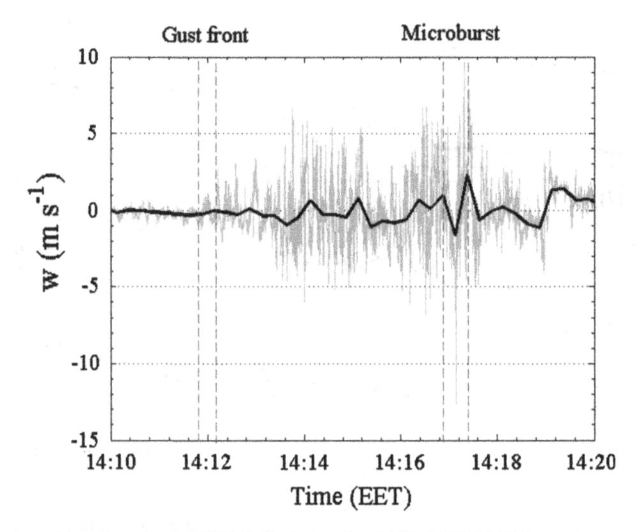

Fig. 6 Time series at location A at SMEAR II station for 1410–1420 EET: instantaneous (grey line) and 15-sec average (black line) vertical wind speed ($\mathrm{m\,s^{-1}}$). The vertical dashed lines represent the passage of the gust front and the microburst, respectively

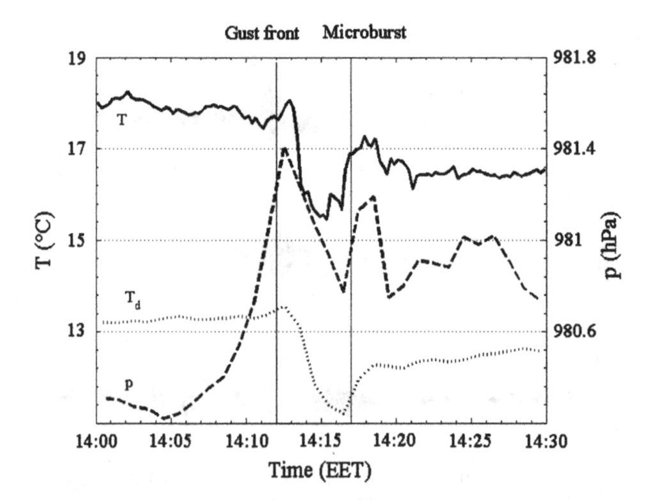

Fig. 7 Time series at location A at SMEAR II station for 1400–1430 EET: 15-sec average temperature at a height of 23 m (°C, solid line), 1-min average dew-point temperature at a height of 23 m (°C, dotted line), and 1-min average pressure at the ground (hPa). The vertical lines represent the arrival of the gust front and microburst, respectively

vertical winds stayed gusty (Figs. 4–5), consistent with meteorological measurements from other microburst studies (e.g., Wakimoto 1982; Fujita 1985, his Fig. 6.42).

The measured wind speed maxima of 21 and 22 $\mathrm{m\,s^{-1}}$ at locations A and B, respectively, are consistent with maximum wind peaks from other studies. Ohno et al. (1994) measured a peak wind speed of 26 $\mathrm{m\,s^{-1}}$, whereas lower wind speed maxima of 15 and 10.5 $\mathrm{m\,s^{-1}}$ were measured by Sherman (1987) and Takayama et al. (1997), respectively. The majority of microbursts measured during the JAWS experiment had wind speed maxima between 12 and 16 $\mathrm{m\,s^{-1}}$ (Wakimoto 1985). A two-peaked wind speed maximum, which was most apparent

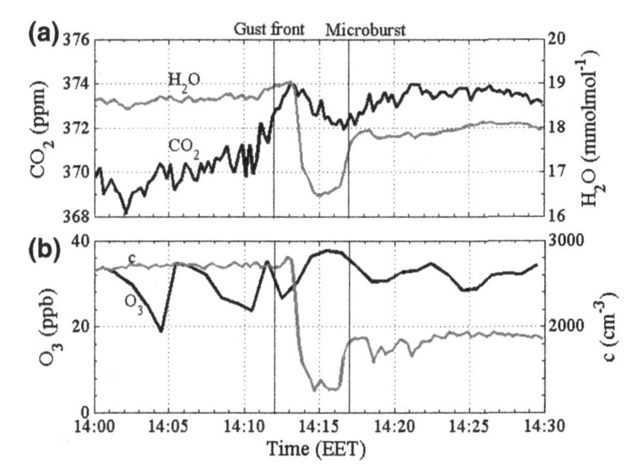

Fig. 8 Time series at location A at SMEAR II station for 1400–1430 EET: (**a**) 15-sec average carbon dioxide (ppm, black line) and water vapour (mmol mol^{-1}, grey line) concentrations, and (**b**) 1-min average ozone (ppb, black line) and 15-sec average total particle (cm^{-3}, grey line) concentrations. The vertical lines represent gust front and microburst, respectively

at location A, is consistent with an observed and modelled wind speed maximum just after the gust front and another maximum close to the downdraft (e.g. Charba 1974; Mitchell and Hovermale 1977).

In addition to the period of downward motion during the microburst, the vertical wind speed showed several brief periods of downward and upward motion between 1412 and 1418 EET (Fig. 6). Mueller and Carbone (1987) showed a similar periodic pattern in vertical wind speeds behind the gust front. Around the time of the strong downward motion during the microburst at 1417 EET, periods of upward motion occurred. These could be caused by vortices formed at the front and rear flanks of the strongest downward motion. Sherman (1987) observed similar behaviour in time series of vertical wind speed from a microburst in Australia.

The forest damage was most probably caused by the powerful microburst around 1417 EET when the wind speeds were strongest (Fig. 4). Based on prior experience in Finland, the small patch of downed trees in the forest would imply the wind speed was about 30 m s^{-1}. Such estimated wind speeds are about 10 m s^{-1} higher than the maximum wind speed values at the above-canopy locations, suggesting that the microburst did not directly strike the observing instrumentation.

The wind direction was similar at both above-canopy measurement locations but had somewhat larger fluctuations at the sub-canopy where enhanced turbulent production occurs (Fig. 4c). In contrast, the horizontal wind speeds were over 10 m s^{-1} smaller, and the vertical wind speed was 10% less at the sub-canopy location than at the above-canopy location.

3.4 Gas and aerosol particle characteristics

The data collected during the microburst at Hyytiälä provide a unique opportunity to observe gas and aerosol properties at ground level during such an extreme event. Carbon dioxide concentrations (CO_2) increased from 370 to 374 ppm at the time of the gust front (Fig. 8a) due to strong mixing of the surface-layer air with the overlying boundary-layer air. When descending cold air arrived at the measurement site and rain started at 1413 EET, the water vapour concentrations decreased from 19 to 16 mmol mol^{-1} in two minutes (Fig. 8a), consistent with the decrease in dew-point temperature (Fig. 7). The effect of the rain scavenging was clear as the total particle concentration decreased from 2800 to 1200 cm^{-3} during the event (Fig. 8b). The ozone concentration did not have a distinct pattern between 1400 and 1430 EET, although the concentration rose during the microburst (Fig. 8b), consistent with the mid-tropospheric source region of the microburst where higher ozone concentrations occur.

Before the microburst, the aerosol number size distribution was dominated by Aitken and accumulation mode particles (Fig. 9a). The total number concentration was of the order of a few thousand, and positive and negative ion concentrations varied between 400 to 1000 (Fig. 9b). Concurrent with the microburst event (1410–1420 EET), the sub-micron aerosol size distribution was dominated by ultra-fine particles (<10 nm in diameter) (Fig. 9a). The total number concentration, together with the negative ion concentrations, increased an order of magnitude at the time of the gust front (Fig. 9b). After the peak, a decrease in total aerosol concentrations occurred, similar to Fig. 8b. The peak in the number concentration did not occur in Fig. 8b since the detection limit of the CPC (approximately 10 nm) is higher than the detection limit of the DMPS. The ion concentration did not decrease due to the rain, but fluctuated rapidly, likely due to the production of intermediate ions during rain (Hirsikko et al. 2007).

Previously, increased number concentration of ultra-fine particles has been measured in relation to clouds. Perry and Hobbs (1994) detected increased concentrations of ultra-fine particles in the clean marine air at the level of the anvil outflow. Also Clarke et al. (1998) observed new particle formation in cloud outflow. Highest formation rates were associated with elevated concentrations of sulphuric acid due to enhanced photochemical activity and lower pre-existing particle surface concentration. Keil and Wendisch (2001) detected ultra-fine particles over western Europe, and observed bursts of newly formed particles in the cloud-free air at cloud top. These observations could explain the increased number concentration of ultra-fine particles in the microburst air. Another possible explanation of concurrent high concentrations of ultra-fine particles and ions is the production of charged particles associated with rain (Laakso et al. 2006), although the aerosol and ion sizers were not fast enough to capture the full variability of concentration during the microburst.

3.5 Turbulence spectra and turbulent kinetic energy (TKE)

The power spectra of u_{rot} and w were calculated for 5- to 10-min periods before, during and after the microburst at location A. Before the microburst (1355–1405 EET), the prevailing wind direction was 314°, the average wind speed was 2.0 m s^{-1} and the atmosphere was stably stratified. During the microburst event (1413–1418 EET), the wind direction was 071° and the average wind speed was 11.1 m s^{-1}. Calculating the thermal stability during the event was not possible due to the nonstationarity of the situation. After the microburst (1425–1435 EET), the wind direction was 206°, the wind speed was 2.8 m s^{-1} and the atmosphere was slightly stable. The power spectra were normalized by the variances of u_{rot} and w, smoothed over logarithmically changing frequency intervals, and plotted as a function of normalized frequency $n = f(\bar{z} - d)/\bar{U}$, where $\bar{z}$ is measurement height, d is the displacement height assumed here as 2/3 the average tree height and $\bar{U}$ is the average wind speed (Fig. 10), with the predicted $-2/3$ power law for the inertial subrange also shown. The turbulence dissipation rates were calculated before, during and after the microburst for the frequency range 0.5–3 Hz according to Eq. 1.

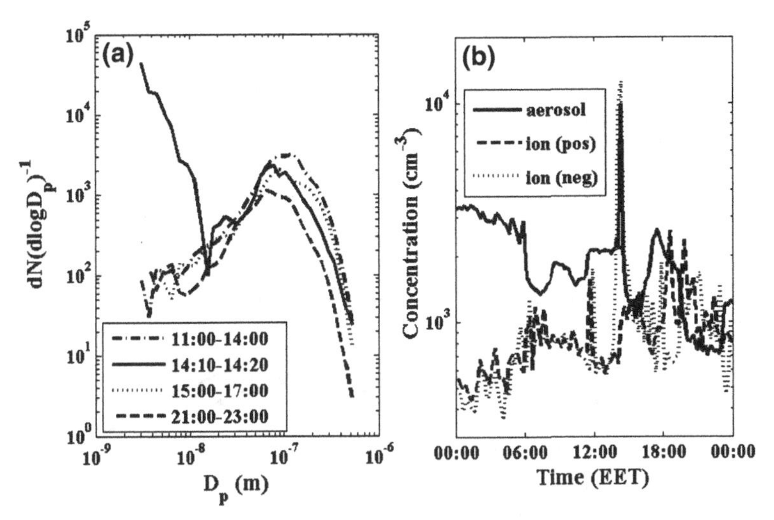

Fig. 9 (**a**) Observed aerosol number size distributions at location A at SMEAR II station before (1100–1400 EET, dashed–dotted line), during (1410–1420 EET, solid line) and after (1500–1700 EET, dotted line; 2100–2300 EET, dashed line) the microburst event. (**b**) Time series at location A at SMEAR II station from 0000 EET 3 July 2004 to 0000 EET 4 July 2004: total aerosol number concentration (cm^{-3}, solid line), positive ion concentration (cm^{-3}, dashed line), and negative ion concentration (cm^{-3}, dotted line)

The power spectra of u_{rot} before and after the microburst showed comparable behaviour (Fig. 10a), both having two maxima around 0.01 and 0.1 Hz. The power spectrum during the event differed from the other spectra in the low-frequency end (below 0.01 Hz), with a faster roll-off at lower frequencies during the event indicating less energy in larger eddies. All spectra followed the $-2/3$ power law in the inertial subrange. Before and after the microburst, the dissipation rates were below $0.01\,\mathrm{m}^2\,\mathrm{s}^{-3}$. During the microburst, the dissipation rate of turbulent energy increased substantially to $0.60 \pm 0.02\,\mathrm{m}^2\,\mathrm{s}^{-3}$. As with the power spectra of u_{rot}, the power spectra of w were very similar in all three cases (Fig. 10b), with a maximum at frequencies 0.05–0.1 Hz and followed the $-2/3$ power law well in the inertial subrange. As with the power spectra of u_{rot}, the greatest differences between these three cases appeared at the low-frequency end, although, in the case of the power spectra of w, the before- and after-microburst periods had less energy than the during-microburst period. The reason for the different behaviours of the spectra of u_{rot} and w at the low-frequency end is not known. Also the reliability of the spectra at the low-frequency end is questionable due to the short calculation periods. Overall, both the spectra of u_{rot} and w followed typical model spectra before, during and after the microburst event.

The turbulent kinetic energy (TKE) at all three measuring locations was calculated for 15-sec time periods during 1400–1430 EET 3 July 2004 (Fig. 11). TKE was less than $0.1\,\mathrm{m}^2\,\mathrm{s}^{-2}$ above the canopy and less than $0.01\,\mathrm{m}^2\,\mathrm{s}^{-2}$ at the sub-canopy before the gust front arrived. About two minutes before the gust front arrived, TKE at all locations increased. By the time of the microburst at 1417 EET, the peak TKE values were greater than $24\,\mathrm{m}^2\,\mathrm{s}^{-2}$ above the canopy and $5\,\mathrm{m}^2\,\mathrm{s}^{-2}$ at the sub-canopy. At all measurement locations, TKE had a two-peaked maximum at the time of the microburst (1416–1418 EET), with lower values at 1417 EET when the highest horizontal values were measured. This implies greater flow fluctuations in the vortices along the front and rear flanks of the strongest downward motion.

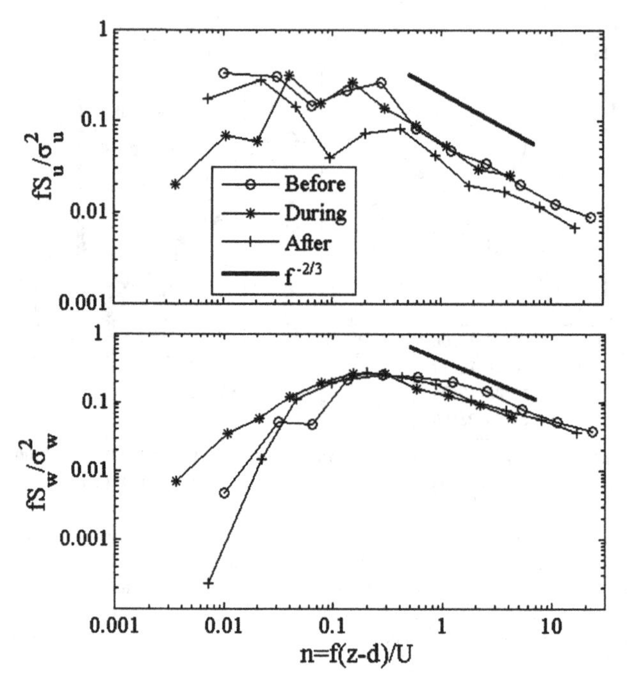

Fig. 10 The turbulence spectra of (**a**) wind u_{rot} component and (**b**) vertical wind velocity w before (1355–1405 EET), during (1413–1418 EET) and after (1425–1435 EET) the event. Black line shows the $-2/3$ power law in the inertial subrange

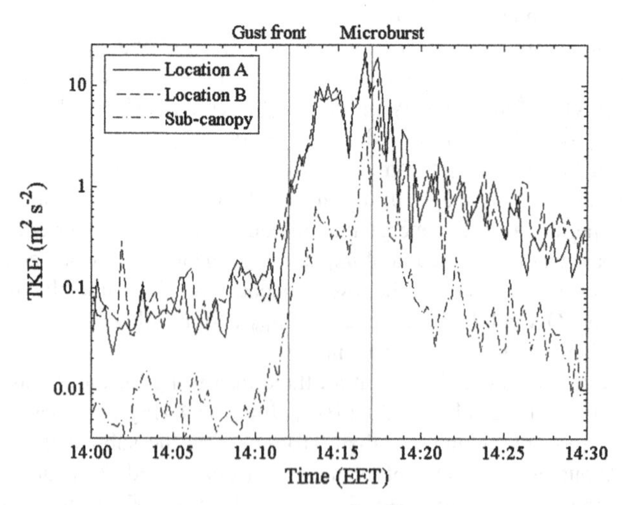

Fig. 11 Time series of turbulent kinetic energy (TKE, m^2 s^{-2}) at SMEAR II station for 1400–1430 EET measured above-canopy at locations A and B and sub-canopy. The vertical lines represent the arrival of the gust front and microburst, respectively

4 Conclusions

Severe thunderstorms mainly occur in Finland during the late summer, and forest damage caused by downbursts or tornadoes takes place almost every summer. The event at Hyytiälä on 3 July 2004 caused forest damage next to the SMEAR II field station, providing a unique opportunity to analyze the micrometeorology of a microburst and its attendant gust front. The leading edge of a squall line was reached Hyytiälä at 1330 EET, and the radar reflectivity factor maximum was located in the middle of the line, contrary to observations of other mesoscale convective systems (e.g., Parker and Johnson 2000). Typical radar signatures of downburst-producing convective systems (bow echo, rear-inflow jet, and rear-inflow notch) were observed just north of Hyytiälä preceding and at the time of the microburst. Also, a possible mesovortex was observed in the radar radial velocity data. The individual convective cells moved to the west or west-south-west. Based on the damage, the wind speed was estimated to be around $30\,\mathrm{m\,s^{-1}}$, $10\,\mathrm{m\,s^{-1}}$ more than the measured wind-speed maxima.

Micrometeorological data above the canopy at a height of 23 m (locations A and B), and sub-canopy at a height of 3 m, were used. The horizontal and vertical wind speeds, wind direction, temperature and pressure showed behaviour typical of a thunderstorm outflow (Charba 1974; Mueller and Carbone 1987). An increased number of ultra-fine particles were measured during the gust front. Carbon dioxide concentrations increased due to the stronger turbulent mixing associated with the gust front, and water vapour and particle concentrations decreased due to cold downdrafts and rain. Strong peaks in horizontal wind speeds ($21–22\,\mathrm{m\,s^{-1}}$ above the canopy and $9\,\mathrm{m\,s^{-1}}$ at sub-canopy) and periods of downward vertical motion occurred five minutes after the arrival of the gust front. Simultaneously, temperature, dew-point temperature, water vapour and pressure increased, suggesting a strong microburst. During the 7-min windy period during the event, the vertical wind speed at location A experienced maxima of upward and downward flow (10 and $15\,\mathrm{m\,s^{-1}}$, respectively), comparable with other microburst studies (e.g., Proctor 1988; Hjelmfelt et al. 1989).

The turbulence spectra of horizontal and vertical wind speeds showed typical model spectral behaviour before, during and after the microburst, with the w spectra following the $-2/3$ power law closer than the u_{rot} spectra in the inertial subrange. The dissipation of turbulent kinetic energy increased during the microburst, with TKE having maxima during the microburst (1416–1418 EET), but values were higher around the measured wind-speed maxima (1417 EET) indicating the front and rear flanks of the strongest downward motion.

Acknowledgements We would like to thank the Academy of Finland and project REBECCA (Helsinki Environment Research Centre) for financial support. Antti Mäkelä, Vesa Nietosvaara and Markus Peura from Finnish Meteorological Institute and Ivan Mammarella, Anne Hirsikko, Heikki Junninen and Samuli Launiainen from University of Helsinki are acknowledged for their help in this study.

References

Aalto P, Hämeri K, Becker E, Weber R, Salm J, Mäkelä JM, Hoell C, O'Dowd C, Karlsson H, Hansson H-C, Väkevä M, Koponen I, Buzorius G, Kulmala M (2001) Physical characterization of aerosol particles during nucleation events. Tellus 53B:344–358

Atkins NT, Wakimoto RM (1991) Wet microburst activity over the southeastern United States: implications for forecasting. Wea Forecast 6:470–482

Atkins NT, Arnott JM, Przybylinski RW, Wolf RA, Ketcham BD (2004) Vortex structure and evolution within bow echoes. Part I: Single-doppler and damage analysis of the 29 June 1998 derecho. Mon Wea Rev 132:2224–2242

Atkins NT, Bouchard CS, Przybylinski RW, Trapp RJ, Schmocker G (2005) Damaging surface wind mechanisms within the 10 June 2003 Saint Louis bow echo during BAMEX. Mon Wea Rev 133:2275–2296

Charba J (1974) Application of gravity current model to analysis of squall-line gust front. Mon Wea Rev 102:140–156

Choi ECC (2004) Field measurement and experimental study of wind speed profile during thunderstorms. J Wind Eng Ind Aerodyn 92:275–290

Clarke AD, Varner JL, Eisele F, Mauldin RL, Tanner D, Litchy M (1998) Particle production in the remote marine atmosphere: cloud outflow and subsidence during ACE 1. J Geophys Res 103:16397–16409

Droegemeier KK, Wilhelmson RB (1987) Numerical simulation of thunderstorm outflow dynamics. Part I: Outflow sensitivity experiments and turbulence dynamics. J Atmos Sci 44:1180–1210

Evans JS, Doswell CA III (2001) Examination of derecho environments using proximity soundings. Wea Forecast 16:329–342

Fujita TT (1979) Objectives, operation and results of project NIMROD. Preprints, 11th conference on severe local Storms, Kansas City, MO, Amer Meteorol Soc 259–266

Fujita TT (1985) The Downbursts—microburst and macroburst. SMRP researcher paper number 210, University of Chicago, Department of the Geophysical Sciences, pp 60–118

Hari P, Kulmala M (2005) Station for measuring ecosystem–atmosphere relations (SMEAR II). Boreal Env Res 10:315–322

Hirsikko A, Bergman T, Laakso L, Dal Maso M, Riipinen I, Hõrrak U, Kulmala M (2007) Identification and classification of the formation of intermediate ions measured in boreal forest. Atmos Chem Phys 7:201–210

Hjelmfelt MR, Orville HD, Roberts RD, Chen JP, Kopp FJ (1989) Observational and numerical study of a microburst line-producing storm. J Atmos Sci 46:2731–2744

Holton JR (1992) A introduction to dynamic meteorology, 4th edn. Academic Press, San Diego, California, pp 120–122

Kaimal JC, Kristensen L (1991) Time series tapering for short data samples. Boundary-Layer Meteorol 57:187–194

Kaimal JC, Finnigan JJ (1994) Atmospheric boundary layer flows. Oxford University Press, New York, USA, pp 234–240

Keil A, Wendisch M (2001) Bursts of Aitken mode and ultrafine particles observed at the top of continental boundary layer clouds. J Aerosol Sci 32:649–660

Laakso L, Hirsikko A, Grönholm T, Kulmala M, Luts A, Parts T-E (2006) Waterfalls as sources of small charged aerosol particles. Atmos Chem Phys Discuss 6:9297–9314

Mitchell KE, Hovermale JB (1977) A numerical investigation of the severe thunderstorm gust front. Mon Wea Rev 105:657–675

Mueller CK, Carbone RE (1987) Dynamics of a thunderstorm outflow. J Atmos Sci 44:1879–1897

Ogura Y, Liou M-T (1980) The structure of a midlatitude squall line: A case study. J Atmos Sci 37:553–567

Ohno H, Suzuki O, Nirasawa H, Yoshizaki M, Hasegawa N, Tanaka Y, Muramatsu Y, Ogura Y (1994) Okayama downbursts on 27 June (1991) downburst identifications and environmental conditions. J Meteorol Soc Japan 72:197–221

Parker MD, Johnson RH (2000) Organizational modes of midlatitude mesoscale convective systems. Mon Wea Rev 128:3413–3436

Perry KD, Hobbs PV (1994) Further evidence for particle nucleation in clear air adjacent to marine cumulus clouds. J Geophys Res 99:22803–22818

Piper M, Lundquist JK (2004) Surface layer turbulence measurements during a frontal passage. J Atmos Sci 61:1768–1780

Proctor FH (1988) Numerical simulations of an isolated microburst. Part I: Dynamics and Structure. J Atmos Sci 45:3137–3160

Przybylinski RW, DeCaire D (1985) Radar signatures associated with the derecho, a type of mesoscale convective system. Preprints, 14th Conference on severe local storms. Indianapolis, IN, Amer Meteorol Soc, pp 228–231

Przybylinski RW (1995) The bow echo: Observations, numerical simulations, and severe weather detection methods. Wea Forecast 10:203–218

Punkka A-J, Teittinen J, Johns RH (2006) Synoptic and mesoscale analysis of a high-latitude derecho-severe thunderstorm outbreak in Finland on 5 July 2002. Wea Forecast 21:752–763

Sherman DJ (1987) The passage of a weak thunderstorm downburst over an instrumented tower. Mon Wea Rev 115:1193–1205

Smull BF, Houze RA Jr. (1987) Rear inflow in squall lines with trailing stratiform precipitation. Mon Wea Rev 115:2869–2889

Takayma H, Niino H, Watanabe S, Sugaya J (1997) Downbursts in the Northerwestern Part of Saitama Prefecture on 8 September 1994. J Meteorol Soc Japan 75:885–905

Trapp RJ, Weisman ML (2003) Low-level mesovortices within squall lines and bow echoes. Part II: their genesis and implications. Mon Wea Rev 131:2804–2823

Vesala T, Haataja J, Aalto P, Altimir N, Buzorius G, Garam E, Hämeri K, Ilvesniemi H, Jokinen V, Keronen P, Lahti T, Markkanen T, Mäkelä JM, Nikinmaa E, Palmroth S, Palva L, Pohja T, Pumpanen J, Rannik Ü, Siivola E, Ylitalo H, Hari P, Kulmala M (1998) Long-term field measurements of atmosphere-surface interactions in boreal forest combining forest ecology, micrometeorology, aerosol Physics and atmospheric chemistry. Trends Heat, Mass Momentum Transfer 4:17–35

Wakimoto RM (1982) The life cycle of thunderstorm gust fronts as viewed with Doppler radar and rawinsonde data. Mon Wea Rev 110:1060–1082

Wakimoto RM (1985) Forecasting dry microburst activity over the high plains. Mon Wea Rev 113:1131–1143

Wakimoto RM (2001) Convectively driven high wind events. Severe Convective Storms, American Meteorological Society, Boston, MA, pp 255–298

Wakimoto RM, Murphey HV, Nester A, Jorgensen DP, Atkins NT (2006) High winds generated by bow echoes. Part I: Overview of the Omaha bow echo 5 July 2003 storm during BAMEX. Mon Wea Rev 134:2793–2812

Weisman ML, Trapp RJ (2003) Low-level mesovortices within squall lines and bow echoes. Part I: Overview and dependence on environmental shear. Mon Wea Rev 131:2779–2803

Wheatley DM, Trapp RJ, Atkins NT (2006) Radar and damage analysis of severe bow echoes observed during BAMEX. Mon Wea Rev 134:791–806

Role of land-surface temperature feedback on model performance for the stable boundary layer

A. A. M. Holtslag · G. J. Steeneveld · B. J. H. van de Wiel

Abstract At present a variety of boundary-layer schemes is in use in numerical models and often a large variation of model results is found. This is clear from model intercomparisons, such as organized within the GEWEX Atmospheric Boundary Layer Study (GABLS). In this paper we analyze how the specification of the land-surface temperature affects the results of a boundary-layer scheme, in particular for stable conditions. As such we use a well established column model of the boundary layer and we vary relevant parameters in the turbulence scheme for stable conditions. By doing so, we can reproduce the outcome for a variety of boundary-layer models. This is illustrated with the original set-up of the second GABLS intercomparison study using prescribed geostrophic winds and land-surface temperatures as inspired by (but not identical to) observations of CASES-99 for a period of more than two diurnal cycles. The model runs are repeated using a surface temperature that is calculated with a simple land-surface scheme. In the latter case, it is found that the range of model results in stable conditions is reduced for the sensible heat fluxes, and the profiles of potential temperature and wind speed. However, in the latter case the modelled surface temperatures are rather different than with the original set-up, which also impacts on near-surface air temperature and wind speed. As such it appears that the model results in stable conditions are strongly influenced by non-linear feedbacks in which the magnitude of the geostrophic wind speed and the related land-surface temperature play an important role.

Keywords Atmosphere-land interaction · Boundary-layer parameter study · GABLS · Model evaluation · Model intercomparison study · Land-surface temperature feedback · Stable boundary layer

---

A. A. M. Holtslag (✉) · G. J. Steeneveld · B. J. H. van de Wiel
Meteorology and Air Quality Section, Wageningen University, Wageningen, The Netherlands
e-mail: Bert.Holtslag@wur.nl

Atmospheric Boundary Layers. A. Baklanov & B. Grisogono (eds.),
doi: 10.1007/978-0-387-74321-9_14, © Springer Science+Business Media B.V. 2007

1 Introduction

In the stable boundary layer over land many small-scale physical processes occur, such as turbulent mixing, radiation divergence, gravity waves (e.g., Mahrt 1999; Holtslag 2006). These processes need to be represented in an effective way in an atmospheric model, and the current understanding of these processes in the stable boundary layer is rather limited (e.g., Delage 1997; Beljaars and Viterbo 1998; Mahrt 1998; Edwards et al. 2006; Steeneveld et al. 2006b). This is relevant for the forecasting of surface and air temperatures, wind speed and direction, the surface fluxes and the boundary-layer depth, and it affects the forecasting of frost and fog episodes (e.g. Clark and Hopwood 2001), and on the dispersion of pollutants and trace gases (e.g. Salmond and McKendry 2005).

To enhance the understanding and to improve the representation of the atmospheric boundary layer in models for weather forecasting, air quality and climate research, frequent model evaluation and intercomparison studies are organized (e.g., Lenderink et al. 2004, Cuxart et al. 2006; Steeneveld et al. 2007). Overall the aim of such studies is to identify strengths and weaknesses of models in comparison with observations (e.g., Poulos et al. 2002) and large-eddy simulations (LES) (e.g. Beare et al. 2006; Kumar et al. 2006).

Usually the intercomparison studies with atmospheric column (1D) models are done with simplified boundary conditions and forcing conditions, such as prescribing a constant geostrophic wind and a prescribed surface temperature (tendency). So far this has also been the approach within the GEWEX Atmospheric Boundary Layer Study (GABLS); see Cuxart et al. (2006) for an overview of the 1D model results for the first GABLS model intercomparison, and Svensson and Holtslag (2006) for the initial results of the second GABLS model intercomparison. Note that evaluation of boundary-layer models with a prescribed surface temperature has been also a typical approach for column models (e.g. Rao and Snodgrass 1979; Delage 1997), as well as large-eddy simulation models (e.g. Beare et al. 2006; Basu et al. 2006).

Instead of prescribing the surface temperature, one may alternatively prescribe the surface sensible heat flux. This has been a useful approach for cases studies over sea and daytime conditions over land (e.g. Lenderink et al. 2004; Kumar et al. 2006), but for nighttime (stable) conditions over land the surface heat flux depends strongly on surface-layer turbulence. Kumar et al. used LES to study the diurnal cycle of the atmospheric boundary layer and they encountered numerical instabilities in stable conditions. At present it is not clear whether this is related to the subgrid closure or to the heat flux boundary condition utilized. In addition, the surface temperature and the surface heat flux are interdependent and are strongly related to the magnitude of the geostrophic wind (e.g. Estournel and Guedalia 1985; Gopalakrishnan et al. 1998; Derbyshire 1999; Delage et al. 2002; van de Wiel et al. 2003; Steeneveld et al. 2006a, b). Thus neither the surface temperature nor the surface heat flux, is a true external boundary condition, at least not for stable conditions (van de Wiel et al. 2007).

It also appears that the results for both the first and second GABLS model intercomparisons show significant variability in the surface fluxes, and the atmospheric wind speed and temperature profiles, despite the relatively simple surface temperature description (and forcing conditions). It is supposed that this is directly related to the different parameterizations of the various models, but it is unknown to what extent the surface temperature boundary condition has an impact on this.

In this study we explore the impact of the surface temperature feedback on the variability of model results. As such, our aim is to investigate to what extent the degree of variability among the model results is influenced by prescribing the surface temperature and not solving for the surface energy balance. The set-up of the second GABLS case is used to study more than two diurnal cycles of the boundary layer over land under clear skies, but we focus on variability created in stable conditions. Some preliminary results of this study were presented by Holtslag et al. (2006).

2 Set-up intercomparison and model description

In the current study we use a first-order closure model and vary the parameters in the turbulence scheme for stable conditions in a reasonable range to mimic the apparent variability among boundary-layer models. Thus, at first, model runs are performed with a prescribed surface temperature as inspired by (but not identical to) the observations in CASES-99 (Poulos et al. 2002) and as described in the GABLS2 case description (Svensson and Holtslag 2006). Second, the model runs are repeated, but using an interactive prognostic heat budget equation for the surface temperature.

For our study we use the coupled land-surface boundary-layer model of Duynkerke (1991) with the extensions of Steeneveld et al. (2006b). The reference model has 50 logarithmically distributed layers with the first atmospheric model level at 2 m. The roughness lengths for heat z_{oh} and momentum z_{om} are given by 3 mm and 30 mm, respectively. Compared to the reference second GABLS study, the surface boundary condition for specific humidity has been altered by introducing a constant canopy resistance of $800\,\mathrm{s\,m^{-1}}$ (to represent the dry conditions during CASES-99, Steeneveld et al. 2006b). Below a brief discussion of the model assumptions is given.

2.1 Turbulence parameterization

The turbulent fluxes of momentum and heat are described by local diffusion for both the surface layer and the SBL. The eddy diffusivity K is given by a first-order closure, which for the whole stable boundary layer can be written as:

$$K_x = \frac{\ell^2}{\phi_m \phi_x} \left| \frac{\partial \vec{V}}{\partial z} \right|. \tag{1}$$

Here subscript x refers to heat (h) or scalar mixing, and subscript m reflects momentum. The length scale ℓ is given by:

$$\frac{1}{l} = \frac{1}{kz} + \frac{1}{\lambda_0}, \tag{2}$$

where $k = 0.4$ is the Von Karman constant, and λ_0 is the asymptotic mixing length (which is infinite in the reference case). Furthermore, the non-dimensional gradients for heat and momentum in stable conditions are given by (Duynkerke 1991):

$$\phi_x(\zeta) = \frac{kz}{X_*} \frac{\partial \overline{X}}{\partial z} = 1 + \beta_x \zeta \left(1 + \frac{\beta_x}{\alpha_x} \zeta\right)^{\alpha_x - 1}. \tag{3}$$

In Eq. 3 $\zeta = z/\Lambda$, where Λ is the local Obukhov length.

For the reference model we use $\beta_m = 5$, $\beta_h = 5$, and $\alpha_m = \alpha_h = 0.8$. Note that this has been validated with observations at Cabauw (Duynkerke 1991), and for CASES-99 (Steeneveld et al. 2006b; Baas et al. 2006). For unstable conditions the original model by Duynkerke (1991) is used for simplicity, although it neglects the impact of non-local mixing by convection (e.g. Holtslag and Moeng 1991).

2.2 Soil and land-surface scheme

In the interactive model runs, the soil temperature evolution is calculated by solving the diffusion equation (using a grid spacing of 10 mm) and the heat flux G through the vegetation is calculated from:

$$G - (1 - f_{veg})K^{\downarrow} = r_g \left(T_{veg} - T_{s0}\right), \tag{4}$$

where $K^{\downarrow}$ is the incoming shortwave radiation, T_{veg} represents the vegetation surface temperature, and T_{s0} is the soil temperature just below the vegetation (at $z = 0\,\mathrm{m}$). As reference values we have $f_{veg} = 0.9$ and $r_g = 5.9\,\mathrm{W\,m^{-2}\,K^{-1}}$, which are consistent with the observations of CASES-99 (Steeneveld et al. 2006b). Initial soil and surface temperatures are also taken from the CASES-99 observations.

Subsequently, the evolution of T_{veg} is computed by solving the surface energy budget for the vegetation layer:

$$C_v \frac{\partial T_{veg}}{\partial t} = Q^* - G - H - L_v E, \tag{5}$$

where C_v is the heat capacity of the vegetation layer per unit of area ($C_v = 2,000\,\mathrm{J\,m^{-2}\,K^{-1}}$, van de Wiel et al. 2003), Q^* is the net radiation, H is the sensible heat flux and $L_v E$ is the latent heat flux. Q^* is calculated by adopting the Garratt and Brost (1981) radiation scheme.

Note that Eqs. 4 and 5 provide a rather strong coupling of the atmosphere to the vegetated land-surface for the current parameter setting (see also Steeneveld et al. 2006b).

2.3 Model parameter settings

To study the impacts of parameter values on the model results, reference runs are made for coupled and uncoupled cases with alternative permutations in some of the parameter settings for stable conditions. The parameter modifications are chosen such that they cover a realistic range in comparison with existing models of the stable boundary layer (such as described in Cuxart et al. 2006). The alternate values of the parameters to be used in the Eqs. 1–3 are:

- $\alpha_\mathrm{m} = \alpha_\mathrm{h} = 0.95$;
- $\beta_\mathrm{m} = \beta_\mathrm{h} = 3$ or 4.7;
- $\lambda_0 = 15, 50, 100, 250\,\mathrm{m}$;
- $\lambda_0 = \varepsilon \frac{u_{*,local}}{N}$, with $\varepsilon = 0.8, 1.3, 2$.

In addition model runs are made:

- with the use of 30 or 20 layers in the model set-up (instead of 50);
- where the height of the first level (z1) is placed at 10 m above the surface rather than 2 m in the reference case;
- with the inclusion of a value for the molecular diffusivity in one of the runs.

In all model runs the roughness length and the canopy resistance are constant (as in Steeneveld et al. 2006b), and the geostrophic wind is taken at a reference value of $9.5\,\mathrm{m\,s^{-1}}$ (as in Svensson and Holtslag 2006). To study the impact of wind speed on the results (e.g., Estournel and Guedalia 1985; Derbyshire 1999; Gopalakrishnan et al. 1998), additional runs are done with a reduced geostrophic wind (see below). Note that it is not our intention to do a full parameter study here. The impact of changing the asymptotic length scale (λ_0) was already discussed for the current model by Steeneveld et al. (2006a).

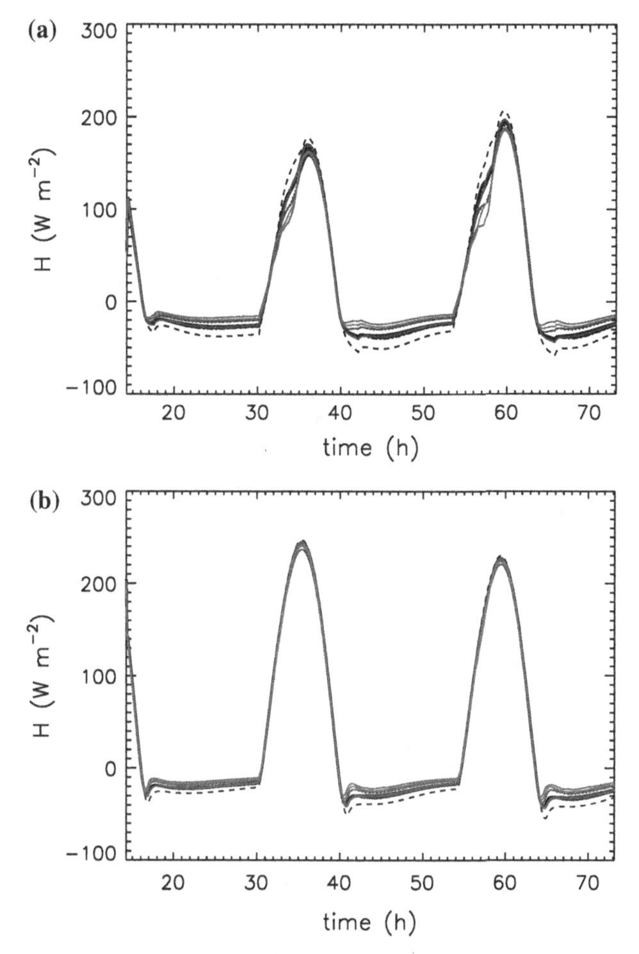

Fig. 1 Time series (in hours commencing at 1400 LT) of model results for the sensible heat flux in a model intercomparison study with (**a**) prescribed surface temperatures, and (**b**) by solving the surface energy budget

3 Results

The model results for all parameter permutations are first presented for the sensible heat flux (Fig. 1), friction velocity (Fig. 2), and boundary-layer height (Fig. 3). The latter is defined as the height where the stress is 5% of its surface value divided by 0.95 (as in Cuxart et al. 2006). In each figure the upper sub-frame of the figure (labelled a) indicates the results achieved with the uncoupled model (using prescribed surface temperature) and in (b) the results are given achieved by solving the energy budget equation. The local starting time in the model runs is 1400 LT on October 22, 1999 (rather than 1600 LT in the GABLS2 runs). The duration of all runs is 59 h (so that the axis of all the figures indicates 14 until 73 h, covering a period of 2.5 diurnal cycles).

Overall the variety of results in the upper frames (Figs. 1a, 2a, 3a) is comparable to the variety within the GABLS2 intercomparison study in stable conditions for the uncoupled models (see also Svensson and Holtslag 2006). Thus we have a range of -10 to $-40\,\mathrm{W\,m^{-2}}$

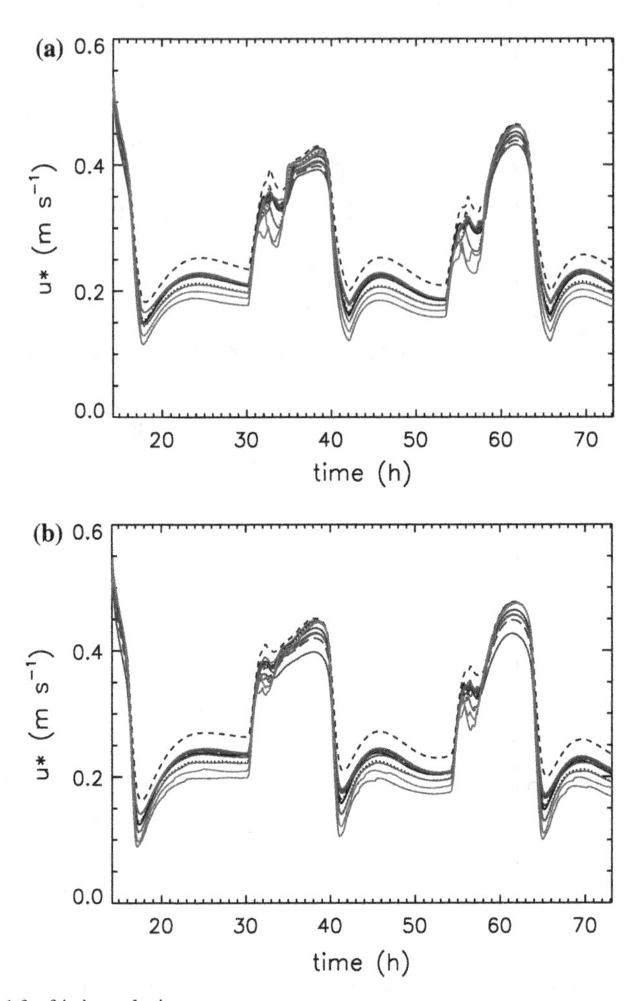

Fig. 2 As Fig. 1 for friction velocity

for the sensible heat flux, a range of 0.19 to 0.25 m s^{-1} for friction velocity and boundary-layer depths varying between 80 and 220 m (all indicated values apply for the variables at the end of the first night e.g. at the indicated time of 30 h in the figures). The variability is a result of the range of parameters chosen above and the impact is apparently sufficient to mimic the different parameterizations for stable conditions in the models used within GABLS2.

Next we repeat all model runs and allow for surface feedback using Eqs. 4 and 5. The results for the coupled model runs are given in the lower frames (Figs, 1b, 2b, 3b). Now we have a range of -10 to -25 W m^{-2} for the sensible heat flux, a range of 0.20–0.26 m s^{-1} for friction velocity and boundary-layer depths between 100 and 270 m (again all indicated values apply for the variables at the end of the first night e.g. at the time of 30 h). Thus it appears that the variety of model results is smaller for the sensible heat flux in the coupled case, in particular. At the same time the variability appears to be somewhat larger for friction velocity and boundary-layer depth, which seems to be related to the larger variability in the near-surface air temperature and wind speed (see Figs. 5 and 6 below).

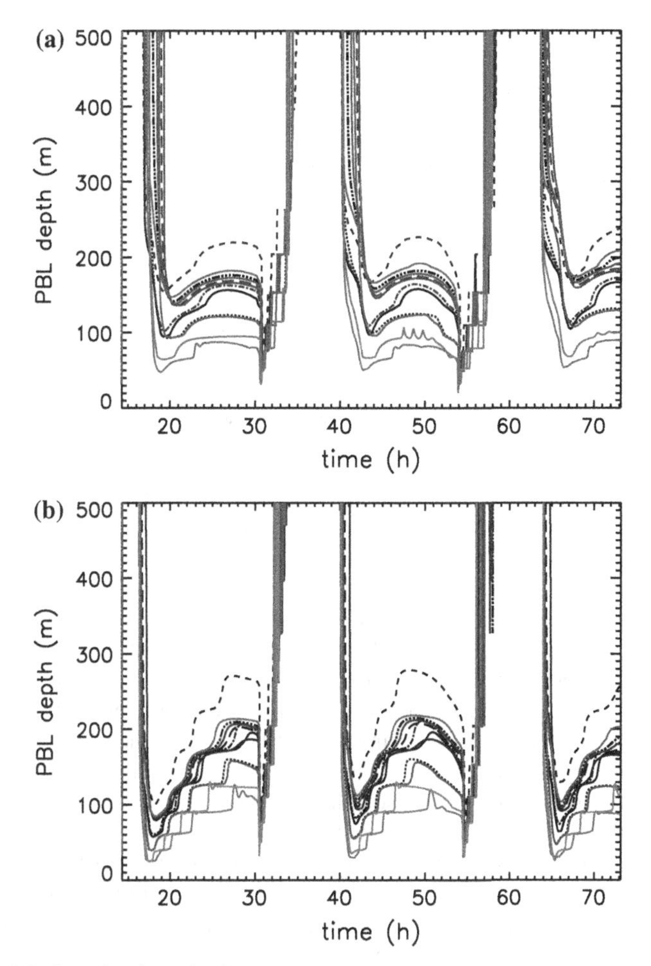

Fig. 3 As Fig. 1 for boundary-layer depth

During daytime the sensible heat fluxes are rather similar for all model runs within one category (either coupled or uncoupled), but the maximum values differ. In addition, due to the coupling the sensible heat fluxes show a smoother behaviour in the morning hours as compared with the uncoupled results (Fig. 1). Thus, surface feedback is influencing the model results and is also able to compensate for some variation in the model parameter values. Note also that the variability in the friction velocities of the first night remains during the morning hours in the uncoupled runs, but not so much in the coupled case.

In Fig. 4 the surface temperatures are given as specified for the uncoupled case (the dash-dotted line), and the temperatures as calculated in the various interactive model runs (various grey lines). It is seen that the latter values are quite different from each other (in particular at night). It is also important to note that the surface temperature by the ensemble of coupled model runs is clearly different from the specified temperature in the uncoupled case. This affects also the absolute values and the range of air temperatures (given at 2 m), and the near-surface wind speeds (given at 10 m). This can be seen in the time series for these variables in Figs. 5 and 6, respectively.

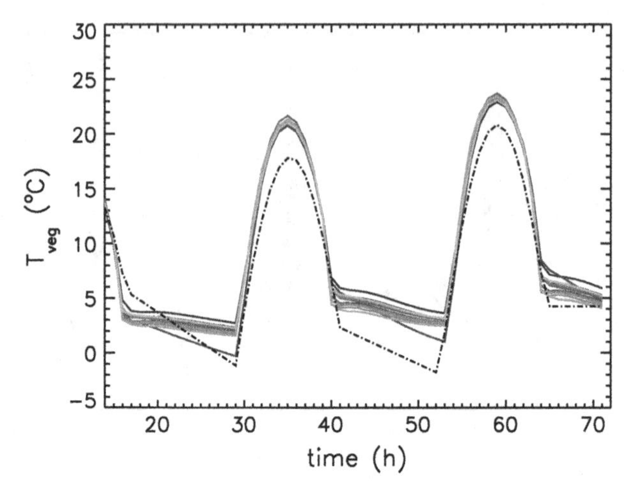

Fig. 4 Time series of modelled surface temperature for coupled runs. *Dash-dotted line*: prescribed surface temperature in the uncoupled case

To further understand this issue, we show in Figs. 7a and 7b the dependence of the sensible heat flux on the potential temperature difference between the surface and the lowest atmospheric model level for given wind speed at that level (at $z = 2$ m). As such we have integrated Eq. 3 for heat and momentum in the surface layer for the reference model parameter settings in stable conditions. In addition, the symbols refer to the outcome of the variety of model runs with perturbed parameter values for the first night, either in the uncoupled case (Fig. 7a) or in the coupled case (Fig. 7b). The indicated lines apply for the reference model and show values for the wind speed at the lowest atmospheric model level (at 2 m). The figures are inspired by earlier works of van de Wiel (2002) and Delage et al. (2002).

In Figs. 7a and 7b it is seen that for lower wind speeds, the curves show a maximum. In fact two regimes can be distinguished, namely the 'well-behaved regime' (at the left-hand side of the maximum) where sensible heat flux is proportional to the potential temperature difference for a given wind speed. This occurs in weakly to moderately stable conditions in which turbulence is sufficiently strong and can maintain itself. At the right-hand side of the maximum, turbulence is suppressed by stability effects so that the exchange decreases and consequently also the sensible heat flux decreases if the potential temperature increases (see also discussion by Holtslag and De Bruin 1988; De Bruin 1994). As such, a positive feedback loop can be established resulting in diminishing turbulence and large temperature gradients.

The results indicate that for a given potential temperature difference and wind speed, a large variation in sensible heat flux can occur due to the different model parameter settings for stable conditions (see Sect. 2). The differences impact clearly on the model results for all model variables as indicated above. Interestingly, it also appears that the results for the uncoupled model runs with perturbed parameter settings (Fig. 7a) show larger absolute values for sensible heat flux for given temperature difference and wind speed than the results of the coupled runs shown in Fig. 7b. In addition, the uncoupled model runs show no maximum for the heat flux, while the coupled model runs do show this. This explains the smaller range of sensible heat fluxes and also the smaller range of variability among model results in the coupled case. The chosen boundary condition has therefore a clear impact on which stability regime is entered. With a prescribed surface temperature condition, the model does not enter the regime with the positive feedback.

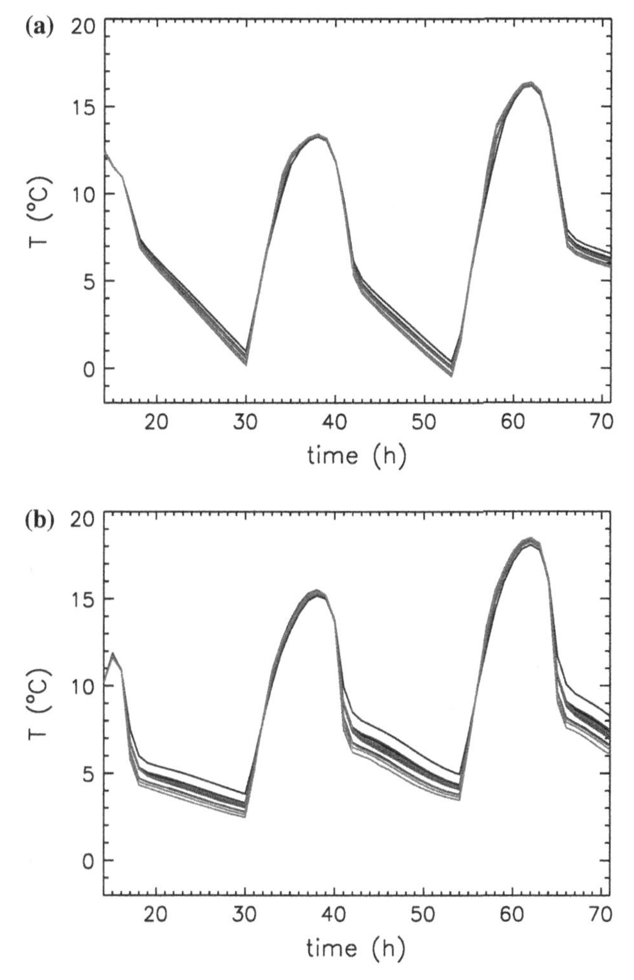

Fig. 5 Time series of the model air temperature at 2 m for (**a**) prescribed surface temperatures, and (**b**) by solving the surface energy budget

In Fig. 7a also a cluster of points is visible at the right-hand side. It appears that this results from the additional model run with lower vertical resolution (see Sect. 2). However, in the coupled case the outcome of the latter run is in more agreement with the other results (Fig. 7b). Overall the findings illustrate that the coupling of the boundary-layer scheme to the land-surface clearly has an impact on the model findings, and this is due to the many feedbacks in this highly nonlinear system (see also McNider et al. 1995).

Forecasted atmospheric profiles for potential temperature and wind speed magnitude after 12 h are given in Figs. 8 and 9 (valid for local nighttime conditions at 0200 on October 23, 1999). Similar range of results is achieved for longer forecasting times during the night. To illustrate the variability in potential temperature and wind speed magnitude, we have calculated the mean square difference (or variance) of the ensemble of model results. Figures 10 and 11 show the outcome of this. Again a distinction is made in uncoupled (upper frames) and coupled cases (lower frames). In all figures the variances are plotted for a height

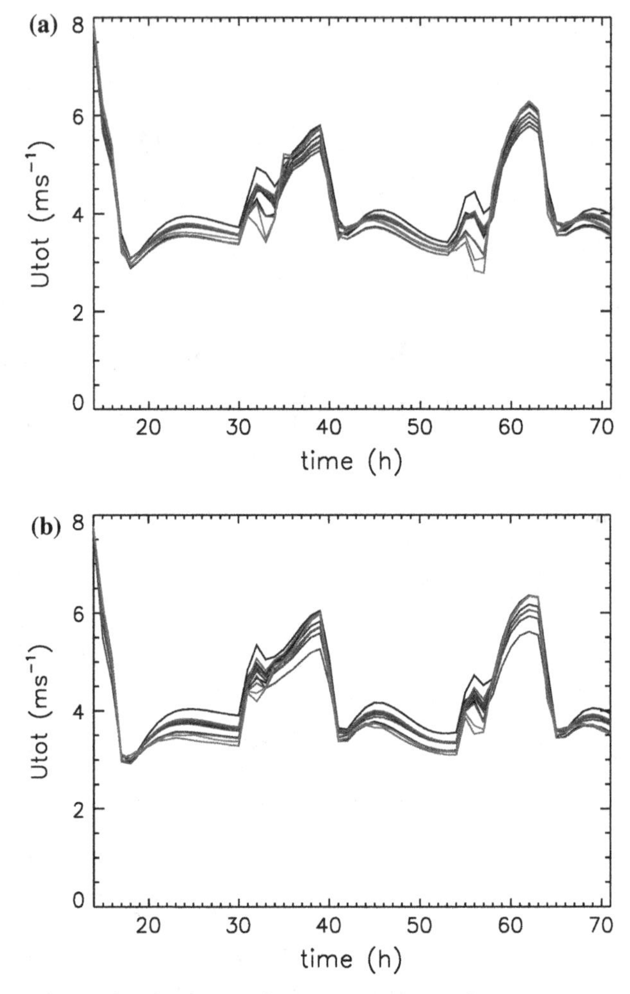

Fig. 6 Time series of the model wind speed at 10 m for (**a**) prescribed surface temperatures, and (**b**) by solving the surface energy budget

up to 300 m and for the complete forecast period. It is clear from these figures that the strongest variability occurs for potential temperature and wind speed in the stable boundary layer at the end of the night in the morning transition hours. This is true for both the coupled and uncoupled cases, although with different magnitudes. During daytime the variability among the models is much less, because of the impact of convective mixing in such conditions.

The variances in the SBL occur over the same depth although with different magnitudes. It is also clear that the variability increases with forecasting time, which is to be expected in this nonlinear system (e.g., McNider et al. 1995). During the second night the maximum variance is 11.2 K^2, while in the first night this is only 4 K^2 (factor 3 smaller) for the uncoupled model runs (Fig. 10a). The variability in the model results is rather different for potential temperature and the wind speed magnitude by comparing their results for the coupled and uncoupled cases. For potential temperature the variability decreases with about a

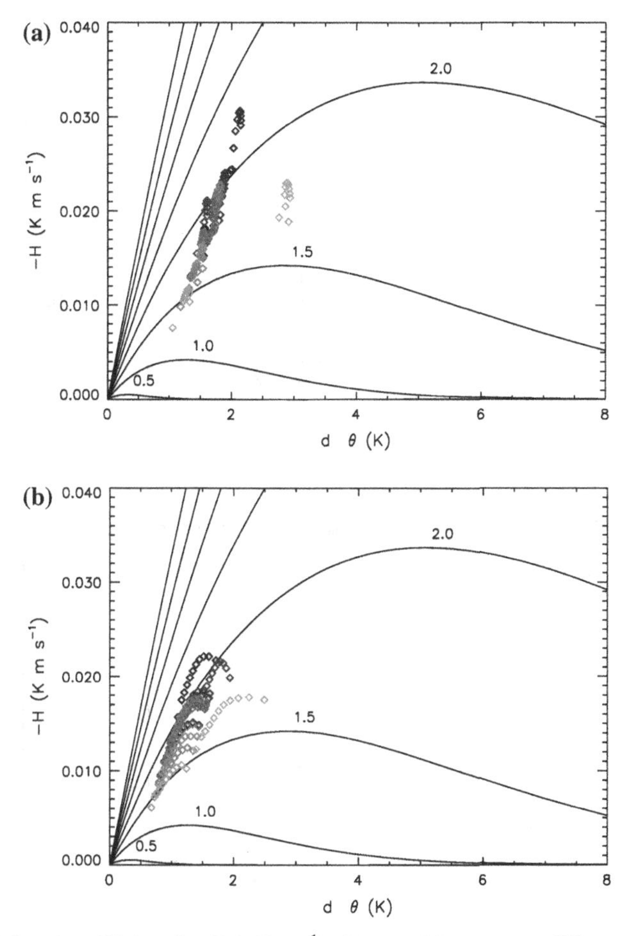

Fig. 7 The variation of sensible heat flux H (in $\mathrm{Kms^{-1}}$) with potential temperature difference near the surface for the reference model and perturbed model results (various symbols) for (**a**) prescribed surface temperatures, and (**b**) by solving the surface energy budget. The *full lines* refer to the wind speed with an interval of $0.5\,\mathrm{m\,s^{-1}}$ at a model height of 2 m

factor of 4 for the coupled case, and for the wind speed magnitude the variance decreases by 30%.

By repeating the model experiments with a lower geostrophic wind of $4.8\,\mathrm{m\,s^{-1}}$ (50% of the reference value), we find overall similar characteristics. However, the magnitudes for the variances of the predicted profiles for potential temperature and for the wind speed are typically smaller in the case of the lower geostrophic wind, both in the coupled and uncoupled model runs (not shown). Thus there is a clear dependence of the model results on both the surface temperature and the geostrophic wind speed, confirming earlier findings by Estournel and Guedalia (1985) and Gopalakrishnan et al. (1998).

From the findings presented, it is apparent that the treatment of the surface temperature boundary condition affects strongly the outcome of the boundary-layer model results and their variety. By repeating the uncoupled model runs with a specified surface temperature as given by the ensemble mean value of the interactive runs, we achieve basically the same

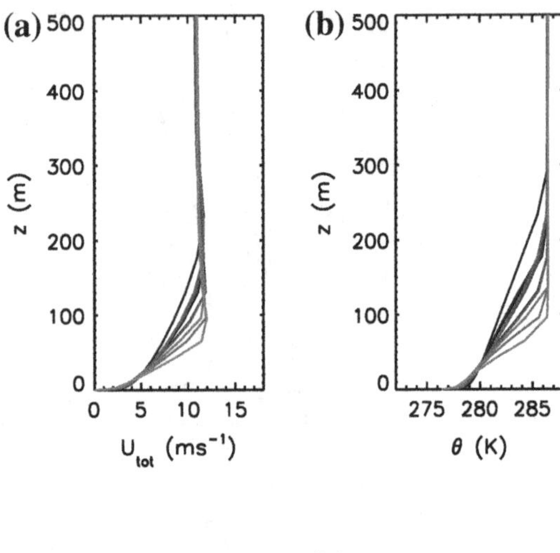

Fig. 8 The profiles of a) wind speed magnitude and b) potential temperature up to 500 m for a 12 h forecast

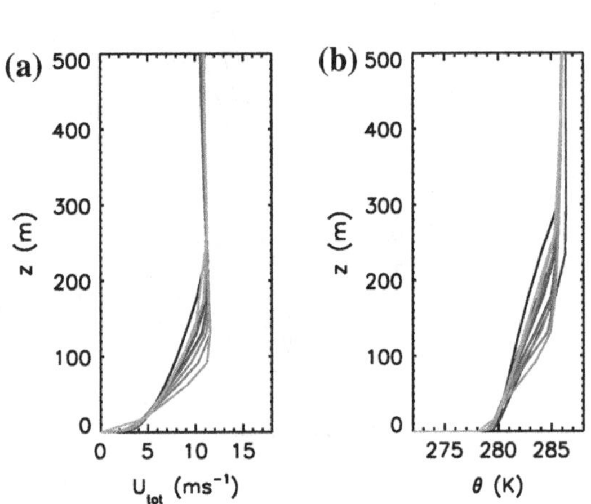

Fig. 9 As Fig. 8 but for the coupled runs

variety of model outputs for the potential temperature and wind as for the coupled cases. This confirms that in model evaluation studies the surface temperature should be taken consistent with the value of the geostrophic wind (although this is likely to be model dependent).

4 Discussion and conclusion

In this paper we have studied the impact of the surface temperature on the variability of results using an atmospheric boundary-layer model. First, it appears that most of the variability seen in the second GABLS model intercomparison case for stable conditions can be reproduced by taking one model and choosing alternative parameter values in a reasonable range. Second, the variety of model results is less when coupled to the land-surface. This is particularly true for the surface sensible heat flux, and the profiles of wind and temperature. However, we find that in the coupled case the realized surface temperatures are clearly different from the specified value of the uncoupled model case study. In addition we find sensitivity of the model results to the magnitude of the geostrophic wind speed.

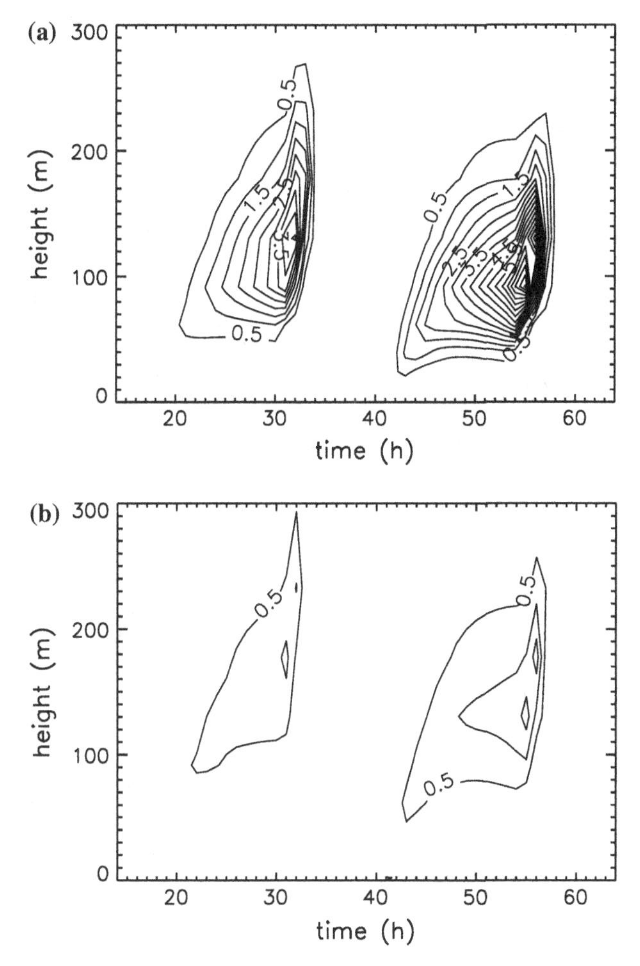

Fig. 10 Contour plot of variance of the predicted potential temperature in a prescribed (**a**) and coupled (**b**) case

From the coupled model results we find that surface feedback can compensate for some of the variety introduced by changing model parameters. Thus the evaluation of boundary-layer models is less critical when coupled to the land-surface, in particular for the nighttime boundary layer over land (see also Holtslag et al. 2006). However, this conclusion seems to depend on the combination of the specified geostrophic wind speed and the surface temperature. In fact these variables are related in the stable boundary layer over land (e.g., Estournel and Guedalia 1985; Derbyshire 1999; Gopalakrishnan et al. 1998 among many other studies).

Steeneveld et al. (2006b) were able to achieve realistic surface temperatures with a coupled model set-up similar to that used here, but by using a more detailed specification of the variation of the geostrophic wind as a function of time. In contrast, in the current study we use a constant geostrophic wind over time as was specified in the GABLS2 model intercomparison case. Then in combination with the specified values for the surface temperature, a larger range of results is found. If a surface temperature is chosen that is consistent with the magnitude of the geostrophic wind then a smaller variation of model results is found.

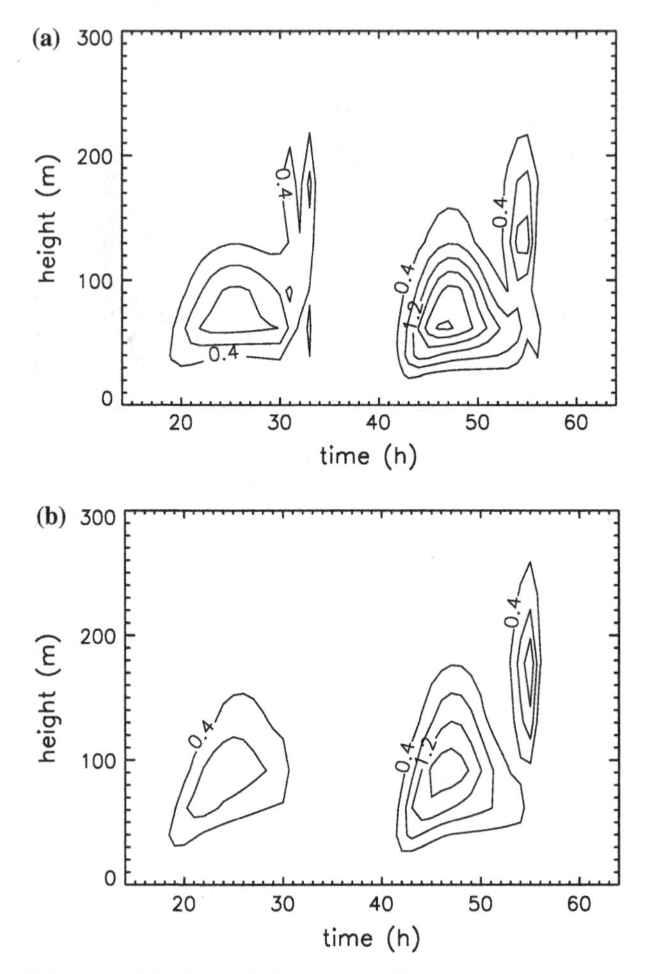

Fig. 11 As Fig. 10 for the model variance of wind speed magnitudes

In conclusion, the intercomparison and evaluation of boundary-layer models are not as simple and straightforward as it may seem. Our results herein indicate that variability among model results in stable conditions is not only related to the different parameterizations, but also to what extent the applied surface temperature forcing and the magnitude of the geostrophic wind are consistent with each other. This conclusion may also be relevant for large-eddy simulation studies (e.g., Beare et al. 2006; Kumar et al. 2006).

Acknowledgements The work was inspired by many discussions with colleagues at our Department and in the GABLS and GLASS/LOCO communities. We acknowledge the suggestions and comments on drafts of this paper by Gunilla Svensson (Stockholm University), Bart van den Hurk (KNMI), Jordi Vila (Wageningen Univ.), Sukanta Basu (Texas Tech Univ.), and the reviewers.

References

Baas P, Steeneveld GJ, van de Wiel BJH, Holtslag AAM (2006) Exploring Self-correlation in flux-gradient relationships for stably stratified conditions. J Atmos Sci 63:3045–3054

Basu S, Porte-Agel F, Foufoula-Georgiou E, Vinuesa JF, Pahlow M (2006) Revisiting the local scaling hypothesis in stably stratified atmospheric boundary-layer turbulence; an integration of field and laboratory measurements with large-eddy simulations. Boundary-Layer Meteorol 119:473–500

Beare R, MacVean M, Holtslag A, Cuxart J, Esau I, Golaz J-C, Jimenez M, Khairoutdinov M, Kosovic B, Lewellen D, Lund T, Lundquist J, McCabe A, Moene A, Noh Y, Raasch S, Sullivan P (2006) An intercomparison of large-eddy simulations of the stable boundary layer. Boundary-Layer Meteorol 118:247–272

Beljaars ACM, Viterbo p (1998) Role of the boundary layer in a numerical weather prediction model. In: Holtslag AAM, Duynkerke PG (eds) Clear and cloudy boundary layers. Royal Netherlands Academy of Arts and Sciences, Amsterdam, 372 pp

Clark PA, Hopwood WP (2001) One-dimensional site-specific forecasting of radiation fog. Part I: model formulation and idealized sensitivity studies. Meteorol Appl 8:279–286

Cuxart J, Holtslag AAM, Beare RJ, Bazile E, Beljaars A, Cheng A, Conangla L, Ek M, Freedman F, Hamdi R, Kerstein A, Kitagawa H, Lenderink G, Lewellen D, Mailhot J, Mauritsen T, Perov V, Schayes G, Steeneveld GJ, Svensson G, Taylor P, Weng W, Wunsch S, Xu K-M (2006) Single-column model intercomparison for a stably stratified atmospheric boundary layer. Boundary-Layer Meteorol 118:273–303

DeBruin HAR (1994) Analytic solutions of the equations governing the temperature fluctuation method. Boundary-Layer Meteorol 68:427–432

Delage Y (1997) Parameterising sub-grid scale vertical transport in atmospheric models under statically stable conditions. Boundary-Layer Meteorol 82:23–48

Delage Y, Bartlett PA, McCaughey JH et al (2002) Study of 'soft' night-time surface-layer decoupling over forest canopies in a land-surface model. Boundary-Layer Meteorol 103:253–276

Derbyshire SH (1999) Boundary layer decoupling over cold surfaces as a physical boundary instability. Boundary-Layer Meteorol 90:297–325

Duynkerke PG (1991) Radiation fog: a comparison of model simulation with detailed observations. Mon Wea Rev 119:324–341

Edwards JM, Beare RJ, Lapworth AJ (2006) Simulation of the observed evening transition and nocturnal boundary layers: single column modelling. Quart J Roy Meteorol Soc 132:61–80

Estournel C, Guedalia D (1985) Influence of geostrophic wind speed on atmospheric nocturnal cooling. J Atmos Sci 42:2695–2698

Garratt JR, Brost RA (1981) Radiative cooling effects within and above the nocturnal boundary layer. J Atmos Sci 38:2730–2746

Gopalakrishnan SG, Sharan M, McNider RT, Singh MP (1998) Study of radiative and turbulent processes in the stable boundary layer under weak wind conditions. J Atmos Sci 55:954–960

Holtslag AAM (2006) GEWEX atmospheric boundary layer study (GABLS) on stable boundary layers. Boundary-Layer Meteorol 118:243–246

Holtslag AAM, De Bruin HAR (1988) Applied modeling of the nighttime surface energy balance over Land. J Appl Meteorol 27:689–704

Holtslag AAM, Moeng CH (1991) Eddy diffusivity and countergradient transport in the convective atmospheric boundary layer. J Atmos Sci 48:1690–1698

Holtslag AAM, Steeneveld GJ, van de Wiel BJH (2006) Exploring variability of model results in the GEWEX atmospheric boundary layer study (GABLS). In: 17th symposium on boundary layers and turbulence, San Diego, USA, 22–25 May. American Meteorological Society, Boston, Paper 8.2. http://ams.confex.com/ams/pdfpapers/110553.pdf

Kumar V, Kleissl J, Meneveau C, Parlange MB (2006) Large-eddy simulation of a diurnal cycle of the atmospheric boundary layer: atmospheric stability and scaling issues. Water Resour Res 42:18

Lenderink G, Siebesma AP, Cheneit S, Ihrons S, Jones CG, Marquet P, Muller F, Olmera D, Calvo J, Sanchez E, Soares PMM (2004) The diurnal cycle of shallow cumulus clouds over land: a single-column model intercomparison study. Quart J Roy Meteorol Soc 130:3339–3364

Mahrt L (1998) Stratified atmospheric boundary layers and breakdown of models. Theo Comp Fluid Phys 11:263–279

Mahrt L (1999) Stratified atmospheric boundary layers. Boundary-Layer Meteorol 90:375–396

McNider RT, England DE, Friedman MJ, Shi X (1995) Predictability of the stable atmospheric boundary layer. J Atmos Sci 52:1602–1614

Poulos GS et al (2002) CASES-99: a comprehensive investigation of the stable nocturnal boundary layer. Bull Amer Meteorol Soc 83:555–581

Rao KS, Snodgrass HF (1979) Some parameterization of the nocturnal boundary layer. Boundary-Layer Meteorol 17:15–28

Salmond JA, McKendry IG (2005) A review of turbulence in the very stable boundary layer and its implications for air quality. Prog Phys Geogr 29:171–188

Steeneveld GJ, van de Wiel BJH, Holtslag AAM (2006a) Modeling the arctic stable boundary layer and its coupling to the surface. Boundary-Layer Meteorol 118:357–378

Steeneveld GJ, van de Wiel BJH, Holtslag AAM (2006b) Modeling the evolution of the atmospheric boundary layer coupled to the land surface for three contrasting nights in CASES-99. J Atmos Sci 63:920–935

Steeneveld GJ, Mauritsen T, de Bruijn EIF, Vilà-Guerau de Arellano J, Svensson G Holtslag AAM (2007) Evaluation of limited area models for the representation of the diurnal cycle and contrasting nights in CASES99. J Appl Meteor Clim (in press)

Svensson G, Holtslag (2006) Single column modeling of the diurnal cycle based on CASES99 data – GABLS second intercomparison project. In: 17th symposium on boundary layers and turbulence, San Diego, USA, 22–25 May. American Meteorological Society, Boston, Paper 8.1. http://ams.confex.com/ams/pdfpapers /110758.pdf

van de Wiel BJH (2002) Intermittency and oscillations in the stable boundary layer over land. PhD thesis. Wageningen University, no. 3319, 129 pp

van de Wiel BJH, Moene AF, Hartogensis OK, de Bruin HAR, Holtslag AAM (2003) Intermittent turbulence and oscillations in the stable boundary layer over land, Part III: a classification for observations during CASES99. J Atmos Sci 60:2509–2522

van de Wiel BJH, Moene AF, Steeneveld GJ, Hartogensis OK, Holtslag AAM (2007) Predicting the collapse of turbulence in stably stratified boundary layers. Flow, Turbulence Combust (in press)

Katabatic flow with Coriolis effect and gradually varying eddy diffusivity

Iva Kavčič · Branko Grisogono

Abstract Katabatic flows over high-latitude long glaciers experience the Coriolis force. A sloped atmospheric boundary-layer (ABL) flow is addressed which partly diffuses upwards, and hence, becomes progressively less local. We present the analytical and numerical solutions for (U, V, θ) depending on (z, t) in the katabatic flow, where U and V are the downslope and cross-slope wind components and θ is the potential temperature perturbation. A Prandtl model that accounts for the Coriolis effect, via f, does not approach a steady state, because V diffuses upwards in time; the rest, i.e., (U, θ), are similar to that in the classic Prandtl model. The V component behaves in a similar manner as the solution to the 1st Stokes (but inhomogeneous) problem. A WKB approach to the problem of the sloped ABL winds is outlined in the light of a modified Ekman-Prandtl model with gradually varying eddy diffusivity $K(z)$. Ideas for parameterizing these high-latitude persistent flows in climate models are revealed.

Keywords Low-level jet · Prandtl model · Strongly stable boundary layer

1 Introduction

Katabatic flows are regular features of the stable atmospheric boundary layer (ABL) over inclined radiatively cooled surfaces. The ubiquitous nature of katabatic flows over e.g., Antarctica and Greenland, not to mention smaller areas such as Iceland,

---

After Wentzel, Kramers and Brillouin, who popularized the method in theoretical physics.

I. Kavčič(✉)
Department of Geophysics, Faculty of Science, University of Zagreb, Horvatovac bb, 10 000 Zagreb, Croatia
e-mail: ivakavc@gfz.hr

B. Grisogono
Department of Geophysics, Faculty of Science, University of Zagreb, Zagreb, Croatia

Atmospheric Boundary Layers. A. Baklanov & B. Grisogono (eds.),

doi: 10.1007/978-0-387-74321-9_15,

and their cumulative effects, implies that the katabatic wind contributes to the general circulation (Parish and Bromwich 1991). Moreover, as katabatic flows may impinge on various coasts (Parmhed et al. 2004; Renfrew and Anderson 2006; Söderberg and Parmhed 2006), they may interact with sea ice and coastal ocean areas. It has been considered that katabatic flows might affect the thermohaline circulation and water mass conversions through the formation of coastal polynyas and the associated strong air–sea interaction (e.g., Gordon and Comiso 1988).

The detailed structure of katabatic flow still remains an important modelling issue (e.g., Weng and Taylor 2003). The stably stratified boundary layer is usually poorly resolved in many numerical models (e.g., Zilitinkevich et al. 2006), i.e., the modelling of katabatic flows is reasonably successful only if a sufficient vertical resolution is used (e.g., Renfrew 2004). A simple model of katabatic flows represents a balance between the negative buoyancy production due to the surface potential temperature deficit and dissipation by turbulent fluxes (e.g., Mahrt 1982; Egger 1990). On long glaciers in higher latitudes the Coriolis force also becomes an important contributor to the katabatic flow balance, deflecting the downslope component and leading to the occurrence of a wind component directed across the slope (Denby 1999; Van den Broeke et al. 2002). Stiperski et al. (2007) extended the Prandtl model by including the Coriolis force in order to be able to cover long polar slopes and the corresponding long-lived strongly stable ABL.

Furthermore, the pure katabatic flow is characterized by a pronounced low-level jet (LLJ) and sharp near-surface vertical temperature gradient (e.g., King et al. 2001; Grisogono and Oerlemans 2001a,b; Van den Broeke et al. 2002). Renfrew (2004) and Renfrew and Anderson (2006) show that significant katabatic flows over Antarctica most often exhibit clearly their LLJ and an anticlockwise backing of the wind with height. The authors suggest that this is due to a decrease in frictional forcing with height through the ABL. Moreover, Renfrew and Anderson (2006) indicate which kind of problems the measurements of katabatic flows may have, e.g., capturing the height of the LLJ that may exist just above a meteorological mast but still below the lowest sodar level. These authors illustrate that even a fine-scale nonhydrostatic numerical weather prediction (NWP) model encounters problems in modelling these widespread flows (to capture the jet-shaped shallow flow a model set-up with high vertical resolution is required), not to mention typical course-grid climate models. Therefore, katabatic flows typically have to be parameterized in large-scale models (e.g., Zilitinkevich et al. 2006), and to this end we further develop the Prandtl model with the Coriolis effect and variable eddy diffusivity.

King et al. (2001) show how sensitive the modelled Antarctic climate is to modifications of ABL parameterizations. Ever increasing resolution of the NWP and various regional models calls for continuous and necessary improvements of current parameterizations (e.g., various corrections to the Obukhov length). There is hardly any horizontal surface over land where the NWP model grid spacing falls below several km; in fact, slopes are typically between 0.5° and 10° to 20°. The surface slope, aside from violating horizontal homogeneity assumption, affects also Monin–Obukhov (MO) scaling as such: MO theory considers only the vertical component of the buoyancy (e.g., Munro and Davies 1978), neglecting its role as the driving force for katabatic flow in the horizontal momentum equation. In this study we revoke a known suggestion that an additional alternative for surface-layer scaling may be invoked — that from the Prandtl model relating to the LLJ height (Munro 1989, 2004; Grisogono and Oerlemans 2001a,b).

We continue the work of Grisogono and Oerlemans (2001a,b) by introducing a gradually varying eddy diffusivity in the analytical model given in Stiperski et al. (2007). The new approximate (and possibly asymptotic) solutions for katabatic boundary-layer flows, obtained by using e.g., the WKB method, may be useful in explaining various measurements (e.g., over the Antarctic), and to lend credibility for a more faithful parameterization of katabatic flows in meteorological and climate models. The paper is organized as follows. In Sect. 2 we present the main findings of Stiperski et al. (2007) as a starting point for introducing the varying eddy diffusivity. In Sect. 3 numerical solutions and approximate WKB solutions are presented. The conclusions are given in Sect. 4.

2 Rotating Prandtl model and solutions for constant eddy diffusivity

The rotating Prandtl model describes a hydrostatic, one-dimensional Boussinesq flow with the effects of the Coriolis force included. As in the classical Prandtl model (Mahrt 1982; Egger 1990; Parmhed et al. 2004), the K-theory is invoked to model the turbulent fluxes. The governing equations of the rotating model are thoroughly derived in Stiperski et al. (2007) under the assumption of a constant eddy thermal diffusivity K_c and a constant turbulent Prandtl number Pr. In the case of non-constant K, the equations for the downslope and cross-slope components of the wind vector (U, V), the potential temperature perturbation θ (total minus the background prescribed potential temperature) and the corresponding boundary conditions are:

$$\frac{\partial U}{\partial t} = g\frac{\theta}{\theta_0}\sin(\alpha) + f\cos(\alpha)V + \mathrm{Pr}\frac{\partial}{\partial z}\left(K\frac{\partial U}{\partial z}\right), \tag{1}$$

$$\frac{\partial V}{\partial t} = -f\cos(\alpha)U + \mathrm{Pr}\frac{\partial}{\partial z}\left(K\frac{\partial V}{\partial z}\right), \tag{2}$$

$$\frac{\partial \theta}{\partial t} = -\gamma\sin(\alpha)U + \frac{\partial}{\partial z}\left(K\frac{\partial \theta}{\partial z}\right), \tag{3}$$

$$\theta(z=0) = C, U(z=0) = V(z=0) = 0, \tag{4}$$

$$\theta(z\to\infty) = U(z\to\infty) = V(z\to\infty) = 0. \tag{5}$$

Here the z axis is not vertical but perpendicular to the surface (x axis) sloped with the negative (clockwise) angle α from the horizontal. The symbols have their usual meaning: θ_0 is a reference potential temperature, f is the Coriolis parameter, g is acceleration due to gravity and $C < 0$ is the constant surface-potential-temperature deficit, applied to an undisturbed atmosphere–surface interface instantaneously at the time $t = 0$. Slope angle α, for which the katabatic wind is successfully treated by the model, typically does not exceed 10°, therefore giving a reasonable assumption of using the constant gradient of the background potential temperature γ in the true vertical (Eq. 3). More about the model derivation can be found in e.g., Denby (1999).

Equations (1) through (5) can be used to describe the "primarily katabatic driven" flow, as selected by the criteria described in Renfrew and Anderson (2002). That is, such flows develop in the stable ABL where the surface radiation balance is a net cooling to space and the mesoscale pressure gradient is small, so that the influence from larger-scale weather systems is reduced. Such "typical" katabatic flow is shallow, with winds aloft decaying with height and rather weak compared to near-surface winds (Renfrew and Anderson 2006).

Before attempting to derive the analytical solutions for U, V and θ let us briefly revisit the main conclusions of Stiperski et al. (2007) for the case of $K(z) = K_c$, as they represent the starting point of discussion for the more general case of varying K.

- The approximate solutions for the steady-state potential temperature perturbation and down-slope velocity component (θ_s and U_s) are analogous to the classical Prandtl model:

$$\theta_s = C\exp\left(-\frac{z}{h_p}\right)\cos(\frac{z}{h_p}), \tag{6}$$

$$U_s = \frac{CK_c\sigma^2}{\gamma\sin(\alpha)}\exp\left(-\frac{z}{h_p}\right)\sin(\frac{z}{h_p}), \tag{7}$$

where $h_p = \sqrt{2}/\sigma$ is the Prandtl layer height,

$$\sigma = \left(\frac{N^2\,\mathrm{Pr}\sin^2(\alpha) + f^2\cos^2(\alpha)}{\mathrm{Pr}^2 K_c^2}\right)^{1/4}, \tag{8}$$

and N is the buoyancy (Brunt-Vaisala) frequency, satisfying $N^2 = \gamma g/\theta_0$. In (6) and (7) θ_s and U_s are the solutions of the 6th-order partial differential equation for each of the unknowns represented by the flow vector $F = (\theta, U, V)$:

$$\frac{d^2}{dz^2}\left(\frac{d^4F}{dz^4} + \sigma^4F\right) = 0. \tag{9}$$

Numerical solutions for U and θ asymptotically approach their steady state values U_s and θ_s after the characteristic time scale for the katabatic flow $T = 2\pi/(Nsin(\alpha))$ (Mahrt 1982; Grisogono 2003).

- Numerical solution for the cross-slope velocity component does not reach the steady state, but diffuses upwards through a several hundred m thick layer. However, the scale analysis carried out in Stiperski et al. (2007) has shown that the changes in V do not exert a significant influence on U and θ, which remain very close to their steady profiles U_s and θ_s. The ratio of the Coriolis term to the buoyancy term in (1) is, for typical katabatic flows, $O(10^{-2})$; hence, it is reasonable to neglect the Coriolis term for the analytical treatment of the simplified problem. Then (1) and (3) become weakly decoupled from (2), which becomes a forced diffusion equation. The analytic solution for V is thus obtained from Eq. (2), with U_s on the right-hand side as its forcing:

$$V_f = \frac{Cf\cot(\alpha)}{\mathrm{Pr}\,\gamma}\left[1 - erf\left(\frac{z}{2\sqrt{tK_c\,\mathrm{Pr}}}\right) - \exp\left(-\frac{z}{h_p}\right)\cos(\frac{z}{h_p})\right]. \tag{10}$$

The above solution holds after time $t > T$ needed for the forcing in (2), via U_s, to approach its steady state.

The derived solutions, together with the results from Grisogono (2003), lead us to the hypothesis that similar behaviour can also be expected in the case of a vertically varying eddy diffusivity. That is, the numerical results for U and θ would approach steady state within $T - 1.5T$, while V would continue to diffuse upwards, only this time with the limitations imposed by the $K(z)$ profile. Thus, V would behave as a solution to the 1st Stokes inhomogeneous problem (e.g., Kundu and Cohen 2002).

3 Solutions for varying eddy diffusivity

3.1 The WKB solutions

For $K = K(z)$, analytical solutions can be derived using the WKB method (Grisogono 1995; Grisogono and Oerlemans 2001a,b). More about the mathematical background of the method can be found in Bender and Orszag (1978). Furthermore, its use for pure katabatic flows is justified in Grisogono and Oerlemans (2002) and Parmhed et al. (2004).

We apply the method with a zero-order solution for θ and U. This approach keeps the balance between the terms with the largest amplitude in Eq. (9), modified for the varying K. Here, the derivatives of K are neglected and only its variations in σ are allowed. Nevertheless, it must be emphasized that, for the WKB method to be valid, the $K(z)$ profile must be either constant or gradually varying with respect to the vertical scale variations of the analytical solution. The latter means not only that $K(z)$ has to be a gradually varying function itself (Grisogono and Oerlemans 2001a), but also that the height of the maximum value of $K(z)$ (hereafter denoted by $K_{\max}$) must be above the LLJ height. In this paper we use the analytical $K(z)$ profile from Grisogono and Oerlemans (2001a,b), and Parmhed et al. (2004):

$$K(z) = K_{\max}\sqrt{e}\frac{z}{h}\exp\left(-\frac{z^2}{2h^2}\right), \tag{11a}$$

$$K_{\max} = 3K_c, \tag{11b}$$

where h is the level where $K_{\max}$ is reached. Here h can be estimated from the fact that the WKB solution for U will always place the LLJ below that calculated via the constant-K solution (Grisogono and Oerlemans 2001a,b). Moreover, the position of the LLJ height in V_f is always higher than in U_s, and also gradually increases in time, reaching ≈ 100 m (Stiperski et al. 2007). Simultaneously, the value of h is limited by the depth of the strongly stable ABL (Grisogono and Oerlemans 2002). The above conditions, together with the conditions imposed by the WKB method, give us a reasonable estimate of $h = 200$ m for the $K(z)$ profile used in the following example (Subsect. 3.2).

Relations between the best choices for K_c and $K_{\max}$ are discussed in Grisogono and Oerlemans (2001a). Here we just adopt the fact that it is reasonable if $K_c \approx 30\%$ of $K_{\max}$, as in Eq. (11b). Of course, other choices are possible depending on specific cases addressed. Further details on estimating $K_{\max}$ and h can be found in Grisogono and Oerlemans (2002) and Parmhed et al. (2004); Parmhed et al. (2005).

As discussed in Sect. 2, following the scale analysis in Stiperski et al. (2007) we neglect the Coriolis term in (1). This enables us to straightforwardly use the zero-order WKB approach for the modified flow vector $F = (\theta, U)$:

$$F_0 \propto \exp\left[-\frac{(1-i)}{\sqrt{2}}\sigma_0 I(z)\right], \tag{12}$$

where

$$I(z) = \int_0^z K(z)^{-1/2}\,dz, \tag{13}$$

and

$$\sigma_0^4 = \frac{N^2 \,\mathrm{Pr}\sin^2(\alpha) + f^2\cos^2(\alpha)}{\mathrm{Pr}^2}. \tag{14}$$

Furthermore, we define:

$$\sigma_{\mathrm{WKB}}(z) = \sigma_0 I(z), \tag{15}$$

which, together with the boundary conditions given in Eqs. (4) and (5), yield the solutions for θ and U:

$$\theta_{\mathrm{WKB}} = C\exp\left(-\frac{\sigma_{\mathrm{WKB}}(z)}{\sqrt{2}}\right)\cos\left(\frac{\sigma_{\mathrm{WKB}}(z)}{\sqrt{2}}\right), \tag{16}$$

$$U_{\mathrm{WKB}} = \frac{C\sigma_0^2}{\gamma\sin(\alpha)}\exp\left(-\frac{\sigma_{\mathrm{WKB}}(z)}{\sqrt{2}}\right)\sin\left(\frac{\sigma_{\mathrm{WKB}}(z)}{\sqrt{2}}\right). \tag{17}$$

As can be seen from previous studies (Grisogono 1995; Grisogono and Oerlemans 2001a,b, 2002) the WKB solutions are structurally similar to the constant-K case. In this study $I(z)$ is evaluated numerically, but it may be calculated also analytically, carefully taking into consideration its often divergent nature that is successfully overcome by the negative exponential in (12), and then in (16) and (17).

Moreover, the WKB solutions approach the constant-K solutions (6) and (7) as $K(z) \rightarrow K_c$. Then, $I(z)$ in (13) becomes $K_c^{-1/2}z$, and $\sigma_{\mathrm{WKB}}(z)/\sqrt{2}$ in (16) and (17) becomes $\sigma z/\sqrt{2} = z/h_p$ (Eqs. 6 and 7). This yields a reasonable assumption that V_f may also be considered as the limit value of the corresponding WKB solution and implies the expansion of the argument of the error function in (10) for the case of variable $K(z)$. That is, $zK_c^{-1/2} \rightarrow I(z)$ in (10), giving us the solution for $V(z,t)$:

$$V_{\mathrm{WKB}} \approx \frac{Cf\cot(\alpha)}{\mathrm{Pr}\,\gamma}\left[1 - erf\left(\frac{I(z)}{2\sqrt{t\,\mathrm{Pr}}}\right) - \exp\left(-\frac{\sigma_{\mathrm{WKB}}(z)}{\sqrt{2}}\right)\cos\left(\frac{\sigma_{\mathrm{WKB}}(z)}{\sqrt{2}}\right)\right]. \tag{18}$$

Again, $t > T$ as in (10). The comparison between the analytical and numerical solutions, as well as comparison with the constant-K case, is given in the following section.

3.2 Comparison with the numerical and constant-K solutions

Following Stiperski et al. (2007), the analytical solutions are verified against the numerical solutions of the time-dependent system (1)–(3) obtained using the simple numerical model from Grisogono (2003). The numerical and WKB solutions for U and $\theta^{\mathrm{tot}} = \theta + \gamma z$ are compared for a case with physical parameters $(f,\alpha,\gamma,Pr,C) = (1.1\times10^{-4}\,\mathrm{s}^{-1}, -4^\circ, 4\times10^{-3}\,\mathrm{K\,m}^{-1}, 1.1, -8^\circ\mathrm{C})$, and the prescribed $K(z)$ from (11a) and (11b). Here θ^{tot} is calculated and plotted without the reference potential temperature θ_0 (to reword, the constant θ_0 is already subtracted from θ^{tot}). From Fig. 1 it can be seen that the numerical solution (dashed) for both U and θ^{tot} are in excellent agreement with the steady state solutions (16) and (17) for $t \geq T$ (solid). Such agreement is expected from the results for the constant-K case described in Stiperski et al. (2007, see their Figure 2).

Figure 2 displays both the WKB and constant-K solutions for U and θ^{tot}, showing the improvement in describing the sharp near-surface gradients in temperature

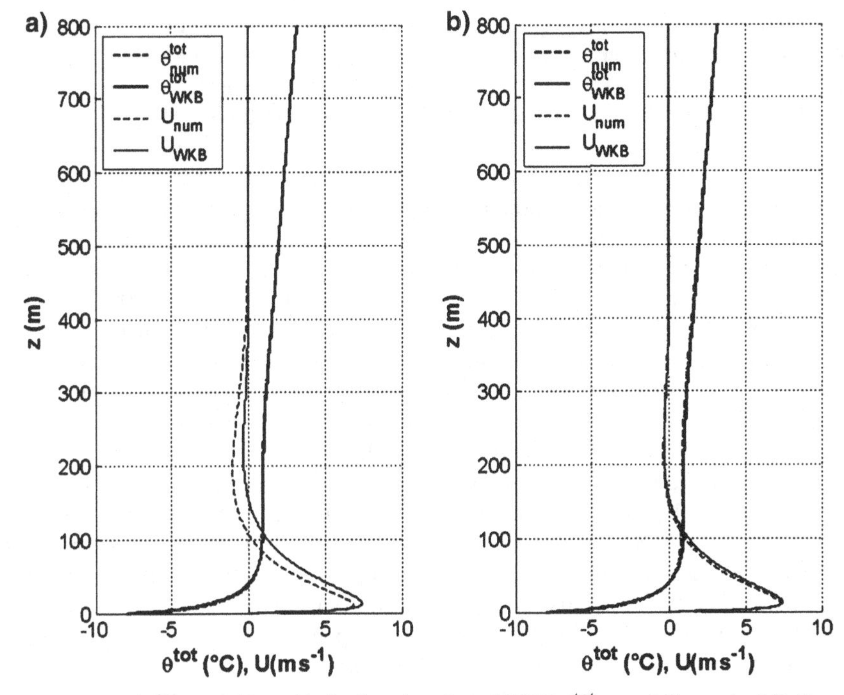

Fig. 1 Numerical $\theta^{\mathrm{tot}}_{\mathrm{num}}$ and U_{num} (dashed) and analytical WKB $\theta^{\mathrm{tot}}_{\mathrm{WKB}}$ and U_{WKB} (solid), Eqs. (16) and (17), solutions for the Prandtl model, at (**a**) $t = T$ and (**b**) $t = 10T, T = 2\pi/(N\sin(\alpha)) \approx 2.1\,\mathrm{h}$. Here $K(z)$ is from (11a) and (11b), with $K_{\max} = 3\,\mathrm{m^2s^{-1}}$ at $h = 200\,\mathrm{m}$; other parameters are $(f, \alpha, \gamma, Pr, C) = (1.1 \times 10^{-4}\,\mathrm{s^{-1}}, -4°, 4 \times 10^{-3}\,\mathrm{K\,m^{-1}}, 1.1, -8°\mathrm{C})$. The numerical model top is at 2,000 m

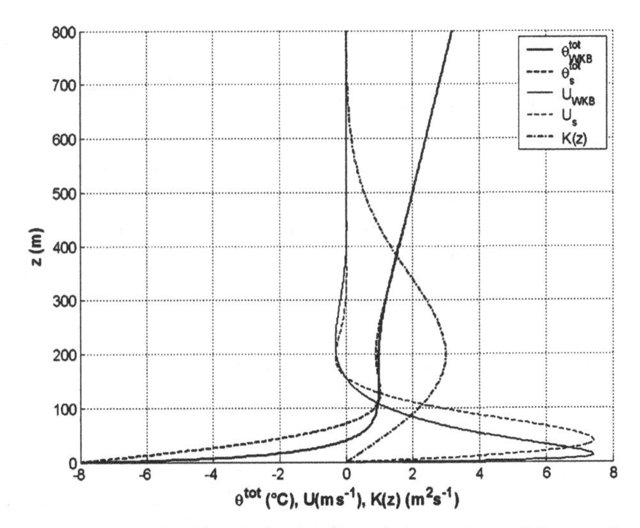

Fig. 2 The prescribed $K(z)$ profile (dot-dashed) and analytic solutions of the rotating Prandtl model for the case of varying (solid) and constant K (dashed). Here $K(z)$ is from Eqs. (11a) and (11b), $K_c = 1\,\mathrm{m^2s^{-1}}$, and θ^{tot}_s and U_s from (6) and (7). The rest as in Fig. 1

and wind that are often observed (Defant 1949; Munro 1989; Egger 1990; Oerlemans 1998; Parmhed et al. 2004). This is also in agreement with the analysis of Grisogono and Oerlemans (2001a,b) for the non-rotating model, and yields the better estimate of both the LLJ height, and surface heat and momentum fluxes. Yet another difference can be seen between U_{WKB} and U_s: both profiles have the return flow around $z \approx 200\,\mathrm{m}$ of similar amplitude, but this layer is thicker for the $K(z)$ case.

The sharper near-surface gradient and the lower LLJ height are also seen for the cross-slope wind component V, when $K(z)$ is employed, Fig. 3. There V_{num} still diffuses upwards but, as expected, its propagation is now limited to the height where the values of $K(z)$ approach zero ($z \approx 800\,\mathrm{m}$, Fig. 2). This leads us to the conclusion that the hypothesis of V influencing the polar vortex after sufficient time imposed by Stiperski et al. (2007) should be more relaxed in this more realistic case. There is another significant difference, i.e., the presence of a secondary bulge in V above the height of K_{max} at $z \approx 400$ or $500\,\mathrm{m}$. As the integration time increases, this bulge strengthens and expands with height, nevertheless obeying the limitations imposed by $K(z)$. The bulge in $V(z,t)$ occurs because of two opposing effects. Both $V(z,t)$, namely V_{num} and V_{WKB}, try to diffuse upwards as in the 1st Stokes problem, which is nicely emulated in Stiperski et al. (2007). However, at progressively higher levels there is less and less $K(z)$ for mixing the V component upward. Hence, $V(z,t)$ finds less and less medium to diffuse through and starts to accumulate below $K(z) \to 0$ level (Fig. 3, black solid line). On the contrary, deep and non-decaying K supports the vertical diffusion of $V(z,t)$ (Fig. 3, grey solid line).

The overall behaviour of V_{num} is very well described with the new approximate WKB solution V_{WKB} from (18), only slightly overestimating the maximum amplitude. Similar behaviour of the analytical solution V_f has also been observed for the constant-K case in Stiperski et al. (2007). The detailed calculation presented here for $V(z,t)$ also explains the behaviour of the V component in Denby (1999), which was not commented there (see his Figure 2e and 5).

Additional remarks on how to estimate the input parameters for this Ekman–Prandtl model type with $K(z)$ can be found in Parmhed et al. (2004); Parmhed et al. (2005). The new analytical solutions $(U,V,\theta)_{\mathrm{WKB}}$, (16), (17) and (18) are not named "asymptotic", which usually holds for the WKB solutions, only because we weakly decoupled (2) so that V_{WKB} does not feed back to the original system (1)–(3). The numerical result shows, as also in Stiperski et al. (2007), that the V effect on the katabatic dynamics is negligible. However, the induced $V(z,t)$ affects the wind direction and the horizontal momentum flux.

4 Conclusions

A better understanding of katabatic flows is necessary for better treatment and parameterization of the coupling between the atmosphere and cool, inclined surfaces (e.g., King et al. 2001; Weng and Taylor 2003). The rotating Prandtl model (Stiperski et al. 2007), although providing the analytical tool for analyzing this coupling, does not hold for the real atmosphere due to the assumption of constant eddy diffusivity. In this work an attempt is made towards a more realistic description of the long-lived katabatic strongly stable ABL through the approach of Grisogono and Oerlemans (2001a,b). There, the asymptotic solutions for the Prandtl model with gradually varying $K(z)$, but without rotation, were obtained using the WKB method. The obtained solutions

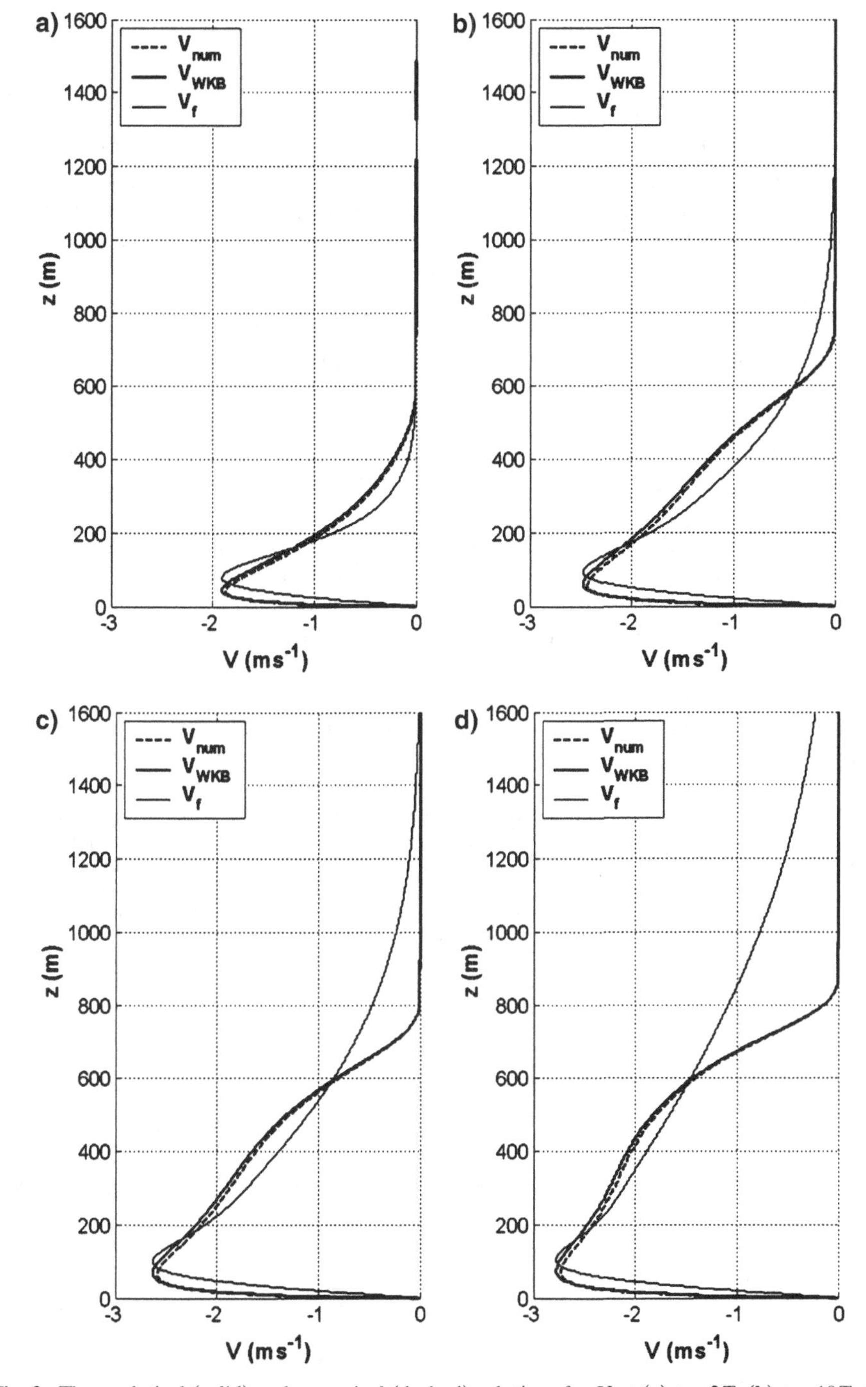

Fig. 3 The analytical (solid) and numerical (dashed) solutions for V at (**a**) $t = 2T$, (**b**) $t = 10T$, (**c**) $t = 20T$ and (**d**) $t = 50T$. The WKB solution V_{WKB} is in (18); the constant K solution, V_f, is given in (10). The rest as in Fig. 1

were verified against the results from the numerical model (Grisogono 2003), and independently against a dataset from Breidamerkurjokull, Iceland (Parmhed et al. 2004). Here, the analytical and numerical solutions for (U, V, θ) depending on (z, t) in the rotating katabatic flow are presented.

As expected, the overall change of the flow vector (U, V, θ) is structurally similar to the constant-K case (Stiperski et al. 2007). Both U and θ reach their steady-state profiles after the typical time scale for simple katabatic flows $T \approx 2\pi/(N\sin(\alpha))$, and V still diffuses upwards in time without a well-defined time scale. Contrary to the constant-K case, the upward propagation of $V(z,t)$ is now limited by the vertically decaying values of $K(z)$ above its maximum. As the result, the elevated bulge in the $V(z,t)$ profile is observed above the weak return flow in U. This feature indicates the trapping of the V momentum at the height where $K(z)$ approaches a zero value, whereas for the constant-K values the V momentum continuously propagates under diffusion in the vertical (Stiperski et al. 2007). For example, if there was pre-existing elevated turbulence, e.g., residual turbulent layer(s), then the katabatic effect could, in principle, still influence the polar vortex after sufficiently long duration of the flow during the polar night.

This study shows that the WKB method of zero-order may be successfully applied to find the approximate analytical solutions for all the model components. The new WKB solution is relatively simple to derive and calculate either by analytical or numerical evaluation of the integral expression (13). The proposed analytical solutions (16), (17) and (18) can be used for studying katabatic flows over long slopes. Together with the introduction of the varying eddy diffusivity profile, the proposed solutions give a more realistic description of sloped surface-flux parameterizations in climate models and data analysis.

Acknowledgements Danijel Belušić is thanked for his insightful comments that helped to improve the manuscript substantially. Constructive criticism from three anonymous reviewers is appreciated. Ivana Stiperski, Dale R. Durran and Peter A. Taylor are thanked for the many fruitful discussions. This study was supported by the Croatian Ministry of Science, Education and Sports under the projects "Numerical methods in geophysical models" (No. 037-1193086-2771, Dept. of Mathematics), and "BORA" (No. 119-1193086-1311, Dept. of Geophysics).

References

Bender CM, Orszag SA (1978) Advanced mathematical methods for scientists and engineers. Mc Graw-Hill, Inc., New York, 593 pp

Defant F (1949) Zur theorie der Hangwinde, nebst Bemerkungen zur Theorie der Bergund Talwinde. Arch Meteor Geophys Biokl Ser A1:421–450

Denby B (1999) Second-order modelling of turbulence in katabatic flows. Boundary-Layer Meteorol 92:67–100

Egger J (1990) Thermally forced flows: theory. In: Blumen W (ed) Atmospheric processes over complex terrain. American Meteorological Society, Boston, MA, pp 43–57

Gordon AL, Comiso JC (1988) Polynyas in the Southern Ocean. Sci Am 258(6):90–97

Grisogono B (1995) A generalized Ekman layer profile within gradually varying eddy diffusivities. Quart J Roy Meteorol Soc 121:445–453

Grisogono B, Oerlemans J (2001a) Katabatic flow: analytic solution for gradually varying eddy diffusivities. J Atmos Sci 58:3349–3354

Grisogono B, Oerlemans J (2001b) A theory for the estimation of surface fluxes in simple katabatic flows. Quart J Roy Meteorol Soc 127:2725–2739

Grisogono B, Oerlemans J (2002) Justifying the WKB approximation in the pure katabatic flows. Tellus 54A:453–463

Grisogono B (2003) Post-onset behaviour of the pure katabatic flow. Boundary-Layer Meteorol 107:157–175
King JC, Conneley WM, Derbyshire SH (2001) Sensitivity of modelled Antarctic climate to surface and boundary-layer flux parameterizations. Quart J Roy Meteorol Soc 127:779–794
Kundu PK, Cohen IM (2002) Fluid mechanics, 2nd ed. Academic Press, San Diego, Calif., London, 730 pp
Mahrt L (1982) Momentum balance of gravity flows. J Atmos Sci 39:2701–2711
Munro DS (1989) Surface roughness and bulk heat transfer on a glacier: comparison with eddy correlation. J Glaciol 35:343–348
Munro DS (2004) Revisiting bulk heat transfer on the Peyto glacier in light of the OG parameterization. J Glaciol 50:590–600
Munro DS, Davies JA (1978) On fitting the log-linear model to wind speed and temperature profiles over a melting glacier. Boundary-Layer Meteorol 15:423–437
Oerlemans J (1998) The atmospheric boundary layer over melting glaciers. In: Holtslag AAM, Duynkerke PG (eds) Clear and cloudy boundary layers. Royal Netherlands Academy of Arts and Sciences, Place, VNE 48, ISBN 90-6984-235-1: 129–153
Parish TR, Bromwich DH (1991) Continental-scale simulation of the Antarctic katabatic wind regime. J Climate 4:135–146
Parmhed O, Oerlemans J, Grisogono B (2004) Describing the surface fluxes in the katabatic flow on Breidamerkurjokull, Iceland. Quart J Roy Meteorol Soc 130:1137–1151
Parmhed O, Kos I, Grisogono B (2005) An improved Ekman layer approximation for smooth eddy diffusivity profiles. Boundary-Layer Meteorol 115:399–407
Renfrew IA, Anderson PS (2002) The surface climatology of an ordinary katabatic wind regime in Coats Land, Antarctica. Tellus 54A:463–484
Renfrew IA (2004) The dynamics of idealized katabatic flow over a moderate slope and ice shelf. Quart J Roy Meteorol Soc 130:1023–1045
Renfrew IA, Anderson PS (2006) Profiles of katabatic flow in summer and winter over coats land, Antarctica. Quart J Roy Meteorol Soc 132:779–882
Söderberg S, Parmhed O (2006) Numerical modelling of katabatic flow over a melting outflow glacier. Boundary-Layer Meteorol 120:509–534
Stiperski I, Kavčič I, Grisogono B, Durran DR (2007) Including Coriolis effects in the Prandtl model for katabatic flow. Quart J Roy Meteorol Soc 133:101–106
Van den Broeke MR, van Lipzig NPM, van Meijgaard E (2002) Momentum budget of the East-Antarctic atmospheric boundary layer: results of a regional climate model. J Atmos Sci 59:3117–3129
Weng W, Taylor PA (2003) On modelling the one-dimensional atmospheric boundary layer. Boundary-Layer Meteorol 107:371–400
Zilitinkevich S, Savijärvi H, Baklanov A, Grisogono B, Myrberg K (2006) Forthcoming meetings on planetary boundary-layer theory, modelling and applications. Boundary-Layer Meteorol 119:591–593

Parameterisation of the planetary boundary layer for diagnostic wind models

Massimiliano Burlando · Emilia Georgieva · Corrado F. Ratto

Abstract The planetary boundary-layer (PBL) parameterization is a key issue for the definition of initial wind flow fields in diagnostic models. However, PBL theories usually treat separately stable, neutral, and convective stability conditions, so that their implementation in diagnostic wind models is not straightforward. In the present paper, an attempt is made to adopt a comprehensive PBL parameterisation, covering stable/neutral and unstable atmospheric conditions, which appears suitable to diagnostic models. This parameterisation is implemented into our diagnostic mass-consistent code. A validation of the consistency between the implemented PBL parameterisations has been checked through an analysis of the sensitivity of the vertical wind profiles to atmospheric stability.

Keywords Applied oriented numerical models · Atmospheric stability · High-resolution surface winds · PBL parameterisation

1 Introduction

Planetary boundary-layer (PBL) parameterisation is a key issue for a range of numerical codes, for applied oriented research. Wind energy studies, dispersion of air pollutants on the local scale, forest fire risk assessment, weather forecasting, weather routing, and wind action on structures are some of the themes, which require the simulation of high-resolution winds in the PBL.

At the Department of Physics, University of Genoa, we have developed, adapted, and applied a number of such codes in recent years, to tackle such type of problems. In particular, our modelling tools include different numerical codes: the operational forecast model BOLAM (Buzzi et al. 1994), the diagnostic mass consistent code WINDS

M. Burlando (✉) · C. F. Ratto
Department of Physics, University of Genoa, Via Dodecaneso 33, Genoa 16146, Italy
e-mail: burlando@fisica.unige.it

E. Georgieva
Institute of Geophysics, Bulgarian Academy of Sciences, Sofia, Bulgaria

Atmospheric Boundary Layers. A. Baklanov & B. Grisogono (eds.),
doi: 10.1007/978-0-387-74321-9_16,

(Ratto et al. 1990), the dispersion package SAFE_AIR II (Canepa and Ratto 2003). In the present paper we outline the current PBL parameterisation in the flow model WINDS, because this model is part also of the dispersion code and because it is used for downscaling of the coarser grid weather forecast model.

The first guess wind field for any diagnostic wind model is crucial for the final results. Indeed, in the absence of sufficient wind observations, a proper PBL parameterisation would be required to reproduce the wind profile starting from wind observations aloft, wind observations at the surface, or to downscale the wind field of the coarser meteorological model. Actually, to the authors' knowledge, PBL height and wind profile parameterisations are usually treated in a different way under different stability conditions. In particular, stable/neutral and unstable atmospheric conditions are usually considered separately. In this context, putting into effect a PBL model, coherent with all atmospheric conditions, and which is suitable for mass-consistent models, is not straightforward, and any parameterisation has to be carefully tested in order to check its compatibility with previously implemented models.

The algorithm implemented in WINDS is based on recent theories for the PBL under stable/neutral conditions (Zilitinkevich et al. 1998; Zilitinkevich and Calanca 2000; Zilitinkevich and Esau 2002), while for the convective case somewhat older parameterisations (Zilitinkevich et al. 1992) are introduced. Different WINDS modules are affected by these parameterisations, i.e. the top-down and bottom-up algorithms for wind profile calculations starting respectively from upper air, or near ground wind data, as well as the internal boundary-layer module that takes into account the abrupt change in the surface roughness and its influence on the wind profile.

The vertical profile of the mean wind components under stable/neutral conditions in a barotropic atmosphere is obtained through the polynomial expressions of Zilitinkevich et al. (1998), which involve the knowledge of four basic PBL parameters: the PBL height, the friction velocity and two parameters of stratification to take into account the stability inside the surface boundary layer as well as the static stability aloft. Recently proposed multi-limit expressions for the equilibrium PBL height (Zilitinkevich et al. 2002) are incorporated in the algorithm.

Under convective conditions instead of expressions for the horizontal wind components, the algorithm implemented is based on the expressions for the vertical wind speed profile and for the wind turn angle (Zilitinkevich et al. 1992; Mironov, 1999). The main problem in our code is the estimation of the convective boundary-layer (CBL) height, since the flow model is diagnostic whereas up-to-date formulations are prognostic. Thus, a simple prognostic CBL expression has been combined with results for the surface heat flux derived from the meteorological pre-processor ABLE (Georgieva et al. 2003).

This paper discusses the sensitivity of the wind profile parameterisations for different input parameters of the code WINDS. Section 2 briefly outlines the mass-consistent models focusing in particular on the initial wind field calculation that strongly depends on the PBL formulations. Section 3 describes the parameterisations adopted in the mass-consistent model WINDS under different stability conditions. In Sect. 4, a sensitivity analysis for the simulated wind profiles is presented. Finally, conclusions and future perspectives are given in Sect. 5.

2 Mass-consistent models

Diagnostic models are simple numerical codes, easy-to-operate and economical with respect to computer resources, widely used to simulate average wind fields, i.e. the deterministic three-dimensional (3D) motion field with time scales from minutes to days and spatial scales from kilometres on. Mass-consistent models, in particular, are a class of diagnostic models that produce three-dimensional wind fields based on the satisfaction of the physical constraints of mass conservation (Sherman 1978).

These codes reconstruct the three-dimensional wind field by means of a two-step procedure. Firstly, wind data are interpolated over the computational domain, transforming the observed wind vectors or other kinds of initial wind data in a three-dimensional "first-guess" wind field; different wind fields can be generated starting from the same initial data depending on the chosen interpolation method, e.g. the PBL parameterisation. Then the interpolated field is adjusted by a minimum possible number of modifications to satisfy the mass conservation constraint, in order to obtain the "final" 3D wind field; mass is conserved over the entire domain, both accounting for flow through the boundaries and imposing mass conservation locally everywhere. A review of mass-consistent models and relative techniques is provided by Ratto et al. (1994).

Mass-consistent codes were originally applied to diagnose atmospheric states from wind observations taken, at a given time, at discrete points in space. Nowadays, a common practice to construct the initial field in the absence of, or in addition to, field measurements is the systematic use of outputs of larger scale and lower resolution numerical weather prediction models, like limited area models (LAM). The diagnostic model is nested into the LAM and it is used to downscale the wind field produced by the prognostic model (see, for instance, Chandrasekar et al. 2003; Furunoa et al. 2004).

An alternative or complementary way of initializing diagnostic codes consists on requiring that the wind field be in barotropic balance with the geostrophic or gradient wind aloft. Following this approach, vertical profiles of the wind velocity, based on similarity-theory formulations, can be used at the initialization step to define the first-guess wind field. This is the case that we will focus on in the present paper, since it does not require a great number of wind measurements but makes wide use of PBL parameterisations.

All the parameters for the sensitivity analysis carried out in this work are produced by the diagnostic flow model Wind-field Interpolation by Non Divergent Schemes (WINDS) (Ratto et al. 1990; Georgieva et al. 2003). During the last decade it has been extensively used for many geophysical and engineering applications, such as dispersion modelling on the local scale (Canepa and Builtjes 2001), wind energy potential evaluations (Burlando et al. 2002) and applied wind engineering problems (Castino et al. 2003). The code has been validated both against wind-tunnel data (Trombetti et al. 1991) and data obtained from field campaigns in coastal mountainous terrains (Canepa et al. 1999).

3 PBL parameterisation

The vertical profiles of the wind velocity, used at the initialization step in WINDS to define the first-guess wind field, are based on similarity-theory formulations. Stable/neutral and convective atmospheric conditions are considered separately in the next sub-sections. The consistency of the two formulations is shown in the sensitivity analysis of Sect. 4.

3.1 The wind profile for the stable and neutral PBL

The formulations for the vertical wind profile under stable/neutral conditions, implemented in the present version of WINDS, have been developed by Zilitinkevich et al. (1998). They represent the wind profile in a barotropic atmosphere in terms of logarithmic plus polynomial functions of dimensionless height z/h (z is the height above the ground and h is the boundary-layer height) with polynomial coefficients depending on boundary-layer governing parameters. Previous wind profile formulations of the same type (Zilitinkevich 1989) account only for the effect of density stratification due to the surface heat flux, while the extended formulations used here, incorporate also the effect of the free flow static stability.

The wind profile expressions are given by:

$$u(\zeta) = \frac{u_*}{\kappa}\left[\ln\frac{\zeta}{\zeta_0} + b_1(\zeta-\zeta_0) + b_2(\zeta-\zeta_0)^2 + b_3(\zeta-\zeta_0)^3\right] \quad (1a)$$

$$v(\zeta) = \frac{u_*}{\kappa}\delta\left[-(\zeta-\zeta_0)\ln(\zeta) + a_1(\zeta-\zeta_0) + a_2(\zeta-\zeta_0)^2 + a_3(\zeta-\zeta_0)^3\right] \quad (1b)$$

where u and v are the components of the horizontal wind velocity along the $x-$ and $y-$axis of a right-hand Cartesian coordinate system with the $x-$axis along the surface stress, $\zeta \equiv z/h$ and $\zeta \equiv z_0/h$ are dimensionless heights, z_0 is the surface roughness length, h is the stable/neutral boundary-layer (SBL) height, $\kappa \approx 0.4$ is the von Kármán constant and $\delta = fh/\kappa u_*$ is the dimensionless "rotation rate parameter" with f being the Coriolis parameter and u_* the surface friction velocity. Note that the vertical component w of the wind speed, directed upward, is assumed to be zero since Eq. 1 holds over flat terrain. Furthermore, this expression has been slightly modified with respect to the original one in Zilitinkevich et al. (1998), in order to give $u(z = z_0) = 0$ and $v(z = z_0) = 0$.

The coefficients $a_1, a_2, a_3, b_1, b_2, b_3$ depend on dimensionless stratification/rotation parameters, functions of the Obukhov length $L = u_*^3/\kappa B_s$, where B_s is the surface buoyancy flux (Zilitinkevich et al. 1998; Mironov 1999).

Based upon Eq. 1 the calculation of the vertical profile of the mean wind speed components involves the knowledge of four basic PBL parameters: the SBL height, the surface friction velocity u_*, and the stability parameters $\mu = u_*/|f|L$ and $\mu_N = N/|f|$ (or the Obukhov length L and the Brunt Väisälä frequency N above the SBL).

The use of N implies some knowledge of the SBL itself, requiring also temperature profile observations not routinely available. According to Zilitinkevich and Esau (2002) at mid latitudes $\mu_N \approx 10^2$ and it does not present strong variations. Thus, it is assumed as a constant in the considered wind flow model.

Traditional theory of the neutral and stable PBL has been recently revised to take into account the effect of the free-flow static stability and baroclinicity on the turbulent transport of momentum and scalars in the boundary layer (Zilitinkevich and Esau 2002, 2003; Zilitinkevich et al. 2002; Hess and Garratt 2002; Hess 2004). According to the new developments, different types of PBL regimes can be distinguished: truly neutral; conventionally neutral; short lived nocturnal and long-lived.

In the WINDS code the SBL height, h, is evaluated according to the expression recently introduced by Zilitinkevich et al. (2002). This expression represents a multi-limit equation for the equilibrium PBL height that covers the above mentioned types of neutral and stable conditions in the atmosphere.

3.2 The wind profile for the unstable PBL

The CBL wind profile formulation implemented in the present version of the code WINDS is given by Zilitinkevich et al. (1992) and Mironov (1999). Instead of expressions for the horizontal wind components, a relation for the modulus of the wind speed, U, in the surface layer and in the layer above it has been derived:

$$U(z) = \begin{cases} \frac{u_*}{\kappa} \ln\left(\frac{z}{z_0}\right) & at\ \kappa z_0 \le \kappa z \le |\xi_u L| \\ \frac{u_*}{\kappa} \left[a_u + C_u \left(\frac{\kappa z}{-L}\right)^{-1/3} + \ln\left(\frac{-L}{\kappa z_0}\right)\right] & at\ |\xi_u L| \le \kappa z \le \kappa h \end{cases} \quad (2)$$

where $\xi_u \approx 0.1$, $a_u \approx 0.7$, and $C_u \approx 1.4$ are dimensionless constants, and h is the convective boundary-layer height. The angle of wind turn in the boundary layer is given by an expression due to Mironov (1999). To calculate the vertical profile of the horizontal wind components, the wind turn angle is assumed to vary linearly with height reaching zero at the top of the boundary layer.

In contrast to the stable and neutral cases, the estimation of the CBL height to be used in WINDS is not straightforward, as the code is diagnostic and the recommended parameterisations, reviewed in Fisher et al. (1998), are prognostic. In order to overcome this problem, a simple prognostic expression for the CBL height has been combined with results deriving from a meteorological pre-processor that simulates atmospheric boundary-layer parameters based on the surface energy balance method. The resulting height, h, of the unstable PBL refers to the maximal CBL height and is calculated by:

$$h^2 = h_0^2 + 2\frac{C_I}{N^2}\int_{t_0}^{t} B_s dt \quad (3)$$

where h_0 is the height of the stable/neutral PBL at time t_0, when convection starts, t is the time of fully developed convection or when the CBL height reaches a maximum, B_s is the buoyancy heat flux at the surface, $C_I = 1.2$ is an empirical coefficient, $N = 10^{-2}\,\mathrm{s}^{-1}$ is the free-flow Brunt Väisälä frequency assumed to have a constant value. Here, h_0 is estimated by the stable/neutral parameterization as $h_0 = 0.125u_*/f$, while B_s is related to the surface sensible heat flux H through:

$$B_s = \frac{\beta}{\rho c_p} H \quad (4)$$

with $\rho \approx 1.23\,\mathrm{kg\,m^{-3}}$ being the air density, $c_p \approx 1004\,\mathrm{J\,kg^{-1}\,K^{-1}}$ is the specific heat at constant pressure, and $\beta \approx 0.03\ \mathrm{m\,s^{-2}\,K^{-1}}$ is the buoyancy parameter. For estimation of the sensible heat flux, as well as the time interval for the integral in Eq. 3, we have applied the meteorological pre-processor ABLE (Georgieva et al. 2001) to a few test regions in Italy.

4 Sensitivity analysis

A sensitivity analysis has been carried out to the aforementioned stable/neutral and unstable PBL parameterisation schemes in order to determine the sensitivity of the model outcomes to changes in the input parameters. Classically, this kind of analysis is used to check if small changes in a parameter result in relatively large changes in the outcomes, so that the results are said to be sensitive to that parameter. In the present context, we will use the sensitivity analysis mainly to check the consistency between the two PBL schemes outlined in Sects. 3.1 and 3.2 Our purpose, therefore, is not to have a complete variance analysis between all the parameters involved in the model, but to test mainly the effect of atmospheric stability on the calculated wind profile and boundary-layer height.

The PBL parameterisation schemes implemented in our code depend mainly on the following three parameters: the wind forcing aloft, G; the roughness length, z_0; and the Obukhov length, L, or alternatively the buoyancy heat flux B_s. Note that the atmospheric boundary-layer height, h, in particular, is strongly dependent on the surface heat flux both in stable/neutral and convective case.

The effects of roughness length and wind forcing on wind velocity profiles are widely known and well documented, so that we will mainly focus the present analysis on the effects of atmospheric stability, through L and B_s, on wind velocity profiles and PBL height. In the following we will assume $z_0 = 0.3$ m, which corresponds to bushland in the area around Genoa, and $G = 10\,\mathrm{m\,s^{-1}}$, for compatibility with most of the atmospheric conditions for this area. Finally, the analysis will be performed in the range $-0.1\,\mathrm{m}^{-1} \leq 1/L \leq 0.1\,\mathrm{m}^{-1}$, which corresponds to Pasquill-Gifford classes from A (very unstable) to G (very stable) and has been indicated as suitable for Italy by Cenedese et al. (1998). For example, the frequencies associated to the aforementioned classes of atmospheric stability, as derived from the measurements by the Italian Air Force during the years 1963–1991 (ENEL-AM 1994), at eight anemometric stations in Liguria Region, northern Italy (the region where Genoa is the main town) vary between the limits reported in Table 1.

Figure 1 shows the vertical profile of the wind velocity normalised with wind at the top of the boundary layer, i.e. the geostrophic wind, for different atmospheric stability conditions. The vertical coordinate is scaled by z_0 (noting that $z_0 = 0.3$ m). Only the lower part of the profiles, up to about 500 m above the ground level (a.g.l.), is shown. Non-neutral profiles are shown within the interval of the inverse Obukhov length $1/L = \pm\, 0.1\,\mathrm{m}^{-1}$.

The differences in the wind velocities for negative values of $1/L$ are more evident close to the surface, for example at $z/z_0 \approx 150$ the wind speeds for diverse instabilities range from 0.68 to 0.81 of the geostrophic wind, while at $z/z_0 = 1{,}500$ the wind speed has values ranging over 0.92–0.94 of the geostrophic wind (G) and is not so sensitive to the grade of the instability. For the stable cases the variability of the wind speed in the height interval $150 < z/z_0 < 1{,}500$ is in the range from $0.57G$ to G presenting also stronger vertical gradients compared to the unstable cases.

The variability of the PBL height for various stability conditions has been also tested. By means of the meteorological pre-processor ABLE the sensible heat fluxes for summer, winter and intermediate periods have been calculated. Negative values of sensible heat flux are not considered, since the multi-limit equation for the boundary-layer height used under stable conditions does not depend explicitly on H. For various conditions of meteorological parameters (including surface temperature and cloud cover) sensible heat flux values are estimated to be in the range 60–310 W m^{-2}. Further on for the tests we have selected summer values in the range $60 \leq H \leq 310\,\mathrm{W\,m^{-2}}$, winter values in the range $60 \leq H \leq 210\,\mathrm{W\,m^{-2}}$ and for the intermediate period values in the range $60 \leq H \leq 260\,\mathrm{W\,m^{-2}}$.

Figure 2 shows the PBL heights as a function of $1/L$ in the interval $\pm\, 0.1\,\mathrm{m}^{-1}$. Under unstable conditions, the heights corresponding to all the three different ranges of sensible heat flux

Table 1 Limits of the frequencies of the Pasquill-Gifford classes of atmospheric stability, as derived from the measurements by the Italian Air Force at eight anemometric stations in Liguria Region, during the years 1963–1991

Stability class	A	B	C	D	E	F+G
Maximum frequency (%)	3.4	14.7	9.0	59.8	21.6	31.2
Minimum frequency (%)	1.6	5.0	4.8	38.6	3.3	9.5

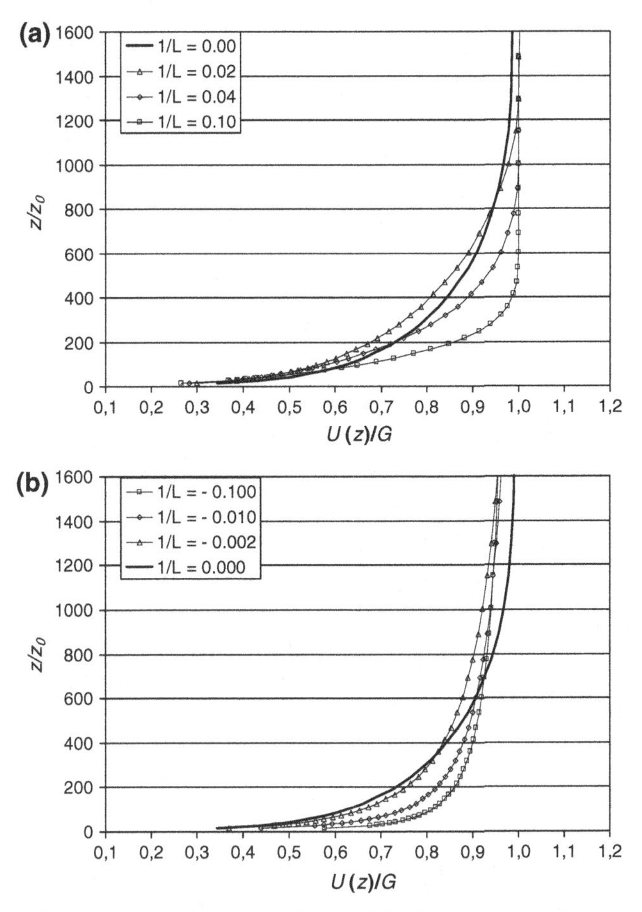

Fig. 1 Vertical profiles of the non-dimensional wind speed, $U(z)/G$, as a function of the atmospheric stability through the inverse of the Obukhov length $1/L$: stable/neutral cases (**a**); unstable/neutral cases (**b**)

values are shown. It is worth noting that under stable/neutral conditions, having assumed μ_N as a constant, the multi-limit equation for the boundary-layer height (Zilitinkevich et al. 2002) is mainly function of $1/L$ through the stability parameter μ, and it is insensitive to the sensible heat flux H. On the contrary, the PBL height under unstable conditions depends directly on H in Eqs. 3–4, so that for a given value of $1/L$ different values of h could be simulated. This effect is more pronounced for strong instabilities. For instance, at $1/L = -0.1\,\mathrm{m}^{-1}$, the simulated PBL height could have values from 3.8 to 4.4 times greater than the neutral PBL height. Under stable conditions, the applied PBL parameterisations give values lower by 15% to 38% with respect to value at $1/L = 0$.

The calculated heights shown in Fig. 2 are in good agreement with the values reported in the literature. For example, Gassmann and Mazzeo (2001) present a model of the evolution of the nocturnal stable boundary-layer height, function of the Obukhov length, surface potential temperature of air, and roughness length. This model, applied to the micrometeorological data of three stations in Argentina at mid-latitudes (30.0°–45.0° N), produced monthly mean PBL heights ranging between 133 and 272 m. As far as unstable atmospheric conditions are concerned, Grimsdell and Angevine (1998) reported values of the CBL height

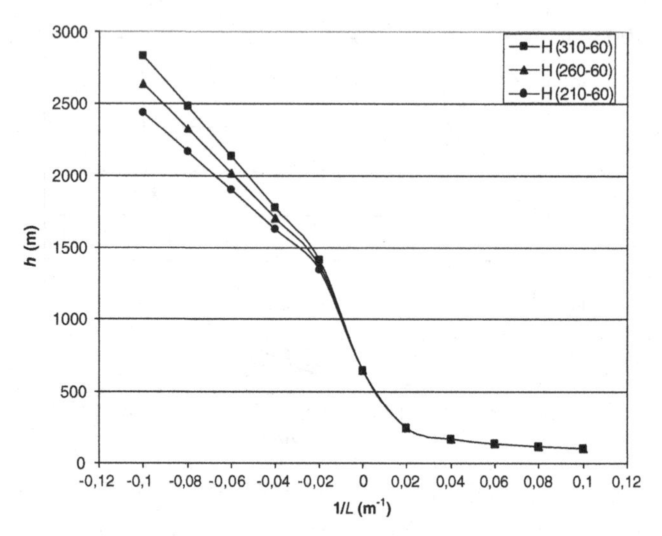

Fig. 2 Boundary-layer heights for different stability conditions

over Illinois (37.5°–42.5° N) evaluated from radiosondes and wind-profiler measurements ranging between a few hundreds metres and 2,000 m, while Lammert and Bösenberg (2006) determined CBL heights over the Baltic Sea, Oklahoma and Germany (35.0°–60.0° N) with laser remote sensing up to around 2,500 m.

5 Conclusions

The sensitivity tests on the PBL parameterisations schemes for stable/neutral and unstable conditions as used in the code WINDS have demonstrated to produce reasonable values for the PBL height and the wind profiles. The choice of some predefined input parameters could be justified on the basis of further comparisons with observations.

It is worth noting that the applied PBL parameterisations impose a strong constraint on the simulated flow, which has to be in a barotropic state, at least with a fairly good approximation and within the simulation volume. Further efforts are being made to use WINDS for downscaling the output of the operational weather forecast model BOLAM in order to produce fine scale wind fields, with due regard to improved PBL parameterisations.

References

Burlando M, Podestà A, Castino F, Ratto CF (2002) The wind map of Italy. In: Proceedings of the world wind energy conference and exhibition, 2–6 July Berlin, Germany, 4pp

Buzzi A, Fantini M, Malguzzi P, Nerozzi F (1994) Validation of a limited area model in case of Mediterranean cyclogenesis: surface fields and precipitation scores. Meteorol Atmos Phys 53:137–153

Canepa E, Builtjes PJH (2001) Methodology of model testing and application to dispersion simulation above complex terrain. Int J Environ Pollut 16:101–115

Canepa E, Ratto CF (2003) SAFE_AIR algorithms to simulate the transport of pollutant elements: a model validation exercise and sensitivity analysis. Environ Model Software 18(4):365–372

Canepa E, Ratto CF, Zannetti P (1999) Calibration of the dispersion code SAFE_AIR using a release in nocturnal low wind conditions, In: Brebbia CA, Jacobson M, Power H (eds) Air Pollution Vol VII. Computational Mechanics Publications, Southampton, UK, pp 3–11

Castino F, Rusca L, Solari G (2003) Wind climate micro-zoning: a pilot application to Liguria Region (North-Western Italy). J Wind Eng Ind Aerod 91:1353–1375

Cenedese A, Leuzzi G, Monti P, Querzoli G (1998) Results from Italian datasets. In: Harmonisation of the pre-processing of meteorological data for atmospheric dispersion models. COST 710 Final Report, Fisher B, Erbrink JJ, Finardi S, Jeannet P, Joffre S, Morselli MG, Pechinger U, Seibert P, Thompson DJ (eds) EUR 18195 EN, Luxemburg, pp 67–83

Chandrasekar A, Philbrick CR, Clark R, Doddridge B, Georgopoulos P (2003) Evaluating the performance of a computationally efficient MM5/CALMET system for developing wind field inputs to air quality models. Atmos Environ 37(23):3267–3276

Fisher B, Erbrink J, Finardi S, Jeannet Ph, Joffre S, Morselli M, Pechinger U, Seibert P, Thompson D (1998) Harmonisation of the preprocessing of meteorological data for atmospheric dispersion models. COST 710 Final report, CEC Publication EUR 18195, Luxemburg, 431 pp

Furunoa A, Teradaa H, Chinoa M, Yamazawab H (2004) Experimental verification for real-time environmental emergency response system: WSPEEDI by European tracer experiment. Atmos Environ 38:6989–6998

Gassmann MI, Mazzeo NA (2001) Nocturnal stable boundary layer height model and its application. Atmos Res 57:247–259

Georgieva E, Canepa E, Ratto CF (2001) The determination of the mixing height: an extended version of the SAFE_AIR dispersion model. In Air pollution modelling and its application. vol 14. Kluwer Academic/Plenum Publisher, New York, pp 719–720

Georgieva E, Canepa E, Bonafè G, Minguzzi E (2003) Evaluation of ABLE mixing height algorithms against observed and modelled data for the Po Valley Italy. Preprint of 26th NATO/CCMS international technical meeting on air pollution modelling and its applications, 26–30 May, Istanbul, Turkey, pp 396–403

Grimsdell AW, Angevine WM (1998) Convective boundary layer height measurement with wind profilers and comparison to cloud base. J Atmos Ocean Technol 15:1331–1338

Hess GD (2004) The neutral, barotropic planetary boundary layer, capped by a low-level inversion. Boundary-Layer Meteorol 110:319–355

Hess GD, Garratt JR (2002) Evaluating models of the neutral, barotropic planetary boundary layer using integral measures: part I. Overview. Boundary-Layer Meteorol 104:333–358

Lammert A, Bösenberg J (2006) Determination of the convective boundary-layer height with laser remote sensing. Boundary-Layer Meteorol 119:159–170

Mironov DV (1999) The SUBMESO meteorological preprocessor. Physical parametrisations and implementation. Laboratoire de Mecanique des Fluides UMR CNRS. Ecole Centrale de Nantes, France

National Electric Energy Supplier (ENEL), and Meteorological Service of Italian Air Force (AM) (1994) Caratteristiche diffusive dei bassi strati dell'atmosfera, Vol. 6 (Liguria)

Ratto CF, Festa R, Nicora O, Mosiello R, Ricci A, Lalas DP, Frumento OA (1990) Wind field numerical simulation: a new user-friendly code. In: European community wind energy conference. Palz W (ed) H. S. Stephens & Associates, Madrid, Spain, pp 130–134

Ratto CF, Festa R, Romeo C, Frumento OA, Galluzzi M (1994) Mass-consistent models for wind fields over complex terrain: the state of the art. Environ Soft. 9:247–268

Sherman CA (1978) A mass-consistent model for wind fields over complex terrain. J Appl Meteorol 17:312–319

Trombetti F, Martano P, Tampieri F (1991) Data sets for studies of flow and dispersion in complex terrain: 1) the "RUSHIL" wind tunnel experiment (flow data) Technical Report, FISBAT-RT-91/1, Bologna, Italy.

Zilitinkevich SS (1989) Velocity profiles, the resistance law and the dissipation rate of mean flow kinetic energy in a neutrally and stably stratified planetary boundary layer. Boundary-Layer Meteorol 46:367–387

Zilitinkevich SS, Calanca P (2000) An extended similarity-theory for the stably stratified atmospheric surface layer. Quart J Roy Meteorol Soc 126:1913–1923

Zilitinkevoch SS, Esau IN (2002) On integral measures of the neutral, barotropic planetary boundary layers. Boundary-Layer Meteorol. 104:371–379

Zilitinkevich SS, Esau IN (2003) The effect of baroclinicity on the depth of neutral and stable planetary boundary layers. Quart J Roy Meteorol Soc 129:3339–3356

Zilitinkevich SS, Fedorovich EE, Shabalova M (1992) Numerical model of a non-steady atmospheric planetary boundary layer, based on similarity theory. Boundary-Layer Meteorol 59:387–411

Zilitinkevich SS, Johansson P-E, Mironov DV, Baklanov A (1998) A similarity-theory model for wind profile and resistance law in stably stratified planetary boundary layers. J Wind Eng Indust Aerodyn 74–76: 209–218

Zilitinkevich SS, Perov VL, King JC (2002) Near-surface turbulent fluxes in stable stratification: calculation techniques for use in general circulation models. Quart J Roy Meterol Soc 128:1571–1587

Participants of the NATO Advanced Research Workshop 'Atmospheric Boundary Layers: Modelling and Applications for Environmental Security' (NATO PBL ARW) in Dubrovnik, Croatia, 18–22 April, 2006

Participants in the photo (from the left to the right):

1. Leif Enger
2. Igor Esau
3. Boris Galperin
4. Robert Bornstein
5. Amela Jeričević
6. Arakel Petrosyan
7. Maja Telišman Prtenjak
8. Leonid Bobylev
9. Vladimir Makin
10. Hannu Savijarvi
11. Sergej Zilitinkevich
12. Matthias Roth
13. Dimiter Syrakov
14. Adolf Ebel
15. Harindra (Joe) Fernando
16. Sergiy Stepanenko
17. Vasily Lykosov
18. Alexander Baklanov
19. Branko Grisogono
20. Søren Larsen
21. Alexander Korablev
22. Georgy Golitsyn
23. Thorsten Mauritsen
24. Danijel Belušić
25. Domenico Anfossi
26. Matteo Buzzi
27. Natan Kleeorin
28. Tov Elperin
29. Igor Granberg
30. Massimiliano Burlando
31. Ivan Mammarella
32. Andrey Grachev
33. Zvjezdana Bencetić Klaić
34. Peter A. Taylor
35. George Djolov
36. Iva Kavčič
37. Igor Rogachevskii
38. Tarmo Soomere
39. Mathias Rotach
40. Silvia Trini Castelli
41. Larry Mahrt
42. Mark Žagar

Participants not in the photo:

1. Helmut Baumert
2. Ivan Čačić
3. Iva Grisogono
4. Sven-Eryk Gryning
5. Bert Holtslag
6. John King
7. Lev Karlin
8. Dmitrii Mironov
9. Pekka Plathan
10. Janis Rimshans
11. Leonid Savelyev
12. Mikhail Sofiev
13. Rein Tamsalu
14. Domingos Viegas
15. Yrjö Viisanen